# Introduction to Holomorphic Functions of Several Variables

Volume II

Local Theory

M. Adams, V. Guillemin, *Measure Theory and Probability*
W. Beckner, A. Calderón, R. Fefferman, P. Jones, *Conference on Harmonic Analysis in Honor of Antoni Zygmund*
G. Chartrand, L. Lesniak, *Graphs & Digraphs, Second Edition*
J. Cochran, *Applied Mathematics: Principles, Techniques, and Applications*
W. Derrick, *Complex Analysis and Applications, Second Edition*
J. Dieudonné, *History of Algebraic Geometry*
R. Dudley, *Real Analysis and Probability*
R. Durrett, *Brownian Motion and Martingales in Analysis*
R. Epstein, W. Carnielli, *Computability: Computable Functions, Logic, and the Foundations of Mathematics*
S. Fisher, *Complex Variables, Second Edition*
A. Garsia, *Topics in Almost Everywhere Convergence*
P. Garrett, *Holomorphic Hilbert Modular Forms*
R. Gunning, *Introduction to Holomorphic Functions of Several Variables*
*Volume I: Function Theory*
*Volume II: Local Theory*
*Volume III: Homological Theory*
J. Kevorkian, *Partial Differential Equations: Analytical Solution Techniques*
R. McKenzie, G. McNulty, W. Taylor, *Algebras, Lattices, Varieties, Volume I*
E. Mendelson, *Introduction to Mathematical Logic, Third Edition*
R. Salem, *Algebraic Numbers and Fourier Analysis* and L. Carleson, *Selected Problems on Exceptional Sets*
R. Stanley, *Enumerative Combinatorics, Volume I*
J. Strikwerda, *Finite Difference Schemes and Partial Differential Equations*
K. Stromberg, *An Introduction to Classical Real Analysis*

# Introduction to Holomorphic Functions of Several Variables

Volume II

Local Theory

**Robert C. Gunning**
*Princeton University*

CRC Press
Taylor & Francis Group
Boca Raton London New York

CRC Press is an imprint of the
Taylor & Francis Group, an **informa** business
A CHAPMAN & HALL BOOK

First published 1999 by Chapman & Hall

Published 2019 by CRC Press
Taylor & Francis Group
6000 Broken Sound Parkway NW, Suite 300
Boca Raton, FL 33487-2742



First issued in paperback 2019



ISBN 13: 978-0-367-45079-3 (pbk)
ISBN 13: 978-0-534-13309-2 (hbk)



**Visit the Taylor & Francis Web site at**
**http://www.taylorandfrancis.com**

**and the CRC Press Web site at**
**http://www.crcpress.com**

# Series Contents

## VOLUME I
## Function Theory



## VOLUME II
## Local Theory

## VOLUME III
## Homological Theory

# Preface

These three volumes together comprise a revised version of the book *Analytic Functions of Several Complex Variables* that Hugo Rossi and I wrote some 20 years ago. The revisions are fairly extensive; indeed, they essentially amount to a complete rewriting. The attempt to include some more recent material, to incorporate a number of corrections and improvements, and to expand and simplify the treatment to make for easier reading have led quite naturally to rather drastic changes. I very much regret that this revision could not have been a joint effort as before; it suffers much in balance, scope, and clarity as a result, and I have no one else to blame for the inevitable errors it contains and the gaps it does not.

These volumes are intended as an extensive introduction to the Oka–Cartan theory of holomorphic functions of several variables and holomorphic varieties, just as was the earlier book, and they cover very much the same range of topics. Great advances have been made in function theory in the past 20 years. To survey all that has been done to a comparable extent would require a few more volumes; that is something I hope to return to later. In the meantime, though, a considerable number of books that treat many of these topics very well indeed are listed in the Bibliography, and some more detailed suggestions for further reading are given in the Outline of the present work that follows.

The Outline is intended as a survey and guide to reading these volumes. It is not likely that the reader will want or need to start at the beginning and read consecutively through to the end. The three volumes cover three somewhat different aspects of the subject and can to a certain extent be read independently. Alternatively, the first sections of each volume can be read as a shorter introduction to the subject.

This revision grew out of courses of lectures on various topics in complex analysis that I have given at Princeton over the years since the first book was written. I should like to express here my sincere gratitude to the many students and colleagues who attended these lectures for the many very helpful comments, suggestions, and corrections they have given me. I should like in particular to thank Jay Belanger for his careful reading of much of the final manuscript. Finally,

I should like to thank all those who have typed various parts, and revisions of parts, and parts of revisions, and revisions of revisions of the manuscript, and especially Maureen Kirkham, who has typed the greater part of the whole project.

*Robert C. Gunning*

# Contents

## II Local Theory

# Series Outline

The prerequisites for reading these books are basically a solid undergraduate training in analysis (especially in the theory of holomorphic functions of a single variable), topology, and algebra. The aim of the books is to introduce the reader to a wide range of topics in the theory of holomorphic functions of several variables, with fairly complete proofs, as preparation either for using this material as part of a general mathematical background or for continuing a more detailed and extensive study of this fascinating and active field of research. The Bibliography at the end lists only other books and general surveys in this and closely related areas; its length is a good indication of just how active a field this really is. An excellent historical discussion of the development of the field can be found in [6].

# Volume I. Function Theory

The first volume deals with holomorphic functions defined in open subsets of the space $\mathbb{C}^n$ of $n$ complex variables, focusing on their properties as ordinary complex-valued functions. **A.** There are various equivalent characterizations of holomorphic functions, the basic and most primitive of which is as those functions that can be represented locally as convergent power series. It is clear from this that a holomorphic function of several variables is holomorphic in each variable separately, when all the others are held fixed. So a great many of the familiar properties of holomorphic functions of a single variable, such as the Cauchy integral formula, can be applied quite easily and directly to obtain corresponding properties of holomorphic functions of several variables. **B.** On the other hand, though quite true, it is surprisingly nontrivial to prove that any function that is holomorphic in each variable separately is actually a holomorphic function of several variables. This result really rests on a closer analysis of the precise domains of convergence of power series of several variables, in the course of which some of the peculiar properties of functions of more than a single variable first become manifest. **C.** The most straightforward properties of differentiable functions of several variables, such as the implicit and inverse function theorems, extend very easily to holomorphic functions; complex manifolds and submanifolds can then be introduced in immediate analogy to the corresponding differentiable notions. These three sections cover the rudiments of the theory of holomorphic functions, the ABC's of the subject.

**D.** There is a natural analogue for holomorphic functions of several variables of Riemann's removable singularities theorem, but much more is true: sufficiently small subsets, such as an isolated point in $\mathbb{C}^n$ when $n > 1$, are automatically removable singularities for holomorphic functions in general, with no hypotheses of boundedness. Several more examples of removable singularities are given; these can be skipped if desired, but they do provide some useful intuitive feel for a property that is peculiarly characteristic of holomorphic functions of several variables. **E.** The differential operator that appears in the analogue for functions of several variables of the Cauchy–Riemann characterization of holomorphic functions can be extended as a linear mapping $\bar{\partial}$ from complex differential forms of bidegree $(p, q)$ to those of bidegree $(p, q + 1)$ and has the property that $\bar{\partial}\bar{\partial} = 0$. Thus if $\phi = \bar{\partial}\psi$, then $\bar{\partial}\phi = 0$, and that raises the question whether, conversely, if $\bar{\partial}\phi = 0$, then there is a differential form $\psi$ such that $\bar{\partial}\psi = \phi$. That is true locally and also for differential forms in such simple domains as polydiscs. What may seem to be a digression here is actually not one; this result has a very nice application to another removable singularities theorem, and this is in turn the model for a deep and powerful general use of the $\bar{\partial}$ operator in complex analysis, to be discussed in section O. **F.** Another application of properties of this differential operator yields an interesting and useful result about the approximation of holomorphic functions in special domains in $\mathbb{C}^n$ by polynomials, in extension of the Runge approximation theorem for functions of a single variable. For several variables, though, there can be such approximation theorems only for somewhat special domains, as indicated by an example.

**G.** The discussion of removable singularities indicates that there are some pairs of open subsets $D \subseteq E \subseteq \mathbb{C}^n$ such that every function holomorphic in $D$ extends to a holomorphic function in $E$. Particularly natural and useful are those subsets $D \subseteq \mathbb{C}^n$ for which there is actually no nontrivial such extension—that is, for which necessarily $E = D$. Such sets are called domains of holomorphy and have a number of special properties that will be considered subsequently. Some alternative characterizations and various examples of domains of holomorphy are provided here. **H.** If $D$ is not a domain of holomorphy, it is but natural to ask for the maximal set $E$ to which all holomorphic functions in $D$ extend. A complication is that the extended functions may be multiple-valued; but all of them can be viewed as single-valued holomorphic functions on a complex manifold spread out as a locally unbranched covering space over an open subset of $\mathbb{C}^n$. In the category of such manifolds, called Riemann domains, there is a unique maximal one to which all holomorphic functions in $D$ extend, the envelope of holomorphy of $D$; it can be described intrinsically in terms of the ring of holomorphic functions on $D$. **I.** An envelope of holomorphy is itself maximal, the analogue of a domain of holomorphy among Riemann domains, virtually by definition. It is possible to extend many of the alternative characterizations of domains of holomorphy to hold on Riemann domains as well, as is done here with elementary but in part quite nontrivial arguments. A more complete and in many ways preferable treatment of this topic is given in section P, after some more machinery is developed, so this section can be omitted altogether by those who are willing to proceed further; it is really inserted just to conclude the discussion for those wishing to go no further in this direction.

The preceding portion is by itself a reasonable introduction to function theory in several variables. It is quite possible for those so inclined to omit the rest of this volume, and proceed to either Volume II or Volume III. On the other hand, those interested in functions as such may wish to plow on through the rest of Volume I, which concentrates on a particularly useful class of nonholomorphic functions.

**J.** Plurisubharmonic functions have played a considerable role in function theory of several variables. They are natural and useful extensions to several variables of the possibly familiar subharmonic functions of a single complex variable. A review of the relevant properties of subharmonic functions is included here for those who may not be that well acquainted with them. **K.** The basic properties of plurisubharmonic functions are discussed in some detail. **L.** Various special or generalized classes of plurisubharmonic functions are useful at some points and so are discussed. It is quite possible, and perhaps at first reading advisable, to skip this section until the need arises in references to it later.

**M.** There is a special class of open subsets of $\mathbf{C}^n$ described in terms of the plurisubharmonic functions in them, the pseudoconvex subsets. These are in some ways analogous to domains of holomorphy. Indeed, it is demonstrated in section P that they are precisely domains of holomorphy, although that is quite a nontrivial result. These sets have a number of special properties and alternative characterizations, all rather simple to demonstrate because of the flexibility and abundance of plurisubharmonic functions. If it is demonstrated that pseudoconvex subsets are really just domains of holomorphy, these special properties and alternative characterizations hold immediately for the latter subsets; that will yield some very useful and powerful results about domains of holomorphy. **N.** Similar arguments hold for Riemann domains, although with some further complications. The discussion here can be omitted altogether by those readers interested in only subsets of $\mathbf{C}^n$ rather than general Riemann domains. **O.** The main technical tool needed to establish the equivalence of pseudoconvexity and holomorphic convexity is the solvability of the $\bar{\partial}$ equation in pseudoconvex domains, which is established here. **P.** The equivalence and some of the immediate consequences are demonstrated quite easily.

**Q.** The preceding results rest eventually on the close relationship between plurisubharmonic and holomorphic functions; some other facets of this relationship are discussed here. **R.** In conclusion, and as an introduction to another vast area of great interest, the special case of domains with smooth boundaries is briefly discussed.

A very quick and convenient survey of various other and more detailed results about pseudoconvex sets can be found in [19]. Powerful tools in studying domains with smooth boundaries are versions of the Cauchy integral formula that involve integrating over the whole boundary; such formulas and their consequences are nicely treated in [48]. Closely related to this is the finer analysis of the $\bar{\partial}$ operator in smoothly bounded domains, in particular in connection with boundary value problems; this is treated in [17] and [25], where more references to the literature can also be found. There are analogues of other classical boundary value problems in function theory of a single variable treated, for instance, in [50] and [51]. There is an extensive theory of holomorphic mappings and differential-geometric methods, as discussed, for instance, in [35].

# Volume II. Local Theory

A holomorphic or meromorphic function of a single variable has the appealingly simple form $z^n$ in suitable local coordinates, for some integer $n$. The zero set of a holomorphic function of one variable or the singular set of a meromorphic function of one variable is an isolated point. The situation is much more complicated and hence much more interesting for functions of several variables. It leads to an extensive collection of results in a quite different direction from that of Volume I, indeed more akin to algebraic geometry than to classical function theory.

**A.** A basic tool for the local study of holomorphic functions of several variables is a convenient canonical form for those functions, the Weierstrass polynomials. From that it is easy to derive some general properties of the ring of local holomorphic functions at a point in $\mathbb{C}^n$. **B.** A holomorphic subvariety of an open subset in $\mathbb{C}^n$ is a subset that can be described locally as the set of common zeros of finitely many holomorphic functions. Such subvarieties are described locally in terms of ideals in the ring of local holomorphic functions, with interesting simple relations between the algebraic and geometric structures. It is convenient to consider some natural equivalence classes of holomorphic subvarieties as abstract holomorphic varieties. **C.** The simplest local form of a holomorphic mapping between holomorphic varieties is a finite branched holomorphic covering; any mapping is of this form in the case of a single variable. The general topological and algebraic properties of such mappings are considered first. **D.** It is demonstrated that every holomorphic variety can be represented locally as a finite branched holomorphic covering of $\mathbb{C}^n$, when it is locally irreducible. This provides a very convenient local description of holomorphic varieties. **E.** This description is used to establish some basic local properties of holomorphic varieties, such as the analytic form of Hilbert's zero theorem.

**F.** There are some very subtle but important properties of holomorphic functions and varieties that are really semilocal rather than local, in the sense that they involve relations between the local rings of holomorphic functions at all points of an open neighborhood of a fixed point. These are perhaps the most complicated results of the local theory, but they play a major role in its development. These results are reformulated as coherence theorems in Volume III; they can really be used without worrying about the proofs, although that approach to mathematics has both dangers and drawbacks and should not be encouraged.

**G.** Perhaps the basic invariant attached to a holomorphic variety is its dimension. This can be characterized either geometrically or algebraically. There are a number of relations between the dimension of a subvariety and the number of holomorphic functions needed to describe the subvariety. A useful digression here is the description of general divisors and divisors of functions.

The results obtained so far in this volume provide an introduction to the local theory of holomorphic functions and varieties, quite sufficient for many purposes. It is possible for those so interested to skip the remainder of this volume and pass to the global theory in Volume III. On the other hand, though, a great deal more can be said about the local theory.

**H.** It is natural to consider holomorphic functions on holomorphic varieties, but it is not altogether trivial to show that a limit of such functions is also holomorphic. That closure property and an extension to the closure of modules of various sorts are useful and important, if somewhat technical, results.

**I.** Another useful local invariant is not the dimension of a variety itself but rather the least value $m$ such that the variety can be realized as a subvariety of $\mathbb{C}^m$; this is known as the tangential or imbedding dimension of the variety. This invariant is strictly larger than the ordinary dimension at singular points and is indeed a measure of the singularity. It can be described in terms of the natural notion of the tangent space to a variety, even at singular points, and is useful in examining differential notions such as the inverse mapping theorem at singular points. **J.** Global sections of the tangent bundle, or holomorphic vector fields, can be introduced on varieties with singularities, and they play a role much like the one they play on manifolds.

**K.** With the additional machinery that has been developed here, it is possible to complete the discussion of holomorphic extensions of holomorphic functions from section D of Volume I. There are analogous results for the holomorphic extension of holomorphic varieties.

**L.** Finite holomorphic mappings were discussed in section C. Local properties of general holomorphic mappings are examined in some detail here. The simplest classes of holomorphic mappings are those for which the dimension of the inverse image of a point is independent of the point; they have relatively simple standard representations, extending those examined earlier for finite holomorphic mappings. **M.** For completeness, and in preparation for the further discussion of local properties of holomorphic mappings, a review of some standard properties of complex projective spaces is given. This digression can be skipped by readers already familiar with such spaces. **N.** The discussion of holomorphic mappings continues with an analysis of proper holomorphic mappings, those for which the inverse image of a compact set is compact. The images of these mappings, and of those of a natural generalization of this class of mappings, are always holomorphic varieties themselves, a useful and nontrivial result. Among these mappings is the special class of monoidal or quadratic transforms, which play a major role in the detailed analysis of singularities of holomorphic varieties.

**O.** Meromorphic functions were earlier considered locally; here they are examined globally, first in open subsets of $\mathbb{C}^n$. **P.** The extension of this discussion to holomorphic functions on varieties is more complicated. **Q.** There is a special class of varieties of great importance here; those for which the bounded meromorphic functions are necessarily holomorphic. They are called normal varieties. **R.** To any holomorphic variety there is naturally associated a unique one-sheeted branched covering that is itself a normal variety. This normalization is a convenient and useful partial step in the classification of singularities of holomorphic varieties.

Further discussion of general properties of holomorphic varieties with singularities and other references to the literature can be found in [1], [16], [23], [28], [43], and [60]. The topological properties of singularities of one-dimensional varieties are a classical topic in algebraic geometry, while the higher-dimensional study was pioneered in [41]. The case of two-dimensional singularities is the next most extensively studied class, as surveyed, for instance, in [37]. For many purposes it is important to study holomorphic varieties with more general rings of holomorphic functions than those considered here, rings with nilpotent elements, reflecting a description of subvarieties by particular families of defining equations; these topics are surveyed nicely in [23].

# Volume III. Homological Theory

Sheaves have proved to be very useful tools in organizing and simplifying various arguments and calculations in function theory, as well as in algebraic geometry, and should be part of the equipment of those working in this area. **A.** The definitions and basic properties of sheaves are developed in a fairly general form, assuming no previous acquaintance; those readers already familiar with sheaves can skip or skim through this section. **B.** Sheaves provide a particularly convenient mechanism for handling the semilocal properties of holomorphic functions and varieties discussed in the preceding volume. The notion of coherence plays a prominent role here.

**C.** Sheaves provide an equally convenient mechanism for handling some of the basic global problems of complex analysis through cohomology theory. Again no previous acquaintance with this theory is presupposed. The algebraic structure underlying cohomology theory is developed first, including some general properties of cohomology theories and the technique of spectral sequences. **D.** The cohomology groups of a space with coefficients in a sheaf are examined in some detail, together with techniques for calculating these groups by using convenient auxiliary sheaves. **E.** There is another technique for calculating these groups, a useful technique in some analytic applications; it involves sections of the sheaf over the sets forming various open covers of the space. **F.** The behavior of sheaves and sheaf cohomology groups under mappings of the base spaces is examined only as necessary for applications here.

**G.** The discussion turns next to the use of sheaf cohomology in complex analysis. A basic technical tool is developed first, a result about matrices of holomorphic functions, a variant of the possibly familiar Riemann–Hilbert problem. **H.** This auxiliary result is used to show that an open polydisc has trivial cohomology groups in all positive dimensions, when the coefficient sheaf is any holomorphic sheaf satisfying the basic semilocal coherence condition. **I.** The vanishing of all of these cohomology groups has a wide range of useful and interesting consequences. In general, a holomorphic variety such as a polydisc for which these cohomology groups vanish is called a Stein variety. Some of the fundamental properties of Stein varieties, including approximation results reminiscent of the Runge theorem, are established first. **J.** The special properties of the Frechet algebra of holomorphic functions on a Stein variety are examined. **K.** The global properties of meromorphic functions on a Stein variety, and their relationship to holomorphic functions, are discussed. L. Alternative characterizations of Stein varieties are established to show that there actually do exist a wide range of interesting such varieties. For instance, an open subset of $\mathbb{C}^n$ is a Stein variety precisely when it is a domain of holomorphy; any subvariety of a Stein variety is itself a Stein variety, and certain increasing unions of Stein varieties are Stein varieties.

The results established up to this point in Volume III provide a good introduction to the use of homological methods in complex analysis, as well as to the properties of Stein varieties, so some readers may be content to stop here.

There are, however, further interesting properties and alternative characterizations of Stein varieties. M. For instance, any holomorphic variety that admits a finite proper holomorphic mapping into a Stein variety is itself a Stein variety. That generalizes the earlier observation that any subvariety of a Stein variety is a Stein variety, and can be used to simplify some of the other characterizations of Stein varieties discussed in the preceding section. N. For another instance, a holomorphic variety is Stein if and only if its normalization is Stein; more generally, if $F: V \rightarrow W$ is a finite branched holomorphic covering, then $V$ is Stein if and only if $W$ is Stein. O. There are in addition various cohomological characterizations of Stein varieties; some of these have applications to the problem of the number of holomorphic functions needed to describe a holomorphic subvariety.

P. One of the basic properties of Stein varieties, as made manifest in the preceding discussion, is that they have a plenitude of global holomorphic functions. Indeed, it is noted here that a general $n$-tuple of holomorphic functions on an $n$-dimensional Stein variety $V$ determines a finite holomorphic mapping from $V$ into $\mathbb{C}^n$. Q. Moreover, a dense set of $(n + 1)$-tuples of holomorphic functions on $V$ determines a proper holomorphic mapping from $V$ to $\mathbb{C}^{n+1}$. R. These results can be combined to yield yet more characterizations of Stein varieties, perhaps the most appealing and satisfactory of all the characterizations. For instance, a holomorphic variety $V$ is Stein precisely when there is a holomorphic homeomorphism from $V$ to a holomorphic subvariety of some space $\mathbb{C}^n$.

A very important topic in the continuation of the discussion of global methods is another solution of the Levi problem using sheaf-theoretic techniques, due to Grauert; this was discussed in the first version of the present book. It is possible to extend the notion of plurisubharmonic functions to general holomorphic varieties as well and approach the Levi problem this way. These topics have not yet appeared very accessibly in textbooks, but a useful survey of the literature can be found in [6]. Other topics are treated in [4] and [24], among other places. There are extension theorems for holomorphic sheaves, analogues of the extension theorems for holomorphic functions and varieties, discussed in [54].

# A

# Local Rings of Holomorphic Functions

The local properties of holomorphic functions of one variable are really quite simple and can essentially be summarized in two assertions: (i) the set of zeros of a nontrivial holomorphic function of one variable is a discrete set of points; and (ii) any holomorphic function $f$ in an open neighborhood of a point $a \in \mathbb{C}$ can be written in the form $f(z) = (z - a)^\nu u(z)$ where $u$ is holomorphic and nonvanishing in an open neighborhood of $a$ and $\nu$ is a nonnegative integer. For holomorphic functions of several variables the situation is much more complicated, but some knowledge of the local properties is essential as a background to any study of holomorphic functions of several variables. To handle these local questions it is convenient to introduce the notion of the germ of a function. Let $A$ be a fixed point of $\mathbb{C}^n$ and let $U$, $V$, $W$ be some open neighborhoods of $A$. For any such neighborhood $U$ consider the set of all complex-valued functions defined on the set $U$, continuous or not; as a purely temporary notation such functions will be denoted by $f_U$, $g_U$, and so on, to emphasize the domain of definition. If $V \subseteq U$, then the restriction of a function $f_U$ to the subset $V$ of its domain of definition will be denoted by $f_U | V$. This is formally a different function, since its domain of definition has been changed.

1. **DEFINITION.** *Two functions $f_U$ and $g_V$ are called* **equivalent at the point** $A \in U \cap V$ *if there is an open neighborhood $W$ of the point $A$ such that $W \subseteq U \cap V$ and $f_U | W = g_V | W$. This is an equivalence relation in the usual sense, and an equivalence class is called a* **germ of a complex-valued function at the point A**.

In general, germs (equivalence classes) of functions will be denoted by boldface letters. Any function $f$ defined in some open neighborhood of a point $A$ belongs to some equivalence class at $A$; this class will be called the **germ** of the function $f$ at the point $A$ and will be denoted by $\mathbf{f}$ or $\mathbf{f}_A$. It should be noted particularly that the germ of a function at a point $A$ depends on the behavior of the function in a full open neighborhood of $A$, not merely on the value of the function at the point $A$. The polynomials $z$ and $z^2$ have the same values at the origin in the complex plane but determine quite different germs at the origin. All the functions in an equivalence class (all functions representing the same germ) at a point $A$ do have the same value at the point $A$, however; that common value will be called the value of the germ at the point $A$.

The set of germs of complex-valued functions can be given the structure of a ring as follows. For any two germs $\mathbf{f}$ and $\mathbf{g}$ at $A$, select representatives $f_U$ and $g_V$, respectively. If $W$ is an open neighborhood of $A$ such that $W \subseteq U \cap V$, the germ of the function $f_U|W + g_V|W$ is defined to be the sum $\mathbf{f} + \mathbf{g}$, and the germ of the function $(f_U|W)\cdot(g_V|W)$ is defined to be the product $\mathbf{f}\cdot\mathbf{g}$. It is a simple matter to verify that these operations are well defined, in the sense of being independent of the choices of representatives, and do impose on the set of all germs of complex-valued functions the structure of a ring. This ring is a commutative ring, with the zero element being the germ of the function that is identically zero and the identity element being the germ of the function that is identically one. A ring here will always be taken to be a ring with an identity element. Identifying the complex constants with constant-valued functions and taking the germs of these functions yield a canonical embedding of $\mathbb{C}$ as a subring of the ring of germs of complex-valued functions, and that introduces on the ring of germs of complex-valued functions the canonical structure of a complex algebra. To be quite explicit it should be mentioned here that throughout this book a **complex algebra** will mean a ring containing the field $\mathbb{C}$ as a subring, with the element $1 \in \mathbb{C}$ being the identity element for the ring. A homomorphism of complex algebras will mean a ring homomorphism that induces the identity mapping on the subfield $\mathbb{C}$.

The construction of germs of functions can also be carried out by considering continuous functions rather than arbitrary complex-valued functions; the resulting ring is then called the **ring of germs of continuous complex-valued functions** at the point $A$ and is denoted by $\mathscr{C}_A$. Similarly the construction can be carried out by considering only holomorphic functions; the resulting ring is called the **ring of germs of holomorphic functions** at the point $A$ and is denoted by $\mathcal{O}_A$. Both of these rings are of course also algebras over the complex numbers. As further notational conventions, the ring of germs of holomorphic functions at the origin $O = (0, \ldots, 0) \in \mathbb{C}^n$ will sometimes be denoted by $\mathcal{O}$ rather than $\mathcal{O}_O$, and when it is necessary or useful to specify the complex dimension of the underlying space $\mathbb{C}^n$, the notation ${}_n\mathcal{O}_A$ will be used. In particular, $\mathcal{O} = \mathcal{O}_O = {}_n\mathcal{O} = {}_n\mathcal{O}_O$. Translation in $\mathbb{C}^n$ clearly establishes an isomorphism between the rings $\mathcal{O}_A$ and $\mathcal{O}$ for any point $A \in \mathbb{C}^n$. This isomorphism preserves all the analytic properties of interest, so the purely local study of holomorphic functions of $n$ complex variables really amounts to a study of the ring ${}_n\mathcal{O}$.

The most elementary properties of the ring ${}_n\mathcal{O}$ are very easily established. First, *the ring ${}_n\mathcal{O}$ is an integral domain*; the product of any two nonzero elements is a nonzero element. To see this, consider any two germs $\mathbf{f}, \mathbf{g} \in {}_n\mathcal{O}$ and let $f, g$ be representative holomorphic functions in a connected open neighborhood $U$ of the origin. If $\mathbf{f}\cdot\mathbf{g} = 0$, then $f(Z)g(Z) = 0$ for all points $Z \in U$. If also $\mathbf{f} \neq 0$, then $f(A) \neq 0$ for some point $A \in U$; hence, by continuity $f(Z) \neq 0$ for all points $Z$ in some open neighborhood $V$ of $A$. But then $g(Z) = 0$ for all $Z \in V$, so by the identity theorem, Theorem I-A3, it follows that $g(Z) = 0$ for all $Z \in U$ and hence $\mathbf{g} = 0$. Since ${}_n\mathcal{O}$ is an integral domain, it has a well-defined quotient field, which is denoted by ${}_n\mathcal{M}$ and is called the **field of germs of meromorphic functions**. A **unit** in the ring ${}_n\mathcal{O}$ is an element $\mathbf{f} \in {}_n\mathcal{O}$ such that the multiplicative inverse $\mathbf{f}^{-1}$ of $\mathbf{f}$ also belongs to ${}_n\mathcal{O}$. It is clear that the units of ${}_n\mathcal{O}$ are precisely the germs that are nonzero at the origin. Indeed if $\mathbf{f}$ is represented by a holomorphic function $f$ in some open neighborhood $U$ of the origin and if $f(0) \neq 0$, then by shrinking the neighborhood $U$ if necessary, it can be assumed that $f(Z) \neq 0$ for all $Z \in U$; but then $1/f$ is holomorphic throughout $U$, and the germ of this function is evidently $\mathbf{f}^{-1}$. An equivalent way of stating this last observation is that the nonunits of ${}_n\mathcal{O}$ consist of those germs that vanish at the origin; the nonunits consequently form an ideal ${}_n\mathfrak{m}$ in the ring ${}_n\mathcal{O}$. Finally, the ring ${}_n\mathcal{O}$ can be described more explicitly as follows.

**2. THEOREM.** *The ring ${}_n\mathcal{O}$ of germs of holomorphic functions of n variables is isomorphic to the ring $\mathbb{C}\{z_1, \ldots, z_n\}$ of convergent complex power series in n variables.*

*Proof.* Each power series convergent in some open neighborhood of the origin represents a holomorphic function in that neighborhood and hence determines a unique germ in ${}_n\mathcal{O}$. Conversely, any representative $f$ of a germ $\mathbf{f} \in {}_n\mathcal{O}$ has a convergent power series expansion at the origin, and any two representatives of the same germ agree in some open neighborhood of the origin and hence determine the same power series expansion. The ring operations both arise from pointwise addition and multiplication of the functions represented, and that is enough to conclude the proof.

For the more detailed study of properties of the rings ${}_n\mathcal{O}$, it is convenient to develop a technique that facilitates the natural inductive step from ${}_{n-1}\mathcal{O}$ to ${}_n\mathcal{O}$. Any representative function $f$ for a germ $\mathbf{f} \in {}_{n-1}\mathcal{O}$ can be viewed as a holomorphic function of $n$ complex variables, but one that does not involve the last variable, and this evidently defines an inclusion ${}_{n-1}\mathcal{O} \subset {}_n\mathcal{O}$. The subring ${}_{n-1}\mathcal{O} \subset {}_n\mathcal{O}$ thus consists of those germs $\mathbf{f} \in {}_n\mathcal{O}$ that are independent of the variable $z_n$. As will be seen, it is useful to split this transition from $n - 1$ to $n$ variables into two steps by introducing an intermediate ring between ${}_{n-1}\mathcal{O}$ and ${}_n\mathcal{O}$. If $\mathbf{a}_0, \ldots, \mathbf{a}_\nu$ are elements of ${}_{n-1}\mathcal{O}$ represented by functions $a_0, \ldots, a_\nu$, then the polynomial expression $a_0 + a_1 z_n + \cdots + a_\nu z_n^\nu$ is a holomorphic function of $n$ variables in an open neighborhood of the origin and hence represents an element of ${}_n\mathcal{O}$. This element can be identified with the formal polynomial $\mathbf{a}_0 + \mathbf{a}_1 z_n + \cdots + \mathbf{a}_\nu z_n^\nu \in {}_{n-1}\mathcal{O}[z_n]$, where as usual ${}_{n-1}\mathcal{O}[z_n]$ denotes the ring of polynomials in $z_n$ with coefficients from the ring ${}_{n-1}\mathcal{O}$. In the pair of inclusions ${}_{n-1}\mathcal{O} \subset {}_{n-1}\mathcal{O}[z_n] \subset {}_n\mathcal{O}$ the first is a simple algebraic extension of rings, while the second extension is basically analytic.

Recall from Definition I-C1 that a function $f$ that is holomorphic in an open neighborhood of the origin in $\mathbb{C}^n$ is said to be regular of order $\nu$ in $z_n$ at the origin if $f(0, \ldots, 0, z_n)$ as a function of $z_n$ alone has a zero of order $\nu$ at $z_n = 0$. A germ $\mathbf{f} \in {}_n\mathcal{O}$ will be said to be **regular of order $\nu$ in $z_n$** if some representative of $\mathbf{f}$ is regular of order $\nu$ in $z_n$ at the origin; all representatives of $\mathbf{f}$ are then also clearly regular of order $\nu$ in $z_n$ at the origin. It follows from Lemma I-C2 that any nonzero germ $\mathbf{f} \in {}_n\mathcal{O}$ can be made regular in $z_n$ by a suitable nonsingular change of coordinates in $\mathbb{C}^n$.

**3. DEFINITION.** *A* **Weierstrass polynomial** *of degree $\nu > 0$ in $z_n$ is an element $\mathbf{h} \in {}_{n-1}\mathcal{O}[z_n]$ of the form*

$$\mathbf{h} = z_n^\nu + \mathbf{a}_1 z_n^{\nu-1} + \cdots + \mathbf{a}_{\nu-1} z_n + \mathbf{a}_\nu \tag{1}$$

*where the coefficients* $\mathbf{a}_j \in {}_{n-1}\mathcal{O}$ *are nonunits,* $1 \leq j \leq \nu$.

If $h$ is a holomorphic function in an open neighborhood of the origin in $\mathbb{C}^n$ and represents a Weierstrass polynomial $\mathbf{h}$ of degree $\nu$ in $z_n$, then $h$ has the form

$$h(Z) = z_n^\nu + a_1(z_1, \ldots, z_{n-1}) z_n^{\nu-1} + \cdots + a_{\nu-1}(z_1, \ldots, z_{n-1}) z_n$$

$$+ a_\nu(z_1, \ldots, z_{n-1}) \tag{2}$$

where $a_j$ are holomorphic functions of $n - 1$ variables near the origin in $\mathbb{C}^{n-1}$ and $a_j(0, \ldots, 0) = 0$ for $1 \leq j \leq \nu$; thus, $h(0, \ldots, 0, z_n) = z_n^\nu$ and $\mathbf{h}$ is regular of order $\nu$ in $z_n$. In a sense a Weierstrass polynomial is the generic form for an element of ${}_n\mathcal{O}$ that is regular in $z_n$. This can be made more precise as follows.

**4. THEOREM (Weierstrass preparation theorem).** *If* $\mathbf{f} \in {}_n\mathcal{O}$ *is regular of order* $\nu$ *in* $z_n$, *then there is a unique Weierstrass polynomial* $\mathbf{h} \in {}_{n-1}\mathcal{O}[z_n]$ *of degree* $\nu$ *in* $z_n$ *such that* $\mathbf{f} = \mathbf{u}\mathbf{h}$ *for some unit* $\mathbf{u} \in {}_n\mathcal{O}$.

*Proof.* The proof is actually just an extension of the proof of the implicit function theorem, Theorem I-C4. Let $f$ be a function that represents the germ $\mathbf{f}$, so that $f$ is regular of order $\nu$ in $z_n$ at the origin. By Lemma I-C3 there is an open polydisc $\Delta(0; R) = \Delta(0; R') \times \Delta(0; r_n) \subseteq \mathbb{C}^{n-1} \times \mathbb{C}$ such that $f$ is holomorphic in an open neighborhood of the closure $\bar{\Delta}(0; R)$ and that for any fixed point $Z' \in \bar{\Delta}(0; R')$ the function $f(Z', z_n)$ as a function of $z_n$ alone has precisely $\nu$ zeros (counting multiplicities) in the disc $\Delta(0; r_n)$ and no zeros on the boundary of that disc. Label these zeros $\phi_1(Z'), \ldots, \phi_\nu(Z')$, with repetitions according to multiplicities. The functions $\phi_j$ need not even be continuous, of course, since they are randomly chosen labelings of these sets of points; the only things that can be said of them are that $\phi_j(0) = 0$ and $|\phi_j(Z')| < r_n$ whenever $Z' \in \Delta(0; R') \subseteq \mathbb{C}^{n-1}$. However, for any fixed point $Z' \in \Delta(0; R')$ it is a familiar result from function theory in one variable that the power sums of the zeros $\phi_j(Z')$ of the function $f(Z', z_n)$ can be represented in the form

$$\sum_{j=1}^{\nu} \phi_j(Z')^r = \frac{1}{2\pi i} \int_{|\zeta| = r_n} \zeta^r \frac{\partial f(Z', \zeta)}{\partial \zeta} \frac{d\zeta}{f(Z', \zeta)} \tag{3}$$

The function $f(Z', \zeta)$ is nonzero whenever $Z' \in \Delta(0; R')$ and $|\zeta| = r_n$, so these power sums are clearly holomorphic functions of $Z'$ in $\Delta(0; R')$. Then set

$$h(Z) = \prod_{j=1}^{\nu} (z_n - \phi_j(Z'))$$

$$= z_n^\nu + a_1(Z') z_n^{\nu-1} + \cdots + a_{\nu-1}(Z') z_n + a_\nu(Z')$$

where the coefficients $a_j(Z')$ are the elementary symmetric functions of the values $\phi_1(Z'), \ldots, \phi_\nu(Z')$ and hence, as is well known, can be expressed as polynomials in the power sums (3). These functions $a_j(Z')$ are consequently holomorphic in $\Delta(0; R')$.

Moreover, $a_j(0) = 0$, since $\phi_j(0) = 0$, so that $h$ represents a Weierstrass polynomial $\mathbf{h} \in {}_{n-1}\mathcal{O}[z_n]$; this is the unique Weierstrass polynomial representing a function that has the same zeros as $f$ in the polydisc $\Delta(0; R)$. To complete the proof of the theorem it is only necessary to show that the quotient function $f/h$ is holomorphic and nowhere vanishing in $(0; R)$. For any fixed point $Z' \in \Delta(0; R') \subset \mathbb{C}^{n-1}$ the quotient $u(Z', z_n) = f(Z', z_n)/h(Z', z_n)$ is by construction a holomorphic and nowhere vanishing function of $z_n$ in $\Delta(0; r_n)$. Since $f$ is uniformly bounded on $\bar{\Delta}(0; R)$ and $h(Z', z_n)$ is bounded away from zero on the compact set $\bar{\Delta}(0; R') \times \partial\Delta(0; r_n)$, it follows further that $u$ is uniformly bounded on $\Delta(0; R') \times \partial\Delta(0; r_n)$; but the maximum modulus theorem for holomorphic functions of one variable then shows that $u$ is uniformly bounded on $\Delta(0; R') \times \Delta(0; r_n) = \Delta(0; R)$. Now the quotient $f/h$ is a holomorphic function in $\Delta(0; R)$ outside the thin set $X = \{Z \in \Delta(0; R) : h(Z) = 0\}$. But since $f/h$ is uniformly bounded in $\Delta(0; R) - X$ as just shown, it follows from the extended Riemann removable singularities theorem, Theorem I-D2, that $f/h$ extends to a holomorphic function throughout $\Delta(0; R)$. This extension must coincide with $u$ by continuity, and that concludes the proof.

It is interesting to note that the implicit function theorem, Theorem I-C4, is an immediate corollary of the Weierstrass preparation theorem. Indeed if $\mathbf{f} \in {}_n\mathcal{O}$ is regular of order one in $z_n$, then by the Weierstrass preparation theorem a representative function $f$ can be written $f(Z) = u(Z)(z_n - a_1(Z'))$, where $\mathbf{u} \in {}_n\mathcal{O}$ is a unit and $\mathbf{a}_1 \in {}_{n-1}\mathcal{O}$ is a nonunit, and this decomposition is unique. Hence, $f(Z) = 0$ for $Z$ in a neighborhood of the origin if and only if $Z = (Z', z_n)$ where $z_n = a_1(Z')$. The Weierstrass polynomials are also useful for the following result.

**5. THEOREM (Weierstrass division theorem).** *If $\mathbf{h} \in {}_{n-1}\mathcal{O}[z_n]$ is a Weierstrass polynomial of degree $\nu$, then any $\mathbf{f} \in {}_n\mathcal{O}$ can be written uniquely in the form $\mathbf{f} = \mathbf{g}\cdot\mathbf{h} + \mathbf{r}$ where $\mathbf{g} \in {}_n\mathcal{O}$ and $\mathbf{r} \in {}_{n-1}\mathcal{O}[z_n]$ is a polynomial in $z_n$ of degree $< \nu$. Moreover, if $\mathbf{f} \in {}_{n-1}\mathcal{O}[z_n]$, then $\mathbf{g} \in {}_{n-1}\mathcal{O}[z_n]$.*

*Proof.* Select functions $f$ and $h$ that are holomorphic in an open neighborhood of some closed polydisc $\bar{\Delta}(0; R)$ and represent the germs $\mathbf{f}$ and $\mathbf{h}$, respectively, and assume that the polydisc is so chosen that for any fixed point $Z' \in \bar{\Delta}'(0; R')$, the function $h(Z', z_n)$ as a function of $z_n$ alone has precisely $\nu$ zeros in $\Delta(0; r_n)$ and no zeros on the boundary of that disc. The function

$$g(Z) = \frac{1}{2\pi i}\int_{|\zeta|=r_n} \frac{f(Z', \zeta)}{h(Z', \zeta)}\frac{d\zeta}{\zeta - z_n}$$

is then clearly holomorphic in $\Delta(0; R)$, and the function $r = f - gh$ is consequently also holomorphic in $\Delta(0; R)$ and has the integral representation

$$r(Z) = \frac{1}{2\pi i}\int_{|\zeta|=r_n}\left[f(Z', \zeta) - h(Z', z_n)\frac{f(Z', \zeta)}{h(Z', \zeta)}\right]\frac{d\zeta}{\zeta - z_n}$$

$$= \frac{1}{2\pi i}\int_{|\zeta|=r_n}\frac{f(Z', \zeta)}{h(Z', \zeta)}\cdot\frac{h(Z', \zeta) - h(Z', z_n)}{\zeta - z_n}\,d\zeta$$

If the Weierstrass polynomial $\mathbf{h}$ has the form (1), then

$$\frac{h(Z',\zeta)-h(Z',z_n)}{\zeta-z_n}=\frac{(\zeta^{\nu}-z_n^{\nu})+a_1(Z')(\zeta^{\nu-1}-z_n^{\nu-1})+\cdots+a_{\nu-1}(Z')(\zeta-z_n)}{\zeta-z_n}$$

and after dividing through by $\zeta - z_n$, this expression is a polynomial in $z_n$ of degree $\nu - 1$. Since this is the only place in the formula for the function $r$ in which the variable $z_n$ appears, it follows that $r$ is necessarily a polynomial in $z_n$ of degree $\leq \nu - 1$. Next, for the uniqueness, if there are two representations $\mathbf{f} = \mathbf{gh} + \mathbf{r} = \mathbf{g}_1\mathbf{h} + \mathbf{r}_1$, then in a common polydisc $\Delta(0; R)$ with the properties as above, the representative functions must satisfy the identity $r(Z) - r_1(Z) = h(Z)[g(Z) - g_1(Z)]$. However, for each fixed $Z' \in \Delta(0; R')$ the function $h(Z', z_n)$ has $\nu$ zeros in the disc $\Delta(0; r_n)$, while the function $r(Z', z_n) - r_1(Z', z_n)$ is a polynomial in $z_n$ of degree $< \nu$. That can be the case only when $r = r_1$, and hence $g = g_1$. Finally, if $\mathbf{f} \in {}_{n-1}\mathcal{O}[z_n]$, then the familiar division theorem for polynomials shows that $\mathbf{f} = \mathbf{g}\cdot\mathbf{h} + \mathbf{r}$ where $\mathbf{g}, \mathbf{r} \in {}_{n-1}\mathcal{O}[z_n]$ and $\mathbf{r}$ is of degree $< \nu$ in $z_n$, and the uniqueness shows that this must coincide with the Weierstrass division formula. That suffices to conclude the proof of the theorem.

These two theorems can be used to derive several more properties of the rings ${}_n\mathcal{O}$. In this discussion a nonunit $\mathbf{f} \in {}_n\mathcal{O}$ is called **reducible over** ${}_n\mathcal{O}$ if it can be written as a product $\mathbf{f} = \mathbf{g}_1\mathbf{g}_2$ where $\mathbf{g}_1$ and $\mathbf{g}_2$ are nonunits of ${}_n\mathcal{O}$, and a nonunit $\mathbf{f} \in {}_n\mathcal{O}$ that is not reducible over ${}_n\mathcal{O}$ is called **irreducible over** ${}_n\mathcal{O}$. Similarly, a nonunit $\mathbf{f} \in {}_{n-1}\mathcal{O}[z_n]$ is called **reducible over** ${}_{n-1}\mathcal{O}[z_n]$ if it can be written as a product $\mathbf{f} = \mathbf{g}_1 \cdot \mathbf{g}_2$ where $\mathbf{g}_1$ and $\mathbf{g}_2$ are nonunits of ${}_{n-1}\mathcal{O}[z_n]$, and a nonunit $\mathbf{f} \in {}_{n-1}\mathcal{O}[z_n]$ that is not reducible over ${}_{n-1}\mathcal{O}[z_n]$ is called **irreducible over** ${}_{n-1}\mathcal{O}[z_n]$. Thus for elements of ${}_{n-1}\mathcal{O}[z_n]$ there are two notions of reducibility or irreducibility, which are in general somewhat different. For example, the units of ${}_{n-1}\mathcal{O}[z_n]$ are merely the units of ${}_{n-1}\mathcal{O}$; so the polynomial $z_2 - 1 \in {}_1\mathcal{O}[z_2]$ is a nonunit of ${}_1\mathcal{O}[z_2]$ but a unit in ${}_2\mathcal{O}$. However, for Weierstrass polynomials these notions are essentially the same, in the following sense.

**6. LEMMA.** *A Weierstrass polynomial* $\mathbf{h} \in {}_{n-1}\mathcal{O}[z_n]$ *is reducible over* ${}_{n-1}\mathcal{O}[z_n]$ *if and only if it is reducible over* ${}_n\mathcal{O}$. *If the Weierstrass polynomial* $\mathbf{h}$ *is reducible, its factors are also Weierstrass polynomials, up to units of* ${}_n\mathcal{O}$.

*Proof.* First if $\mathbf{h}$ is reducible over ${}_{n-1}\mathcal{O}[z_n]$, then $\mathbf{h} = \mathbf{g}_1 \cdot \mathbf{g}_2$ where $\mathbf{g}_j$ are nonunits of ${}_{n-1}\mathcal{O}[z_n]$. If, say, $\mathbf{g}_1$ is a unit in ${}_n\mathcal{O}$, then write $\mathbf{g}_2 = \mathbf{g}_1^{-1} \cdot \mathbf{h} + 0$ and apply the Weierstrass division theorem. It follows that $\mathbf{g}_1^{-1} \in {}_{n-1}\mathcal{O}[z_n]$, but that means that $\mathbf{g}_1$ is a unit in ${}_{n-1}\mathcal{O}[z_n]$, a contradiction, so $\mathbf{h}$ must be reducible over ${}_n\mathcal{O}$. Next if $\mathbf{h}$ is reducible over ${}_n\mathcal{O}$, then $\mathbf{h} = \mathbf{g}_1 \cdot \mathbf{g}_2$ where $\mathbf{g}_j$ are nonunits of ${}_n\mathcal{O}$. The germ $\mathbf{h}$ is regular in $z_n$ so both $\mathbf{g}_1$ and $\mathbf{g}_2$ must also be regular in $z_n$, and by the Weierstrass preparation theorem $\mathbf{g}_j = \mathbf{u}_j\mathbf{h}_j$ where $\mathbf{u}_j \in {}_n\mathcal{O}$ are units and $\mathbf{h}_j \in {}_{n-1}\mathcal{O}[z_n]$ are Weierstrass polynomials. Thus, $\mathbf{h} = \mathbf{u}_1\mathbf{u}_2\mathbf{h}_1\mathbf{h}_2$; but since $\mathbf{h}_1\mathbf{h}_2$ is also a Weierstrass polynomial, it follows from the uniqueness in the Weierstrass preparation theorem that $\mathbf{u}_1\mathbf{u}_2 = 1$ and $\mathbf{h} = \mathbf{h}_1 \cdot \mathbf{h}_2$. Hence, $\mathbf{h}$ is reducible over ${}_{n-1}\mathcal{O}[z_n]$ and the factors $\mathbf{h}_1$ and $\mathbf{h}_2$ are also Weierstrass polynomials. That completes the proof.

**7. THEOREM.** *The ring ${}_n\mathcal{O}$ is a unique factorization domain.*

*Proof.* The assertions of the theorem are that ${}_n\mathcal{O}$ is an integral domain with an identity element and that every nonunit of ${}_n\mathcal{O}$ can be written, uniquely up to the order of factors and to units, as a finite product of irreducible factors. The proof will be by induction on the dimension $n$. When $n = 0$, then ${}_n\mathcal{O} = \mathbb{C}$, and a field is trivially a unique factorization domain. For the inductive step assume that ${}_{n-1}\mathcal{O}$ is a unique factorization domain. Any element $\mathbf{f} \in {}_n\mathcal{O}$ can be made regular in $z_n$ by a suitable nonsingular linear change of coordinates in $\mathbb{C}^n$, and then by the Weierstrass preparation theorem $\mathbf{f} = \mathbf{uh}$, where $\mathbf{u} \in {}_n\mathcal{O}$ is a unit and $\mathbf{h} \in {}_{n-1}\mathcal{O}[Z_n]$ is a Weierstrass polynomial. It is a familiar algebraic result, Gauss's theorem, that if ${}_{n-1}\mathcal{O}$ is a unique factorization domain, then so is ${}_{n-1}\mathcal{O}[Z_n]$. Hence, $h$ can be written, uniquely up to the order of factors and to units in ${}_{n-1}\mathcal{O}[z_n]$, as a product of irreducible polynomials. It follows immediately from Lemma 6 that this provides a factorization in ${}_n\mathcal{O}$, unique up to the order of factors and to units in ${}_n\mathcal{O}$, thus concluding the proof.

**8. THEOREM.** *The ring ${}_n\mathcal{O}$ is a Noetherian ring.*

*Proof.* The assertion of the theorem is that ${}_n\mathcal{O}$ is a commutative ring with an identity element, and that every ideal in ${}_n\mathcal{O}$ has a finite basis. The proof of this too will be by induction on the dimension $n$. Again when $n = 0$, then ${}_n\mathcal{O} = \mathbb{C}$, and a field is trivially a Noetherian ring. For the inductive step assume that ${}_{n-1}\mathcal{O}$ is a Noetherian ring, and consider a nontrivial ideal $\mathfrak{A} \subseteq {}_n\mathcal{O}$. Choose any nonzero element $\mathbf{g} \in \mathfrak{A}$. This element can be made regular in $z_n$ by a suitable nonsingular linear change of coordinates in $\mathbb{C}^n$, and it follows from the Weierstrass preparation theorem that, after multiplying by a unit in ${}_n\mathcal{O}$, it can be assumed that $\mathbf{g} \in \mathfrak{A} \cap {}_{n-1}\mathcal{O}[z_n]$ is a Weierstrass polynomial. It is a familiar algebraic result, Hilbert's basis theorem, that if ${}_{n-1}\mathcal{O}$ is a Noetherian ring, then so is ${}_{n-1}\mathcal{O}[z_n]$; hence, the ideal $\mathfrak{A} \cap {}_{n-1}\mathcal{O}[z_n]$ in ${}_{n-1}\mathcal{O}[z_n]$ has a finite basis $\mathbf{g}_1, \ldots, \mathbf{g}_v$. For any element $\mathbf{f} \in \mathfrak{A}$, it follows from the Weierstrass division theorem that $\mathbf{f} = \mathbf{g} \cdot \mathbf{h} + \mathbf{r}$ for some element $\mathbf{h} \in {}_n\mathcal{O}$, where $\mathbf{r} \in {}_{n-1}\mathcal{O}[z_n]$. But since clearly $\mathbf{r} \in \mathfrak{A}$, it further follows that $\mathbf{r} = \sum_j \mathbf{g}_j \mathbf{h}_j$ for some elements $\mathbf{h}_j \in {}_{n-1}\mathcal{O}[z_n]$, and hence $\mathbf{f} = \mathbf{gh} + \sum_j \mathbf{g}_j \mathbf{h}_j$. Thus the elements $\mathbf{g}, \mathbf{g}_1, \ldots, \mathbf{g}_v$ generate the ideal $\mathfrak{A}$, and that suffices to conclude the proof of the theorem. It may be noted incidentally that actually $\mathbf{g}_1, \ldots, \mathbf{g}_v$ generate the ideal $\mathfrak{A}$.

Combining the preceding with an earlier observation leads to the result that the ring ${}_n\mathcal{O}$ is a **local ring**; that is to say, ${}_n\mathcal{O}$ is a Noetherian ring with the property that the nonunits form an ideal. The ideal of nonunits is the unique **maximal ideal** ${}_n\mathfrak{m} \subseteq {}_n\mathcal{O}$, consisting of all germs in ${}_n\mathcal{O}$ that vanish at the origin. The residue class ring ${}_n\mathcal{O}/{}_n\mathfrak{m}$ is naturally isomorphic to the field $\mathbb{C}$ of complex numbers, under the isomorphism induced by the homomorphism $\rho: {}_n\mathcal{O} \to \mathbb{C}$ that associates to any germ $\mathbf{f} \in {}_n\mathcal{O}$ the value $f(0) \in \mathbb{C}$ where $f$ is any representative of $\mathbf{f}$. That the ring ${}_n\mathcal{O}$ is a local ring is perhaps the basic general property needed in the present discussion, although of course the more detailed structural properties of this ring are also of great interest. One property of arbitrary local rings that will be used repeatedly in the sequel is the following.

**9. LEMMA (Nakayama's lemma).** *If $\mathcal{O}$ is a local ring with maximal ideal $\mathfrak{m}$ and if $\mathcal{M}$ and $\mathcal{N} \subseteq \mathcal{M}$ are modules over the ring $\mathcal{O}$ such that $\mathcal{M}/\mathcal{N}$ is finitely generated and $\mathcal{M} = \mathcal{N} + \mathfrak{m}\cdot\mathcal{M}$, then $\mathcal{N} = \mathcal{M}$.*

*Proof.* Pass to the quotient module $\mathcal{L} = \mathcal{M}/\mathcal{N}$, where it suffices to show that if $\mathcal{L}$ is a finitely generated module over the ring $\mathcal{O}$ and if $\mathcal{L} = \mathfrak{m}\cdot\mathcal{L}$, then necessarily $\mathcal{L} = 0$. If $\mathcal{L} \neq 0$, choose a minimal set of generators $X_1, \ldots, X_r$ for the module $\mathcal{L}$; thus, the elements $X_1, \ldots, X_r$ generate $\mathcal{L}$, but no proper subset of these $r$ elements serves to generate $\mathcal{L}$. Since $\mathcal{L} = \mathfrak{m}\cdot\mathcal{L}$, there must be elements $m_j \in \mathfrak{m}$ such that $X_r = \sum_{j=1}^{r} m_j X_j$—hence, such that $(1 - m_r)X_r = \sum_{j=1}^{r-1} m_j X_j$. Since $m_r \in \mathfrak{m}$ necessarily $1 - m_r \notin \mathfrak{m}$, so that $1 - m_r$ is invertible in the ring $\mathcal{O}$. But then $X_r = \sum_{j=1}^{r-1} m_j(1 - m_r)^{-1}X_j$, so that $X_1, \ldots, X_{r-1}$ actually generate the module $\mathcal{L}$, a contradiction. That suffices to conclude the proof.

Note that the proof of the preceding lemma did not really require that the local ring $\mathcal{O}$ be Noetherian; the only finiteness assumption was that the quotient module $\mathcal{M}/\mathcal{N}$ be finitely generated. If $\mathcal{O}$ is a Noetherian local ring, then all ideals in $\mathcal{O}$ are finitely generated $\mathcal{O}$-modules, so Nakayama's lemma shows that if $\mathfrak{A}$ and $\mathfrak{B}$ are any two ideals in $\mathcal{O}$ such that $\mathfrak{B} \subseteq \mathfrak{A}$ and $\mathfrak{A} = \mathfrak{B} + \mathfrak{m}\cdot\mathfrak{A}$, then necessarily $\mathfrak{B} = \mathfrak{A}$. As an application note that the intersection $\mathfrak{A} = \bigcap_{\nu=1}^{\infty} \mathfrak{m}^{\nu}$ is an ideal in $\mathcal{O}$ with the property that $\mathfrak{A} = \mathfrak{m}\cdot\mathfrak{A} = 0 + \mathfrak{m}\cdot\mathfrak{A}$ and consequently $\mathfrak{A} = 0$; similarly, of course, $\bigcap_{\nu=1}^{\infty} (\mathfrak{A} + \mathfrak{m}^{\nu}) = \mathfrak{A}$ for any ideal $\mathfrak{A}$ in $\mathcal{O}$.

In considering finitely generated modules over the rings ${}_n\mathcal{O}$, the simplest module over ${}_n\mathcal{O}$ generated by $\mu$ elements is the **free module ${}_n\mathcal{O}^{\mu}$ of rank $\mu$**, the direct sum of $\mu$ copies of the ring ${}_n\mathcal{O}$ with the usual module operations. An element $\mathbf{F} \in {}_n\mathcal{O}^{\mu}$ is a $\mu$-tuple $\mathbf{F} = (\mathbf{f}_1, \ldots, \mathbf{f}_{\mu})$ of elements $\mathbf{f}_j \in {}_n\mathcal{O}$. Addition is defined to be addition of the separate components, and the module operation is defined by $\mathbf{f}\cdot\mathbf{F} = (\mathbf{f}\mathbf{f}_1, \ldots, \mathbf{f}\mathbf{f}_{\mu})$. The $\mu$ canonical generators of the module ${}_n\mathcal{O}^{\mu}$ are the elements $\mathbf{E}_j = (0, \ldots, 0, 1, 0, \ldots, 0)$, where the entry 1 is in the $j$th place, for an element $\mathbf{F} \in {}_n\mathcal{O}^{\mu}$ as above can be written uniquely in the form $\mathbf{F} = \sum_{j=1}^{\mu} \mathbf{f}_j\mathbf{E}_j$. If $\phi\colon {}_n\mathcal{O}^{\mu} \to {}_n\mathcal{O}^{\nu}$ is a module homomorphism, the images of the canonical generators $\mathbf{E}_j$ of ${}_n\mathcal{O}^{\mu}$ are some elements

$$\phi(\mathbf{E}_j) = (\boldsymbol{\phi}_{j1}, \ldots, \boldsymbol{\phi}_{j\nu}) \in {}_n\mathcal{O}^{\nu}$$

The homomorphism $\phi$ is uniquely determined by the elements $\boldsymbol{\phi}_{ij} \in {}_n\mathcal{O}$, and any $\mu\nu$ elements of ${}_n\mathcal{O}$ determine a homomorphism $\phi$; for if $\mathbf{F} \in {}_n\mathcal{O}^{\nu}$ is an element as above, then since $\phi$ is a module homomorphism necessarily

$$\phi(\mathbf{F}) = \phi\left(\sum_j \mathbf{f}_j\mathbf{E}_j\right) = \sum_j \mathbf{f}_j\phi(\mathbf{E}_j)$$

$$= \sum_j (\mathbf{f}_j\boldsymbol{\phi}_{j1}, \ldots, \mathbf{f}_j\boldsymbol{\phi}_{j\nu})$$

There is thus a natural one-to-one correspondence between module homomorphisms $\phi\colon {}_n\mathcal{O}^{\mu} \to {}_n\mathcal{O}^{\nu}$ and $\mu \times \nu$ matrices $\{\boldsymbol{\phi}_{ij}\}$ of elements of ${}_n\mathcal{O}$. Quite arbitrary finitely generated modules over the ring ${}_n\mathcal{O}$ can then conveniently be described in terms of free modules and homomorphisms of free modules. Indeed, if $\mathcal{S}$ is an ${}_n\mathcal{O}$-module generated by some elements $s_1, \ldots, s_\mu$, then the homomorphism $\sigma\colon {}_n\mathcal{O}^\mu \to \mathcal{S}$ defined by $\sigma(\mathbf{f}_1, \ldots, \mathbf{f}_\mu) = \sum_j \mathbf{f}_j s_j$ is a surjective homomorphism—that is, has as image the entire module $\mathcal{S}$. Since ${}_n\mathcal{O}$ is a Noetherian ring, the kernel of the homomorphism $\sigma$ is a finitely generated submodule of ${}_n\mathcal{O}^\mu$ and the same construction can be applied to this module, and so on.

A customary and convenient terminology for describing constructions such as this last one involves the notion of an exact sequence of ${}_n\mathcal{O}$-modules. A sequence of ${}_n\mathcal{O}$-modules and module homomorphisms of the form

$$\cdots \to \mathcal{S}_{j+1} \xrightarrow{\sigma_{j+1}} \mathcal{S}_j \xrightarrow{\sigma_j} \mathcal{S}_{j-1} \xrightarrow{\sigma_{j-1}} \cdots \tag{4}$$

is said to be **exact at** $\mathcal{S}_j$ if the image of $\sigma_{j+1}$ is precisely the kernel of $\sigma_j$, and the whole sequence is said to be **exact** if it is exact at each module in the sequence. Denote the zero module by 0; an exact sequence of the form

$$0 \to \mathcal{R} \xrightarrow{\rho} \mathcal{S} \xrightarrow{\sigma} \mathcal{T} \to 0 \tag{5}$$

is called a **short exact sequence**. The exactness of this sequence merely amounts to the assertions that $\rho$ is injective (has zero kernel), that $\sigma$ is surjective (has as image all of $\mathcal{T}$), and that the image $\rho(\mathcal{R}) \subseteq \mathcal{S}$ is precisely the kernel of $\sigma$. Thus, (5) is an exact sequence if $\mathcal{R}$ is naturally isomorphic to its image $\rho(\mathcal{R}) \subseteq \mathcal{S}$ and $\mathcal{T}$ is isomorphic to the quotient module $\mathcal{S}/\rho(\mathcal{R})$ under the homomorphism $\sigma$. Note that any exact sequence of the form (4) can be written as a collection of short exact sequences

$$0 \to \mathcal{R}_j \xrightarrow{i} \mathcal{S}_j \xrightarrow{\sigma_j} \mathcal{R}_{j-1} \to 0$$

where $\mathcal{R}_j \subseteq \mathcal{S}_j$ is the kernel of $\sigma_j$ and $i$ is the natural inclusion homomorphism. Now for any finitely generated ${}_n\mathcal{O}$-module $\mathcal{S}$ there is as noted in the preceding paragraph a surjective homomorphism $\sigma\colon {}_n\mathcal{O}^\nu \to \mathcal{S}$. Let $\mathcal{S}_1 \subseteq {}_n\mathcal{O}^\nu$ be the kernel of $\sigma$, so there is thus an exact sequence of ${}_n\mathcal{O}$-modules of the form

$$0 \to \mathcal{S}_1 \to {}_n\mathcal{O}^\nu \xrightarrow{\sigma} \mathcal{S} \to 0$$

Since $\mathcal{S}_1$ is also finitely generated, being a submodule of a finitely generated module over the Noetherian ring ${}_n\mathcal{O}$, the same argument shows that there is also an exact sequence of ${}_n\mathcal{O}$-modules of the form

$$0 \to \mathcal{S}_2 \to {}_n\mathcal{O}^{\nu_1} \xrightarrow{\sigma_1} \mathcal{S}_1 \to 0$$

and so on. This collection of exact sequences can be rewritten as the single exact sequence

$$\cdots \to {}_n\mathcal{O}^{\nu_2} \xrightarrow{\sigma_2} {}_n\mathcal{O}^{\nu_1} \xrightarrow{\sigma_1} {}_n\mathcal{O}^{\nu} \xrightarrow{\sigma} \mathcal{S} \to 0 \tag{6}$$

**10. DEFINITION.** *A* **free resolution** *of a finitely generated ${}_n\mathcal{O}$-module $\mathcal{S}$ is an exact sequence of ${}_n\mathcal{O}$-modules of the form* (6).

The preceding observations can thus be rephrased as the assertion that any finitely generated ${}_n\mathcal{O}$-module admits a free resolution. The exactness of the sequence (4) at $\mathcal{S}_j$ is also described in an older terminology by saying that $\sigma_{j+1}$ is a **syzygy** for the homomorphism $\sigma_j$, and an exact sequence of the form (6) is called a **chain of syzygies** for the homomorphism $\sigma$. The preceding observations can be further rephrased in this terminology as the assertion that any surjective homomorphism $\sigma\colon {}_n\mathcal{O}^v \to \mathcal{S}$ admits a chain of syzygies. These results are, of course, rather trivial consequences of the result that ${}_n\mathcal{O}$ is a Noetherian ring. A deeper result, resting on deeper structural properties of the ring ${}_n\mathcal{O}$, is the following.

**11. THEOREM (Hilbert's syzygy theorem).** *In any free resolution* (6) *the kernel of the homomorphism $\sigma_{n-1}$ is a free ${}_n\mathcal{O}$-module.*

*Proof.* To simplify the notation, $\mathcal{O}$ will be used in place of ${}_n\mathcal{O}$ throughout this proof, with the dimension $n$ being fixed. Let $\mathfrak{m}_j \subseteq \mathcal{O}$ be the ideal generated by the germs $\mathbf{z}_1, \ldots, \mathbf{z}_j$ of the first $j$ coordinate functions in $\mathbb{C}^n$ for $1 \leqq j \leqq n$. Thus, $\mathfrak{m}_n$ is the maximal ideal of the local ring $\mathcal{O}$. Furthermore, let $\mathcal{S}_k \subseteq \mathcal{O}^{v_k}$ be the kernel of the homomorphism $\sigma_k$ for $k = 0, 1, \ldots$, where $\sigma_0 = \sigma$. The first step in the proof is to demonstrate that

$$\mathcal{S}_k \cap \mathfrak{m}_j \mathcal{O}^{v_k} = \mathfrak{m}_j \mathcal{S}_k \quad \text{whenever} \quad 1 \leqq j \leqq k \tag{7}$$

It is obvious that $\mathfrak{m}_j \mathcal{S}_k \subseteq \mathcal{S}_k \cap \mathfrak{m}_j \mathcal{O}^{v_k}$, so it is only necessary to show that $\mathcal{S}_k \cap \mathfrak{m}_j \mathcal{O}^{v_k} \subseteq \mathfrak{m}_j \mathcal{S}_k$ whenever $1 \leqq j \leqq k$, and that will be proved by induction on the index $j$. First consider the case $j = 1$, and suppose that $\mathbf{F} \in \mathcal{S}_k \cap \mathfrak{m}_1 \mathcal{O}^{v_k}$ for some $k \geqq 1$; thus, $\sigma_k(\mathbf{F}) = 0$ and $\mathbf{F} = \mathbf{z}_1 \mathbf{G}$ for some element $\mathbf{G} \in \mathcal{O}^{v_k}$. Now since $0 = \sigma_k(\mathbf{F}) = \mathbf{z}_1 \sigma_k(\mathbf{G}) \in \mathcal{O}^{v_{k-1}}$ and $\mathcal{O}$ is an integral domain, it follows that $\sigma_k(\mathbf{G}) = 0$ and hence that $\mathbf{G} \in \mathcal{S}_k$, and then $\mathbf{F} = \mathbf{z}_1 \mathbf{G} \in \mathfrak{m}_1 \mathcal{S}_k$ as desired. Next make the inductive hypothesis that (7) holds for some index $j \geqq 1$ and all $k \geqq j$, and suppose that $\mathbf{F} \in \mathcal{S}_k \cap \mathfrak{m}_{j+1} \mathcal{O}^{v_k}$ for some $k \geqq j + 1$. Thus, $\sigma_k(\mathbf{F}) = 0$ and

$$\mathbf{F} = \mathbf{z}_1 \mathbf{G}_1 + \cdots + \mathbf{z}_{j+1} \mathbf{G}_{j+1} \tag{8}$$

for some elements $\mathbf{G}_1, \ldots, \mathbf{G}_{j+1}$ in $\mathcal{O}^{v_k}$. Since $\sigma_k(\mathbf{F}) = 0$, it follows from (8) that

$$\mathbf{z}_{j+1} \sigma_k(\mathbf{G}_{j+1}) = -\mathbf{z}_1 \sigma_k(\mathbf{G}_1) - \cdots - \mathbf{z}_j \sigma_k(\mathbf{G}_j) \in \mathcal{O}^{v_{k-1}}$$

Thus, each monomial in the power series expansion of the components of $\mathbf{z}_{j+1} \sigma_k(\mathbf{G}_{j+1})$ must be divisible by one of the germs $\mathbf{z}_1, \ldots, \mathbf{z}_j$, so the same must hold for the components of $\sigma_k(\mathbf{G}_{j+1})$ itself. Hence,

$$\sigma_k(\mathbf{G}_{j+1}) = \mathbf{z}_1 \mathbf{G}'_1 + \cdots + \mathbf{z}_j \mathbf{G}'_j$$

for some elements $\mathbf{G}'_1, \ldots, \mathbf{G}'_j$ in $\mathcal{O}^{v_{k-1}}$. Recall that $\sigma_{k-1}\sigma_k = 0$, so this implies that

$\sigma_k(\mathbf{G}_{j+1}) \in \mathscr{S}_{k-1} \cap \mathfrak{m}_j \mathcal{O}^{\nu_{k-1}}$, and since $k-1 \geqq j$, it then follows from the inductive hypothesis that $\sigma_k(\mathbf{G}_{j+1}) \in \mathfrak{m}_j \mathscr{S}_{k-1} = \mathfrak{m}_j \sigma_k(\mathcal{O}^{\nu_k})$ and hence that

$$\sigma_k(\mathbf{G}_{j+1}) = \mathbf{z}_1 \sigma_k(\mathbf{H}_1) + \cdots + \mathbf{z}_j \sigma_k(\mathbf{H}_j)$$

for some elements $\mathbf{H}_1, \ldots, \mathbf{H}_j$ in $\mathcal{O}^{\nu_k}$. Now define

$$\mathbf{H}_{j+1} = \mathbf{G}_{j+1} - \mathbf{z}_1 \mathbf{H}_1 - \cdots - \mathbf{z}_j \mathbf{H}_j \in \mathcal{O}^{\nu_k}$$

and note that $\sigma_k(\mathbf{H}_{j+1}) = 0$ so that $\mathbf{H}_{j+1} \in \mathscr{S}_k$. Combine this with (8) and note also that

$$\mathbf{F} - \mathbf{z}_{j+1}\mathbf{H}_{j+1} = \mathbf{z}_1(\mathbf{G}_1 + \mathbf{z}_{j+1}\mathbf{H}_1) + \cdots + \mathbf{z}_j(\mathbf{G}_j + \mathbf{z}_{j+1}\mathbf{H}_j)$$

and hence that $\mathbf{F} - \mathbf{z}_{j+1}\mathbf{H}_{j+1} \in \mathscr{S}_k \cap \mathfrak{m}_j \mathcal{O}^{\nu_k}$. But after applying the inductive hypothesis once more, it follows that $\mathbf{F} - \mathbf{z}_{j+1}\mathbf{H}_{j+1} \in \mathfrak{m}_j \mathscr{S}_k$; hence, $\mathbf{F} \in \mathfrak{m}_{j+1}\mathscr{S}_k$ as desired.

The proof of the theorem itself now follows easily from the appropriate application of (7). For this purpose select a minimal set of generators $\mathbf{F}_1, \ldots, \mathbf{F}_r$ for the module $\mathscr{S}_{n-1}$. Thus, $\mathbf{F}_1, \ldots, \mathbf{F}_r$ generate the module $\mathscr{S}_{n-1}$, but no proper subset of these $r$ elements generates that module. Introduce the module homomorphism $\rho: \mathcal{O}^r \to \mathcal{O}^{\nu_{n-1}}$ defined by $\rho(\mathbf{f}_1, \ldots, \mathbf{f}_r) = \sum_j \mathbf{f}_j \mathbf{F}_j$. The image of the homomorphism $\rho$ is precisely $\mathscr{S}_{n-1}$, and to conclude the proof it is only necessary to show that the kernel $\mathscr{K}$ of this homomorphism is zero. Note first that $\mathscr{K} \subseteq \mathfrak{m}_n \mathcal{O}^r$. Indeed, if $(\mathbf{f}_1, \ldots, \mathbf{f}_r) \in \mathscr{K}$ then $\sum_j \mathbf{f}_j \mathbf{F}_j = 0$, and if one of the components $\mathbf{f}_j$ were a unit in $\mathcal{O}$ then the corresponding generator $\mathbf{F}_j$ could be written as a linear combination of the other generators with coefficients from the ring $\mathcal{O}$, contradicting the assumption that $\mathbf{F}_1, \ldots, \mathbf{F}_r$ is a minimal set of generators. Note next that replacing $\sigma_n$ by $\sigma$ in (6) leads to another free resolution

$$\mathcal{O}^r \xrightarrow{\rho} \mathcal{O}^{\nu_{n-1}} \xrightarrow{\sigma_{n-1}} \cdots \xrightarrow{\sigma_1} \mathcal{O}^{\nu} \to \mathscr{S} \to 0$$

After applying (7) to this resolution for the case $j = k = n$, it follows that $\mathscr{K} \cap \mathfrak{m}_n \mathcal{O}^r = \mathfrak{m}_n \mathscr{K}$. Since $\mathscr{K} \subseteq \mathfrak{m}_n \mathcal{O}^r$, this reduces to the identity $\mathscr{K} = \mathfrak{m}_n \mathscr{K}$, and it then follows from Nakayama's lemma that $\mathscr{K} = 0$ as desired, to conclude the proof.

**12. COROLLARY.** *For any finitely generated ${}_n\mathcal{O}$-module $\mathscr{S}$ there is a free resolution of the form*

$$0 \to {}_n\mathcal{O}^{\nu_n} \xrightarrow{\rho_n} {}_n\mathcal{O}^{\nu_{n-1}} \to \cdots \to {}_n\mathcal{O}^{\nu_1} \xrightarrow{\rho_1} {}_n\mathcal{O}^{\nu} \xrightarrow{\rho} \mathscr{S} \to 0$$

*Proof.* Since any finitely generated ${}_n\mathcal{O}$-module $\mathscr{S}$ admits some free resolution of any length, this is an immediate consequence of the preceding theorem.

# B

# Holomorphic Varieties and Subvarieties

After having established some algebraic properties of the local rings of holomorphic functions, the next topic is the study of what might be considered the geometric properties of these local rings. The geometry in question is not so much the geometry of the local rings themselves as the geometry of the sets of common zeros of ideals in these local rings. The basic notion is the following.

**1. DEFINITION.** *A* **holomorphic subvariety** *$V$ of an open set $U \subseteq \mathbb{C}^n$ is a subset $V \subseteq U$ with the property that for each point $A \in U$ there exist an open neighborhood $U_A$ of $A$ in $U$ and finitely many holomorphic functions $f_{A1}, f_{A2}, \ldots$ in $U_A$ such that*

$$V \cap U_A = \{Z \in U_A : f_{A1}(Z) = f_{A2}(Z) = \cdots = 0\}$$

The empty set and the entire set $U$ itself are trivial examples of holomorphic subvarieties of $U$. It is clear from the definition that any holomorphic subvariety is a closed subset of $U$. Furthermore, if it is assumed that $V$ is a closed subset of $U$, then the condition that $V$ be a holomorphic subvariety of $U$ can be rephrased as the condition that for each point $A \in V$, rather than for each point $A \in U$, there exist an open neighborhood $U_A$ and finitely many holomorphic functions in $U_A$ such that $V \cap U_A$ is the set of common zeros of these functions. A proper holomorphic subvariety is, of course, a thin set in $U$; hence, a proper holomorphic subvariety $V \subset U$ is a nowhere dense subset of $U$, and $U - V$ is connected if $U$ is connected. Any complex submanifold of $U$ is, of course, a holomorphic subvariety of $U$, but when $n > 1$ there are holomorphic subvarieties $V \subset U \subseteq \mathbb{C}^n$ that are not complex submanifolds. It is easy to see that the union $V_1 \cup V_2$ and the intersection $V_1 \cap V_2$ of any two holomorphic subvarieties $V_1, V_2$ of $U$ are also holomorphic subvarieties of $U$; for if $f_j, g_k$ are finitely many holomorphic functions in $U_A$ such that $V_1 \cap U_A = \{Z \in U_A : f_j(Z) = 0 \text{ for all } j\}$ and $V_2 \cap U_A = \{Z \in U_A : g_k(A) = 0 \text{ for all } k\}$, then

$$(V_1 \cup V_2) \cap U_A = \{Z \in U_A : f_j(Z)g_k(Z) = 0 \text{ for all } j \text{ and } k\}$$

$$(V_1 \cap V_2) \cap U_A = \{Z \in U_A : f_j(Z) = 0 \text{ for all } j \text{ and } g_k(Z) = 0 \text{ for all } k\}$$

This section is primarily concerned with the local properties of holomorphic subvarieties, and for this purpose it is convenient to introduce the following notion.

**2. DEFINITION.** *Holomorphic subvarieties $V_1$, $V_2$ of open neighborhoods $U_1$, $U_2$ of a point $A \in \mathbb{C}^n$ are called* **equivalent at the point** *$A$ if there is an open neighborhood $W$ of the point $A$ such that $W \subset U_1 \cap U_2$ and $V_1 \cap W = V_2 \cap W$. This is an equivalence relation in the usual sense, and an equivalence class is called a* **germ of a holomorphic subvariety at the point** *$A$.*

In general, germs (equivalence classes) of holomorphic subvarieties will be denoted by boldface letters. Any holomorphic subvariety $V$ in an open neighborhood of a point $A \in \mathbb{C}^n$ belongs to some equivalence class at $A$. This class will be called the **germ** of the holomorphic subvariety $V$ at the point $A$ and will be denoted by $\mathbf{V}$ or $\mathbf{V}_A$. If $A \notin V$, the germ $\mathbf{V}_A$ is just the germ of the empty set at the point $A$. It is of course possible to introduce the notion of the germ of an arbitrary subset, not just of a holomorphic subvariety. It should be noted that the usual set-theoretic operations can be introduced on germs of subsets by applying these operations to any representatives. Since the union and intersection of two holomorphic subvarieties are again holomorphic subvarieties, it is thus possible to speak of the union $\mathbf{V}_1 \cup \mathbf{V}_2$ and the intersection $\mathbf{V}_1 \cap \mathbf{V}_2$ of two germs of holomorphic subvarieties at a point $A \in \mathbb{C}^n$, both of which are again germs of holomorphic subvarieties at the point $A$. It is evident that biholomorphic mappings in $\mathbb{C}^n$ preserve essentially all interesting local properties of holomorphic subvarieties. To make this more precise, two germs $\mathbf{V}_1$, $\mathbf{V}_2$ of holomorphic subvarieties at points $A_1$, $A_2$ in $\mathbb{C}^n$ are called **equivalent germs of holomorphic subvarieties** if for some open neighborhoods $U_1$, $U_2$ of the respective points $A_1$, $A_2$ there exist holomorphic subvarieties $V_1 \subseteq U_1$, $V_2 \subseteq U_2$ representing the germs $\mathbf{V}_1$, $\mathbf{V}_2$ and a biholomorphic mapping $F\colon U_1 \to U_2$ such that $F(A_1) = A_2$ and $F(V_1) = V_2$. This is evidently an equivalence relation in the usual sense, and the local theory of holomorphic subvarieties really deals with these equivalence classes. However, in the general discussion of germs of holomorphic subvarieties it is probably less confusing to ignore this further equivalence relation most of the time, letting it remain understood though that these equivalence classes are the basic entities of interest here. Note, however, that for the local theory it is really sufficient merely to consider germs of holomorphic subvarieties at a single point in $\mathbb{C}^n$—say, the origin.

**3. DEFINITION.** *To each germ $\mathbf{V}$ of a holomorphic subvariety at the origin in $\mathbb{C}^n$ there is canonically associated a subset* id $\mathbf{V} \subseteq {}_n\mathcal{O}$*, called the* **ideal of the germ** $\mathbf{V}$ *and defined by* id $\mathbf{V} = \{\mathbf{f} \in {}_n\mathcal{O} :$ *for some open neighborhood $U$ of the origin there exist a holomorphic subvariety $V \subseteq U$ representing $\mathbf{V}$ and a function $f \in \mathcal{O}_U$ representing $\mathbf{f}$ such that $f(Z) = 0$ whenever $Z \in V\}$. To each ideal $\mathfrak{A} \subseteq {}_n\mathcal{O}$ there is canonically associated a germ* loc $\mathfrak{A}$ *of a holomorphic subvariety at the origin, called the* **locus of the ideal** $\mathfrak{A}$, *and defined as the germ represented by the holomorphic subvariety $V \subseteq U$ where $U$ is an open neighborhood of the origin in $\mathbb{C}^n$, $f_1, f_2, \ldots$ are finitely many holomorphic functions in $U$ such that the germs $\mathbf{f}_1, \mathbf{f}_2, \ldots$ in ${}_n\mathcal{O}$ generate $\mathfrak{A}$, and $V = \{Z \in U : f_1(Z) = f_2(Z) = \cdots = 0\}$.*

It is clear that the ideal of a germ of a holomorphic subvariety at the origin in $\mathbf{C}^n$ is a well-defined ideal in the local ring ${}_n\mathcal{O}$, independent of the choices of representatives. Recall that ${}_n\mathcal{O}$ is a Noetherian ring, so it is also clear that the locus of an ideal in ${}_n\mathcal{O}$ is a well-defined germ of a holomorphic subvariety at the origin in $\mathbf{C}^n$, independent of the various choices of representatives. Note that if $\mathbf{V}$ is the germ of the empty set, then id $\mathbf{V} = {}_n\mathcal{O}$, while if $\mathbf{V}$ is the germ of the full set $\mathbf{C}^n$, then id $\mathbf{V} = 0$. Conversely, loc ${}_n\mathcal{O}$ is the germ of the empty set and loc 0 is the germ of the full set $\mathbf{C}^n$.

**4. THEOREM.** *The following relations hold between germs of holomorphic subvarieties at the origin in $\mathbf{C}^n$ and ideals in the local ring ${}_n\mathcal{O}$:*

(i) $\mathbf{V}_1 \subseteq \mathbf{V}_2$ *implies that* id $\mathbf{V}_1 \supseteq$ id $\mathbf{V}_2$.
(ii) $\mathfrak{A}_1 \subseteq \mathfrak{A}_2$ *implies that* loc $\mathfrak{A}_1 \supseteq$ loc $\mathfrak{A}_2$.
(iii) $\mathbf{V} =$ loc id $\mathbf{V}$
(iv) $\mathfrak{A} \subseteq$ id loc $\mathfrak{A}$ *but equality does not necessarily hold.*
(v) $\mathbf{V}_1 = \mathbf{V}_2$ *if and only if* id $\mathbf{V}_1 =$ id $\mathbf{V}_2$.
(vi) $\mathfrak{A}_1 = \mathfrak{A}_2$ *implies that* loc $\mathfrak{A}_1 =$ loc $\mathfrak{A}_2$ *but not conversely.*
(vii) $\text{id}(\mathbf{V}_1 \cup \mathbf{V}_2) = (\text{id } \mathbf{V}_1) \cap (\text{id } \mathbf{V}_2) \supseteq (\text{id } \mathbf{V}_1) \cdot (\text{id } \mathbf{V}_2)$, *but the inclusion is not necessarily an equality.*
(viii) $\text{id}(\mathbf{V}_1 \cap \mathbf{V}_2) \supseteq (\text{id } \mathbf{V}_1) + (\text{id } \mathbf{V}_2)$ *but equality does not necessarily hold.*
(ix) $\text{loc}(\mathfrak{A}_1 \cdot \mathfrak{A}_2) = \text{loc}(\mathfrak{A}_1 \cap \mathfrak{A}_2) = (\text{loc } \mathfrak{A}_1) \cup (\text{loc } \mathfrak{A}_2)$
(x) $\text{loc}(\mathfrak{A}_1 + \mathfrak{A}_2) = (\text{loc } \mathfrak{A}_1) \cap (\text{loc } \mathfrak{A}_2)$

*Proof.* (i) and (ii) These results are quite obvious, so nothing further needs to be said about either.

(iii) It is obvious that $\mathbf{V} \subseteq$ loc id $\mathbf{V}$. On the other hand, there are finitely many analytic functions $f_j$ in some open neighborhood $U$ of the origin in $\mathbf{C}^n$ such that $\mathbf{V}$ is represented by the holomorphic subvariety $V = \{Z \in U : f_j(Z) = 0 \text{ for all } j\}$. The germs $\mathbf{f}_j$ generate an ideal $\mathfrak{A} \subseteq$ id $\mathbf{V}$ since $\mathbf{f}_j \in$ id $\mathbf{V}$; hence, from (ii) it follows that loc id $\mathbf{V} \subseteq$ loc $\mathfrak{A} = \mathbf{V}$.

(iv) Again it is obvious that $\mathfrak{A} \subseteq$ id loc $\mathfrak{A}$. However even for holomorphic functions of one variable this is not always an equality; for example, the ideal $\mathfrak{A} \subseteq {}_1\mathcal{O}$ generated by $z^2$ has the same locus as the ideal ${}_1\mathfrak{m}$ generated by $z$, and $\mathfrak{A} \subset {}_1\mathfrak{m}$ but $\mathfrak{A} \neq {}_1\mathfrak{m}$.

(v) This is an immediate consequence of (i) and (iii).

(vi) The example in the proof of part (iv) shows that loc $\mathfrak{A}_1 =$ loc $\mathfrak{A}_2$ does not imply $\mathfrak{A}_1 = \mathfrak{A}_2$.

(vii) The only part not really obvious is the assertion that the inclusion is not necessarily an equality. If $V_1 = \{Z \in \mathbf{C}^2 : z_1(z_1 - z_2) = 0\}$ while $V_2 = \{Z \in \mathbf{C}^2 : z_2(z_1 - z_2) = 0\}$, then

$$z_1 z_2(z_1 - z_2) \in (\text{id } \mathbf{V}_1) \cap (\text{id } \mathbf{V}_2) \subseteq {}_2\mathcal{O}_0, \quad \text{but} \quad z_1 z_2(z_1 - z_2) \notin (\text{id } \mathbf{V}_1) \cdot (\text{id } \mathbf{V}_2) \subseteq {}_2\mathcal{O}_0.$$

(viii) Again the only part not really obvious is the assertion that equality does not necessarily hold. If $V_1 = \{Z \in \mathbf{C}^2 : z_1 - z_2^2 = 0\}$ while $V_2 = \{Z \in \mathbf{C}^2 : z_1 = 0\}$, then $V_1 \cap V_2 = 0$ and hence $\text{id}(\mathbf{V}_1 \cap \mathbf{V}_2) = {}_2\mathfrak{m} \subseteq {}_2\mathcal{O}_0$; but id $\mathbf{V}_1 +$ id $\mathbf{V}_2$ is the ideal in ${}_2\mathcal{O}_0$ generated by $z_1$ and $z_2^2$.

(ix) and (x) These results are fairly obvious, and the proof is thereby concluded.

There is a certain lack of symmetry in the preceding theorem, but that is inherent in the subject; a germ of a holomorphic subvariety is completely determined by its ideal, but distinct ideals may have the same locus. This lack of symmetry can be alleviated by determining precisely which ideals $\mathfrak{A}$ in ${}_n\mathcal{O}$ are of the form $\mathfrak{A} = \text{loc}\,\mathbf{V}$ for some germ of a holomorphic subvariety $\mathbf{V}$, as will be done in section E after some more machinery is developed. But this class of ideals is not preserved by the standard algebraic operations, so the lack of symmetry cannot be altogether overcome.

**5. DEFINITION.** *A germ* $\mathbf{V}$ *of a holomorphic subvariety is* **reducible** *if* $\mathbf{V} = \mathbf{V}_1 \cup \mathbf{V}_2$ *where* $\mathbf{V}_1$ *and* $\mathbf{V}_2$ *are germs of holomorphic subvarieties and* $\mathbf{V}_j \neq \mathbf{V}$. *A germ that is not reducible is* **irreducible**.

**6. THEOREM.** *A germ* $\mathbf{V}$ *of a holomorphic subvariety is irreducible if and only if* id $\mathbf{V}$ *is prime.*

*Proof.* If $\mathbf{V} = \mathbf{V}_1 \cup \mathbf{V}_2$ where $\mathbf{V}_j \neq \mathbf{V}$, then $\text{id}\,\mathbf{V} \subset \text{id}\,\mathbf{V}_j$ and this is a proper containment. There must thus exist elements $f_j \in \text{id}\,\mathbf{V}_j$ for which $f_j \notin \text{id}\,\mathbf{V}$; but then $f_1 f_2 \in \text{id}\,\mathbf{V}_1 \cap \text{id}\,\mathbf{V}_2 = \text{id}\,\mathbf{V}$, and thus the ideal id $\mathbf{V}$ is not prime. On the other hand, if id $\mathbf{V}$ is not prime, there exist germs $\mathbf{f}_1$ and $\mathbf{f}_2$ in ${}_n\mathcal{O}$ not contained in id $\mathbf{V}$ but such that $\mathbf{f}_1 \cdot \mathbf{f}_2 \in \text{id}\,\mathbf{V}$. Choose representatives $f_1, f_2, V$ in some open subset of $\mathbb{C}^n$, so the subvarieties $V_j = \{Z \in V : f_j(Z) = 0\}$ represent germs $\mathbf{V}_1$ and $\mathbf{V}_2$ that are properly contained in $\mathbf{V}$. Since $f_1(Z) f_2(Z) = 0$ for any point $Z \in V$, that point must be contained either in $V_1$ or in $V_2$; thus, $\mathbf{V} = \mathbf{V}_1 \cup \mathbf{V}_2$ and $\mathbf{V}$ is not irreducible. That concludes the proof.

**7. THEOREM.** *Any germ* $\mathbf{V}$ *of a holomorphic subvariety can be written as a finite union* $\mathbf{V} = \mathbf{V}_1 \cup \mathbf{V}_2 \cup \cdots$, *where* $\mathbf{V}_j$ *are irreducible germs,* $\mathbf{V}_j \neq \mathbf{V}$, *and* $\mathbf{V}_j \not\subseteq \mathbf{V}_k$ *for* $j \neq k$; *the germs* $\mathbf{V}_j$ *are uniquely determined up to order.*

*Proof.* First suppose that $\mathbf{V}$ cannot be written as a finite union of irreducible germs of subvarieties. Then $\mathbf{V}$ is itself reducible, so that $\mathbf{V} = \mathbf{V}_1 \cup \mathbf{V}_2$ where $\mathbf{V}_j$ are properly contained in $\mathbf{V}$. At least one of the subvarieties $\mathbf{V}_j$ must also have the property that it cannot be written as a finite union of irreducible germs of subvarieties. If $\mathbf{V}_1$ has this property, then $\mathbf{V}_1$ is reducible, so that $\mathbf{V}_1 = \mathbf{V}_{11} \cup \mathbf{V}_{12}$ where $\mathbf{V}_{1j}$ are properly contained in $\mathbf{V}_1$. The argument can evidently be continued, yielding an infinite sequence of germs of holomorphic subvarieties

$$\mathbf{V} \supset \mathbf{V}_1 \supset \mathbf{V}_{11} \supset \mathbf{V}_{111} \supset \cdots$$

for which each containment is proper. Then

$$\text{id}\,\mathbf{V} \subset \text{id}\,\mathbf{V}_1 \subset \text{id}\,\mathbf{V}_{11} \subset \text{id}\,\mathbf{V}_{111} \subset \cdots$$

is an infinite sequence of ideals in ${}_n\mathcal{O}$ and each containment is also proper; but that is impossible since ${}_n\mathcal{O}$ is Noetherian. Thus every germ $\mathbf{V}$ can be written as a finite union of irreducible germs $\mathbf{V} = \mathbf{V}_1 \cup \mathbf{V}_2 \cup \cdots$, and by deleting any redundant terms, it can be assumed that $\mathbf{V}_j \neq \mathbf{V}$ and that $\mathbf{V}_j \not\subseteq \mathbf{V}_k$ for $j \neq k$. If $\mathbf{V} = \mathbf{V}'_1 \cup \mathbf{V}'_2 \cup \cdots$ is another irredundant representation of $V$ as a finite union of irreducible germs, then

$$\mathbf{V}_j = \mathbf{V}_j \cap \mathbf{V} = (\mathbf{V}_j \cap \mathbf{V}'_1) \cup (\mathbf{V}_j \cap \mathbf{V}'_2) \cup \cdots$$

Since $\mathbf{V}_j$ is irreducible necessarily $\mathbf{V}_j = \mathbf{V}_j \cap \mathbf{V}'_{f(j)}$ for some index $f(j)$, and hence $\mathbf{V}_j \subseteq \mathbf{V}'_{f(j)}$. The same argument applied to $\mathbf{V}'_k$ then shows that $\mathbf{V}'_k \subseteq \mathbf{V}_{g(k)}$ for some index $g(k)$. Now $\mathbf{V}_j \subseteq \mathbf{V}'_{f(j)} \subseteq \mathbf{V}_{g(f(j))}$, and since the representation is assumed irredundant, necessarily $g(f(j)) = j$ and $\mathbf{V}_j = \mathbf{V}'_{f(j)}$. Similarly, of course, $f(g(k)) = k$, so the two representations differ merely in the order of the terms. That suffices to conclude the proof.

When a germ $\mathbf{V}$ of a holomorphic subvariety is written in this way as an irredundant finite union $\mathbf{V} = \mathbf{V}_1 \cup \mathbf{V}_2 \cup \cdots$ of irreducible germs, the germs $\mathbf{V}_j$ are called the **irreducible components** or **irreducible branches** of $V$. These notions extend readily to the global case as well. A holomorphic subvariety $V$ of an open subset $U \subseteq \mathbb{C}^n$ is called **reducible** if $V = V_1 \cup V_2$ where $V_1$ and $V_2$ are holomorphic subvarieties of $U$ and $V_j \neq V$; a holomorphic subvariety that is not reducible is called **irreducible**. Obviously a holomorphic subvariety $V$ of an open subset $U \subseteq \mathbb{C}^n$ can be reducible although the germ of $V$ at each point $A \in U$ is irreducible, and $V$ can be irreducible although the germ of $V$ at some point $A \in U$ is reducible. It is also obvious that a holomorphic subvariety $V$ cannot necessarily be written as a finite union of irreducible subvarieties, although it is evident from Theorem 7 that $V$ can be written as a union of at most countably many irreducible subvarieties. These irreducible subvarieties are also uniquely determined up to order, as in Theorem 7, and are called the **irreducible components** or **irreducible branches** of $V$.

**8. DEFINITION.** *If $V$ is a holomorphic subvariety of an open subset of $\mathbb{C}^n$, a* **holomorphic function** *on $V$ is a complex-valued function $f$ defined on the point set $V$, with the property that for each point $A \in V$ there exist an open neighborhood $U_A$ of the point $A$ in the ambient space $\mathbb{C}^n$ and a holomorphic function $f_A$ in $U_A$ such that $f|U_A \cap V = f_A|U_A \cap V$.*

Thus the holomorphic functions on a holomorphic subvariety $V$ are those functions that can be extended at least locally to holomorphic functions in open subsets of the ambient space $\mathbb{C}^n$. Nothing is assumed about the compatibility of these various local extensions, so in particular it is not assumed that a holomorphic function on $V$ is the restriction to $V$ of a holomorphic function defined in an open neighborhood of the whole set $V$ in $\mathbb{C}^n$. The problem of the existence of global extensions is a nontrivial one, which will be discussed in some detail later. Note that the holomorphic functions on $V$ are necessarily continuous functions on $V$ and form a ring under pointwise addition and multiplication. This ring of functions will be denoted by ${}_V\mathcal{O}_V$. The ring ${}_V\mathcal{O}_V$ contains the complex numbers in a canonical manner, by identifying $\mathbb{C}$ with the set of constant functions on $V$; that provides on the ring ${}_V\mathcal{O}_V$ the canonical structure of an algebra over the complex numbers. Any open subset $W$ of $V$ is of course another holomorphic subvariety and so has its algebra ${}_W\mathcal{O}_W$ of holomorphic functions. But it is also natural to view these functions as the holomorphic functions on the open subset $W$ of the original subvariety $V$, and to denote this algebra by ${}_V\mathcal{O}_W$. For the local study of the properties of holomorphic functions on a holomorphic subvariety $V$ it is possible to introduce the algebra of germs of holomorphic functions at a point of $V$, paralleling the treatment of the local theory in $\mathbb{C}^n$ as discussed in the preceding section. Thus two holomorphic functions $f_1 \in {}_V\mathcal{O}_{W_1}$ and $f_2 \in {}_V\mathcal{O}_{W_2}$ in open neighborhoods $W_1$, $W_2$ of a point $A$ in $V$ are called **equivalent at the point** $A$ if there is another open neighborhood $W$ of the point $A$ in $V$ such that $W \subseteq W_1 \cap W_2$ and $f_1|W = f_2|W$. This is evidently an equivalence relation in the usual sense, and an equivalence class is called a **germ of a holomorphic function** at the point $A$ in $V$. The set of all these germs has the natural structure of a ring, or indeed of an algebra over the complex numbers. Note that this ring or algebra really depends only on the germ of the holomorphic subvariety $V$ at the point $A$.

**9. DEFINITION.** *The ring of germs of holomorphic functions at a point A of a holomorphic subvariety V is called the* **local ring** *of the subvariety V at the point A or of the germ* $\mathbf{V}_A$, *and is denoted by* ${}_V\mathcal{O}_A$ *or* ${}_{\mathbf{V}_A}\mathcal{O}$.

**10. THEOREM.** *If V is a holomorphic subvariety of an open subset in* $\mathbb{C}^n$ *and A is any point of V, then* ${}_V\mathcal{O}_A \cong {}_n\mathcal{O}_A/\text{id}\,\mathbf{V}_A$. *The ring* ${}_V\mathcal{O}_A$ *is a Noetherian local ring.*

*Proof.* It is clear that any germ $\mathbf{f}_V \in {}_V\mathcal{O}_A$ can be viewed as the restriction to $V$ of some germ $\mathbf{f} \in {}_n\mathcal{O}_A$. The restriction mapping $\mathbf{f} \to \mathbf{f}|V$ is thus a surjective ring or algebra homomorphism, and since its kernel is obviously the ideal id $\mathbf{V}_A \subseteq {}_n\mathcal{O}_A$, it follows that ${}_V\mathcal{O}_A \cong {}_n\mathcal{O}_A/\text{id}\,\mathbf{V}_A$. That ${}_V\mathcal{O}_A$ is a Noetherian local ring is an immediate consequence of this isomorphism, since ${}_n\mathcal{O}_A$ is a Noetherian local ring. The nonunits in ${}_V\mathcal{O}_A$ are precisely the germs of holomorphic functions that are zero at the point $A$. That suffices to conclude the proof.

The preceding theorem shows that the terminology introduced here is reasonably consistent, at least in the sense that the local ring ${}_V\mathcal{O}_A$ of a holomorphic variety $V$ at a point $A \in V$ is a local ring in the usual sense. The unique maximal ideal of the local ring ${}_V\mathcal{O}_A$ is the ideal ${}_V\mathfrak{m}_A \subseteq {}_V\mathcal{O}_A$ consisting of all germs of holomorphic functions vanishing at the point $A$. The quotient field ${}_V\mathcal{O}_A/{}_V\mathfrak{m}_A$ can be identified canonically with the complex numbers by the mapping that associates to each germ $\mathbf{f} \in {}_V\mathcal{O}_A$ its value $\mathbf{f}(A)$ at the point $A$. These local rings play a major role in the study of local properties of holomorphic subvarieties, as will become quite apparent as the discussion progresses; they play the same role on a subvariety $V$ as does the local ring ${}_n\mathcal{O}$ in the space $\mathbb{C}^n$. For example, it is possible to introduce the notion of a holomorphic subvariety $W$ of the holomorphic subvariety $V$ by the obvious extension of Definition 1. That actually leads to nothing new, since it is clear that such a subvariety $W$ merely corresponds to a holomorphic subvariety in $\mathbb{C}^n$ contained in the holomorphic subvariety $V$. However, there is in extension of Definition

3 a corresponding correspondence between germs of holomorphic subvarieties of a holomorphic subvariety $V$ at a point $A \in V$ and ideals in the local ring ${}_V\mathcal{O}_A$, and for this correspondence the results listed in Theorems 4, 6, and 10 continue to hold. That the algebraic properties of these local rings reflect properties of the subvarieties is well illustrated by the following result.

**11. THEOREM.** *If* $\mathbf{V}$ *is the germ of a holomorphic subvariety of* $\mathbb{C}^n$, *the local ring* ${}_{\mathbf{V}}\mathcal{O}$ *is an integral domain precisely when the germ* $\mathbf{V}$ *is irreducible.*

*Proof.* If $\mathbf{V}$ is the germ at the origin of a holomorphic subvariety $V$ of an open neighborhood of the origin in $\mathbb{C}^n$, then ${}_{\mathbf{V}}\mathcal{O} = {}_n\mathcal{O}/\text{id}\,\mathbf{V}$ by Theorem 10, so ${}_{\mathbf{V}}\mathcal{O}$ is an integral domain precisely when the ideal $\text{id}\,\mathbf{V} \subseteq {}_n\mathcal{O}$ is a prime ideal; but $\text{id}\,\mathbf{V}$ is prime precisely when the germ $\mathbf{V}$ is irreducible by Theorem 6, and that concludes the proof.

That the local ring ${}_{\mathbf{V}}\mathcal{O}$ of a germ $\mathbf{V}$ of a holomorphic subvariety is not necessarily an integral domain means that the notion of a meromorphic function on a general holomorphic subvariety is somewhat more complicated than the corresponding notion in $\mathbb{C}^n$, but the complications are actually rather minor. Recall that for any ring $\mathcal{O}$ it is possible to introduce the total quotient ring $\mathcal{M}$ of $\mathcal{O}$. $\mathcal{M}$ is the ring of all formal quotients $f/g$, where $f$ and $g$ are elements of $\mathcal{O}$ such that $g$ is not a zero divisor, with the usual notion of equivalence and the usual definition of the ring operations. On any germ $\mathbf{V}$ of a holomorphic subvariety the **ring of germs of meromorphic functions** ${}_{\mathbf{V}}\mathcal{M}$ is defined to be the total quotient ring of the local ring ${}_{\mathbf{V}}\mathcal{O}$; thus, the germs of meromorphic functions form a field only when $\mathbf{V}$ is an irreducible germ of a holomorphic subvariety. This rather algebraic construction is quite reasonable analytically. If $\mathbf{V}$ is a reducible germ of a holomorphic subvariety and so can be written as a nontrivial finite union of germs of subvarieties $\mathbf{V} = \mathbf{V}_1 \cup \cdots \cup \mathbf{V}_n$, then the elements $\mathbf{g} \in {}_{\mathbf{V}}\mathcal{O}$ that are not zero divisors are precisely those germs of functions that do not vanish identically on any irreducible component $\mathbf{V}_j$; for if $\mathbf{g}'\mathbf{g}'' = 0$ in $\mathbf{V}_j$ but neither factor is zero, then $\mathbf{V}_j$ can be written as the union of the germs of the proper holomorphic subvarieties $V_j' = \{Z \in V_j : g'(Z) = 0\}$ and $V_j'' = \{Z \in V_j : g''(Z) = 0\}$. If $\mathbf{g}$ is not a zero divisor, it does not vanish identically on any irreducible component $\mathbf{V}_j$, so the quotient function $\mathbf{f}/\mathbf{g}$ can be defined at least somewhere on $\mathbf{V}_j$. This construction will be examined in more detail later.

**Examples.** The problem of determining the precise criteria that some of the other properties of the local rings ${}_n\mathcal{O}$ also hold for the local rings ${}_{\mathbf{V}}\mathcal{O}$ of the germs of holomorphic subvarieties is generally a rather difficult one, with few results as simple as that of Theorem 10. However, it is worthwhile to show by examples some of the properties of ${}_n\mathcal{O}$ that do not always extend to the rings ${}_{\mathbf{V}}\mathcal{O}$.

First, *the local rings* ${}_{\mathbf{V}}\mathcal{O}$ *are not necessarily unique factorization rings*, even when they are integral domains. For instance, consider the holomorphic function $f(z_1, z_2) = z_2^2 - z_1^3$ of two complex variables and the holomorphic subvariety $V = \{(z_1, z_2) \in \mathbb{C}^2 : f(z_1, z_2) = 0\}$. Note that the germ $\mathbf{f} \in {}_n\mathcal{O}_0$ at the origin is irreducible. Indeed, since $\mathbf{f}$ can be viewed as a Weierstrass polynomial of degree two in ${}_1\mathcal{O}[z_2]$, if it were reducible then by Lemma A6 its factors could be assumed to be Weierstrass polynomials as well; there could be only two factors, each necessarily of degree one, leading to a factorization of the form $z_2^2 - z_1^3 = (z_2 - a_1(z_1))(z_2 - a_2(z_1))$ where $\mathbf{a}_1, \mathbf{a}_2 \in {}_1\mathcal{O}$. It would then follow that $a_1 + a_2 = 0$ and $a_1 a_2 = -z_1^3$ and hence that $\mathbf{a}_1^2 = \mathbf{z}_1^3$; but $a_1^2$ has even total order at the origin while $z_1^3$ has odd total order, a contradiction. Since $\mathbf{f}$ is an irreducible element in the unique factorization domain ${}_2\mathcal{O}$, it generates a prime ideal in ${}_2\mathcal{O}$, and it is easy to see that this ideal is precisely $\text{id}\,\mathbf{V} \subseteq {}_2\mathcal{O}$. Indeed if $g$ is a holomorphic function in an open neighborhood of the origin in $\mathbb{C}^2$ and vanishes on $V$ near the origin, then apply the Weierstrass division theorem to write $\mathbf{g} = \mathbf{fh} + \mathbf{r}$, where $\mathbf{r} \in {}_1\mathcal{O}[z_2]$ is a polynomial of degree at most one and $\mathbf{r}$ also vanishes on $\mathbf{V}$. Here $\mathbf{r}$ can be represented by a holomorphic function of the form $r(z_1, z_2) = a_0(z_1)z_2 + a_1(z_1)$ in some open neighborhood of the origin. Now for each $z_1 \neq 0$ near the origin there are two distinct values $z_2$ such that $(z_1, z_2) \in V$. But $r(z_1, z_2)$ can vanish at these two points only when $r(z_1, z_2) \equiv 0$, and hence $\mathbf{g}$ is in the ideal generated by $\mathbf{f}$ as desired. The ideal $\text{id}\,\mathbf{V} \subseteq {}_2\mathcal{O}$ is thus a prime ideal, so $\mathbf{V}$ is an irreducible germ of a holomorphic subvariety and ${}_\mathbf{V}\mathcal{O} = {}_2\mathcal{O}/\text{id}\,\mathbf{V}$ is an integral domain. Consider then the holomorphic functions $f_1 = z_1 | V$ and $f_2 = z_2 | V$ on $V$, for which $f_1^3 = f_2^2$. If the germs $\mathbf{f}_j \in {}_\mathbf{V}\mathcal{O}$ are irreducible—indeed, if only one of them is irreducible—it follows immediately that ${}_\mathbf{V}\mathcal{O}$ is not a unique factorization domain. To see that $\mathbf{f}_2$ is irreducible, suppose to the contrary that $\mathbf{f}_2 = \mathbf{g}_1\mathbf{g}_2$, and choose holomorphic functions $g_j$ in an open neighborhood of the origin in $\mathbb{C}^2$ such that $g_j | V$ represent the germs $\mathbf{g}_j$. Note that $g_1 g_2 - z_2$ vanishes on $V$ and hence represents an element in $\text{id}\,\mathbf{V}$, an element divisible by $f$. Now if $g_1 g_2 = z_2 + (z_2^2 - z_1^3)h$ for some holomorphic function $h$ in an open neighborhood of the origin in $\mathbb{C}^2$, then the total order of the product $g_1 g_2$ at the origin must be one. But $g_j$ are nonunits so each has order at least one; hence, the order of the product $g_1 g_2$ is at least two, a contradiction. Thus $\mathbf{f}_2$ is irreducible as desired.

Next, *the Hilbert syzygy theorem does not necessarily hold in the local rings* ${}_\mathbf{V}\mathcal{O}$, in the strong sense that there can be infinitely long exact sequences of ${}_\mathbf{V}\mathcal{O}$-modules of the form

$$\cdots \rightarrow {}_\mathbf{V}\mathcal{O}^{\nu_2} \xrightarrow{\sigma_2} {}_\mathbf{V}\mathcal{O}^{\nu_1} \xrightarrow{\sigma_1} {}_\mathbf{V}\mathcal{O}^{\nu} \xrightarrow{\sigma_0} \mathcal{S} \rightarrow 0$$

with the property that for no value $j$ is the kernel of the homomorphism $\sigma_j$ a free ${}_\mathbf{V}\mathcal{O}$-module. For instance, consider again the holomorphic subvariety $V = \{(z_1, z_2) \in \mathbb{C}^2 : z_1^3 - z_2^2 = 0\}$ of the preceding example. The holomorphic mapping $\phi: \mathbb{C}^1 \rightarrow \mathbb{C}^2$ that takes a point $t \in \mathbb{C}^1$ to the point $(t^2, t^3) \in \mathbb{C}^2$ is clearly a one-to-one mapping between $\mathbb{C}^1$ and the subvariety $V \subseteq \mathbb{C}^2$. It induces an injective homomorphism $\phi^*: {}_\mathbf{V}\mathcal{O} \rightarrow {}_1\mathcal{O}$ and hence can be viewed as imbedding the local ring ${}_\mathbf{V}\mathcal{O}$ of the germ of the subvariety $V$ at the origin as a subring $\mathcal{R} = \phi^*({}_\mathbf{V}\mathcal{O}) \subseteq {}_1\mathcal{O}$. Indeed, if $h$ is a holomorphic function in an open neighborhood of the origin in $\mathbb{C}^2$, then the image under the homomorphism $\phi^*$ of the germ of the function $h | V$ is the germ of the holomorphic function $h(t^2, t^3)$; thus, the subring $\mathcal{R} \subseteq {}_1\mathcal{O}$ consists of all convergent power series that can be written in the form $h(t^2, t^3)$, where $h$ is a convergent power series in two variables. If $h(z_1, z_2) = \sum h_{j_1 j_2} z_1^{j_1} z_2^{j_2}$, then $h(t^2, t^3) = h_{00} + h_{10}t^2 + h_{01}t^3 + \cdots$, so that the power series $h(t^2, t^3)$ contains no linear term. On the other hand, it is easy to see that any power series in ${}_1\mathcal{O}$ containing no linear term can be written in the form $h(t^2, t^3)$; indeed, if $g(t) \in {}_1\mathcal{O}$ is such a power series the terms of an even order in $g$ can be viewed as of the form $h_1(t^2)$, the terms of an odd order congruent to 0 modulo 3 can be viewed as of the form $h_2(t^3)$, the terms of an odd order congruent to 1 modulo 3 (all of which are necessarily of order strictly greater than one) can be viewed as of the form $t^4h_3(t^3)$, and the terms of an odd order congruent to 2 modulo 3 can be viewed as of the form $t^2h_4(t^3)$. Thus, $\mathscr{R} \subseteq {}_1\mathcal{O}$ is the subring consisting of all convergent power series in one variable that contain no linear term. The maximal ideal ${}_V\mathfrak{m}$ in the local ring ${}_V\mathcal{O}$ corresponds to the ideal $\mathfrak{m} \subseteq \mathscr{R}$ consisting of those power series in $\mathscr{R}$ that have no constant terms. This ideal $\mathfrak{m}$ cannot be generated by a single element; indeed, if $m(t) = \sum_j m_j t^j \in \mathfrak{m}$ and $g(t) = \sum_j g_j t^j \in \mathscr{R}$, then $g(t)m(t) = g_0 m_2 t^2 + g_0 m_3 t^3 + \cdots$, so not both $t^2$ and $t^3$ can be represented as such a product for a fixed element $m(t)$. This shows that $\mathfrak{m}$ is not a free module over the ring $\mathscr{R}$, since the only ideals that can be free modules are obviously those ideals generated by a single element. On the other hand, $\mathfrak{m}$ is generated by the two elements $t^2$ and $t^3$, since the germs at the origin of the coordinate functions $z_1$ and $z_2$ generate the maximal ideal in ${}_2\mathcal{O}$. There is thus a surjective module homomorphism $\sigma_0: \mathscr{R}^2 \to \mathfrak{m}$, defined by $\sigma_0(f(t), g(t)) = t^2 f(t) + t^3 g(t)$. If $(f(t), g(t))$ is in the kernel of $\sigma_0$, then write $f(t) = \sum_j f_j t^j$ and $g(t) = \sum_j g_j t^j$, and it follows from the identity $t^2 f(t) + t^3 g(t) = 0$ that $f_0 = g_0 = f_2 = 0$ and $f_{j+1} = -g_j$ whenever $j \geqq 2$. Thus, equivalently $(f(t), g(t))$ is in the kernel of $\sigma_0$ precisely when $g(t) \in \mathfrak{m}$ and $f(t) = -tg(t)$; hence, the kernel of $\sigma_0$ is evidently isomorphic to $\mathfrak{m}$. The same argument can then be repeated, yielding an exact sequence of $\mathscr{R}$-modules

$$\cdots \to \mathscr{R}^2 \xrightarrow{\sigma_1} \mathscr{R}^2 \xrightarrow{\sigma_0} \mathfrak{m} \to 0$$

in which the kernel of the homomorphism $\sigma_j$ is isomorphic to $\mathfrak{m}$ and is thus not a free module for any index $j \geqq 0$, as desired. It should perhaps also be mentioned, although it will not be demonstrated here, that something more is actually true. For any free resolution of this module, not just for the particular resolution chosen here, the kernels of the homomorphisms $\sigma_j$ are not free modules.

To turn next to the investigation of the extent to which the properties of the local ring ${}_V\mathcal{O}$ reflect the properties of the germ $\mathbf{V}$ of a holomorphic subvariety, it is useful to introduce a natural extension of Definition 8. Note that if $V$, $W$ are holomorphic subvarieties of open subsets $U_V \subseteq \mathbb{C}^m$, $U_W \subseteq \mathbb{C}^n$, then any mapping $F: V \to W$ is completely determined by its $n$ coordinate functions. Write $F(Z) = (f_1(Z), \ldots, f_n(Z))$ for any point $Z \in V$, where the coordinate functions $f_j$ are well-defined complex-valued functions on the point set $V$. It is evident that the mapping $F$ is continuous precisely when its coordinate functions $f_j$ are continuous functions on $V$.

**12. DEFINITION.** *A mapping $F: V \to W$ between holomorphic subvarieties $V$, $W$ of open subsets $U_V \subseteq \mathbb{C}^m$, $U_W \subseteq \mathbb{C}^n$ is called a* **holomorphic mapping** *if its coordinate functions are holomorphic functions on $V$. The mapping $F$ is called a* **biholomorphic mapping** *if it is holomorphic and has a holomorphic inverse.*

A holomorphic mapping $F: V \to W$ between two holomorphic subvarieties is of course a continuous mapping, so whenever $f$ is a continuous complex-valued function on $W$, the composition $f \circ F$ is a continuous complex-valued function on $V$. Thus, $F$ induces a natural mapping $F^*: {}_W\mathscr{C}_W \to {}_V\mathscr{C}_V$ from the algebra ${}_W\mathscr{C}_W$ of continuous complex-valued functions on $W$ into the algebra ${}_V\mathscr{C}_V$ of continuous complex-valued functions on $V$, by defining $F^*(f) = f \circ F$ for any $f \in {}_W\mathscr{C}_W$. It is obvious that the mapping $F^*$ is a homomorphism of complex algebras. If the mapping $F$ takes a point $A \in V$ to a point $F(A) = B \in W$, then $F$ induces locally a corresponding algebra homomorphism $F^*: {}_W\mathscr{C}_B \to {}_V\mathscr{C}_A$ from the algebra ${}_W\mathscr{C}_B$ of germs of continuous complex-valued functions at the point $B$ in $W$ into the algebra ${}_V\mathscr{C}_A$ of germs of continuous complex-valued functions at the point $A$ in $V$. By restricting these homomorphisms to the subalgebras ${}_W\mathcal{O}_W \subseteq {}_W\mathscr{C}_W$ and ${}_W\mathcal{O}_B \subseteq {}_W\mathscr{C}_B$ of holomorphic functions and their germs, there result homomorphisms $F^*: {}_W\mathcal{O}_W \to {}_V\mathscr{C}_V$ and $F^*: {}_W\mathcal{O}_B \to {}_V\mathscr{C}_A$. As might be expected, the images of these homomorphisms are contained in ${}_V\mathcal{O}_V \subseteq {}_V\mathscr{C}_V$ and ${}_V\mathcal{O}_A \subseteq {}_V\mathscr{C}_A$, so that there actually arise the homomorphisms $F^*: {}_W\mathcal{O}_W \to {}_V\mathcal{O}_V$ and $F^*: {}_W\mathcal{O}_B \to {}_V\mathcal{O}_A$; but even more is true.

**13. THEOREM.** *If $V, W$ are holomorphic subvarieties of open subsets $U_V \subseteq \mathbb{C}^m$, $U_W \subseteq \mathbb{C}^n$ and $F: V \to W$ is a continuous mapping taking a point $A \in V$ to a point $F(A) = B \in W$, then $F$ is holomorphic in an open neighborhood of $A$ in $V$ if and only if $F^*({}_W\mathcal{O}_B) \subseteq {}_V\mathcal{O}_A$.*

*Proof.* If $F$ is holomorphic in an open neighborhood of $A$ in $V$, then the coordinate functions of $F$ are holomorphic in that neighborhood and so can be extended to holomorphic functions in a full open neighborhood $U_A$ of the point $A$ in the ambient space $\mathbb{C}^n$. That provides a holomorphic mapping $\tilde{F}: U_A \to \mathbb{C}^n$ such that $\tilde{F}|U_A \cap V = F|U_A \cap V$. Now if $\mathbf{f} \in {}_W\mathcal{O}_B$, then there is a holomorphic function $\tilde{f}$ in an open neighborhood $U_B$ of the point $B$ in the ambient space $\mathbb{C}^n$ such that $\tilde{f}|U_B \cap W$ represents the germ $\mathbf{f}$. But then $F^*(\tilde{f}|U_B \cap W) = \tilde{F}^*(\tilde{f})|U_A \cap V$ represents the germ $F^*(\mathbf{f})$ and hence $F^*(\mathbf{f}) \in {}_V\mathcal{O}_A$ as desired. On the other hand, if $F^*({}_W\mathcal{O}_B) \subseteq {}_V\mathcal{O}_A$ and if $f_j$ are the coordinate functions of the mapping $F$, then $f_j = F^*(w_j)$ where $w_j \in {}_W\mathcal{O}_W$ are the restrictions to $W$ of the coordinate functions in the ambient space $\mathbb{C}^n$. Thus, $\mathbf{f}_j = F^*(\mathbf{w}_j) \in F^*({}_W\mathcal{O}_B) \subseteq {}_V\mathcal{O}_A$, which means that the functions $f_j$ are holomorphic near $A$ and hence that the mapping $F$ is also holomorphic near $A$ as desired. That suffices to complete the proof.

If $F: V \to W$ is a holomorphic mapping between holomorphic subvarieties and $F(A) = B \in W$ for some point $A \in V$, then there results the homomorphism $F^*: {}_W\mathcal{O}_B \to {}_V\mathcal{O}_A$ of complex algebras. This purely algebraic entity already serves to determine the mapping $F$ near $A$ as a consequence of the following extension of the preceding theorem.

**14. THEOREM.** *If $V, W$ are holomorphic subvarieties of open subsets $U_V \subseteq \mathbb{C}^m$, $U_W \subseteq \mathbb{C}^n$ and if $\phi: {}_W\mathcal{O}_B \to {}_V\mathcal{O}_A$ is an algebra homomorphism between the local rings of these subvarieties at points $A \in V, B \in W$, then there is a unique holomorphic mapping $F: V \to W$ in some open neighborhood of $A$ such that $F(A) = B$ and $F$ induces the homomorphism $\phi$.*

*Proof.* Any ring homomorphism $\phi: {}_W\mathcal{O}_B \to {}_V\mathcal{O}_A$ clearly maps units in ${}_W\mathcal{O}_B$ to units in ${}_V\mathcal{O}_A$. An algebra homomorphism also maps nonunits in ${}_W\mathcal{O}_B$ to nonunits in ${}_V\mathcal{O}_A$, or equivalently satisfies $\phi({}_W\mathfrak{m}_B) \subseteq {}_V\mathfrak{m}_A$. To see this, suppose to the contrary that there is an element $\mathbf{f} \in {}_W\mathfrak{m}_B$ such that $\phi(\mathbf{f}) \notin {}_V\mathfrak{m}_A$. Thus, $\mathbf{f}$ is the germ of a holomorphic function on $W$ vanishing at $B$, while $\phi(\mathbf{f})$ is the germ of a holomorphic function on $V$ taking some value $c \neq 0$ at $A$. The element $\mathbf{f} - c \in {}_W\mathcal{O}_B$ is then a unit in ${}_W\mathcal{O}_B$, so $\phi(\mathbf{f} - c) = \phi(\mathbf{f}) - c$ is necessarily a unit in ${}_V\mathcal{O}_A$; but that is a contradiction, since $\phi(\mathbf{f}) - c$ vanishes at $A$. It thus follows that $\phi({}_W\mathfrak{m}_B) \subseteq {}_V\mathfrak{m}_A$, and as a consequence of this actually $\phi({}_W\mathfrak{m}_B^\nu) \subseteq {}_V\mathfrak{m}_A^\nu$ for every positive integer $\nu$.

If $w_1, \ldots, w_n$ are the coordinate functions in $\mathbb{C}^n$, the restrictions $\mathbf{w}_j | W$ of the germs $\mathbf{w}_j \in {}_n\mathcal{O}_B$ are elements of the local ring ${}_W\mathcal{O}_B$, and there exist some holomorphic functions $f_j$ in an open neighborhood of the point $A$ in $\mathbb{C}^m$ such that $\phi(\mathbf{w}_j | W) = \mathbf{f}_j | V$ where $\mathbf{f}_j \in {}_m\mathcal{O}_A$. The functions $f_j$ can be taken as the components of a holomorphic mapping $F$ from an open neighborhood of the point $A$ in $\mathbb{C}^m$ into the space $\mathbb{C}^n$. Since $\mathbf{w}_j | W - w_j(B) \in {}_W\mathfrak{m}_B$, it follows from the observations made in the preceding paragraph that $\mathbf{f}_j | V - w_j(B) = \phi(\mathbf{w}_j | W - w_j(B)) \in {}_V\mathfrak{m}_A$ and hence that $f_j(A) = w_j(B)$. Thus, the mapping $F$ has the property that $F(A) = B$. Now the mapping $F$ induces an algebra homomorphism $F^*: {}_n\mathcal{O}_B \to {}_m\mathcal{O}_A$, and the composition of this homomorphism and the restriction to $V$ is a homomorphism $\tilde{F}^*: {}_n\mathcal{O}_B \to {}_V\mathcal{O}_A$. On the other hand, the composition of the restriction to $W$ and the homomorphism $\phi$ is also a homomorphism $\tilde{\phi}: {}_n\mathcal{O}_B \to {}_V\mathcal{O}_A$. Note that $\tilde{F}^*(\mathbf{w}_j) = \mathbf{f}_j | V = \phi(\mathbf{w}_j | W) = \phi(\mathbf{w}_j)$, so the homomorphisms $\tilde{F}^*$ and $\tilde{\phi}$ agree on the germs of the coordinate functions $w_j$ and they must then of course agree on any polynomial in the variables $w_j$. Note further that for any positive integer $\nu$, an element $\mathbf{h} \in {}_n\mathcal{O}_B$ can be written as the sum $\mathbf{h} = \mathbf{h}' + \mathbf{h}''$, where $\mathbf{h}'$ is a polynomial in the variables $w_j$ and $\mathbf{h}'' \in {}_n\mathfrak{m}_B^\nu$; upon recalling the observation of the preceding paragraph it follows that $\tilde{F}^*(\mathbf{h}) - \tilde{\phi}(\mathbf{h}) = \tilde{F}^*(\mathbf{h}'') - \tilde{\phi}(\mathbf{h}'') \in {}_V\mathfrak{m}_A^\nu$. Since this holds for every positive integer $\nu$, it follows that $\tilde{F}^*(\mathbf{h}) - \tilde{\phi}(\mathbf{h}) \in \bigcap_\nu {}_V\mathfrak{m}_A^\nu$; but since ${}_V\mathcal{O}_A$ is a Noetherian local ring, it follows from Nakayama's lemma, Lemma A9, that $\bigcap_\nu {}_V\mathfrak{m}_A^\nu = 0$ and hence $\tilde{F}^* = \tilde{\phi}$ as homomorphisms. Now since $\text{id}\, \mathbf{W} \subseteq {}_n\mathcal{O}_B$ is in the kernel of $\tilde{\phi}$ by the definition of that homomorphism and since $\tilde{F}^* = \tilde{\phi}$, it follows that id $W$ is also in the kernel of $\tilde{F}^*$. Therefore whenever $\mathbf{f} \in \text{id}\, \mathbf{W}$ necessarily $0 = \tilde{F}^*(\mathbf{f}) = \mathbf{f} \circ \mathbf{F} | V$, or equivalently $f$ vanishes on $F(V)$ in an open neighborhood of $B$. Hence after restricting the mapping $F$ to a sufficiently small open neighborhood of $A$, it must be the case that $F(V) \subseteq W$, so that $F$ defines a holomorphic mapping ${}_VF: V \to W$. For any germ ${}_W\mathbf{f} \in {}_W\mathcal{O}_B$ choose a germ $\mathbf{f} \in {}_n\mathcal{O}_B$ such that $\mathbf{f} | W = {}_W\mathbf{f}$ and note that $F^*({}_W\mathbf{f}) = \tilde{F}^*(\mathbf{f}) = \tilde{\phi}(\mathbf{f}) = \phi({}_W\mathbf{f})$; thus, $\phi$ is just the homomorphism induced by the holomorphic mapping ${}_VF$. The only extent to which the holomorphic mapping $F$ itself is not uniquely determined is merely the choice of different holomorphic functions $f_j$ near $A$ such that $\mathbf{f}_j | V = \phi(\mathbf{w}_j | W)$. But any such choices yield the same germ of the holomorphic mapping ${}_VF$ upon restriction to $V$, and thus ${}_VF$ is uniquely determined. That suffices to conclude the proof.

It should be pointed out that in the preceding theorem it is really not sufficient merely to assume that $\phi$ is a homomorphism of rings. The argument used in the proof required that the homomorphism $\phi$ be the identity on the canonical subrings $\mathbb{C} \subseteq {}_V\mathcal{O}_A$, and indeed any ring homomorphism induced by a holomorphic mapping is necessarily the identity on the canonical subrings of constant functions. However, there are ring homomorphisms that do not have this property and hence are not induced by holomorphic mappings. The homomorphism $\phi\colon {}_1\mathcal{O} \to {}_1\mathcal{O}$ which sends a germ $\mathbf{f} \in {}_1\mathcal{O}$ having the power series expansion $f(z) = \sum_j c_j z^j$ to the germ $\mathbf{g} = \phi(\mathbf{f}) \in {}_1\mathcal{O}$ having the power series expansion $g(z) = \sum_j \bar{c}_j z^j$, where $\bar{c}_j$ is the complex conjugate of $c_j$, is perhaps the simplest example. The preceding theorem does show, however, that the local rings ${}_V\mathcal{O}$, viewed as complex algebras, really do determine the germs of holomorphic subvarieties up to biholomorphic mapping.

**15. COROLLARY.** *Two germs of holomorphic subvarieties $\mathbf{V}_A$ and $\mathbf{W}_B$ at points $A$ and $B$ in $\mathbb{C}^n$ are equivalent germs of holomorphic subvarieties if and only if there is an algebra isomorphism $\phi\colon {}_n\mathcal{O}_A \to {}_n\mathcal{O}_B$ such that $\phi(\text{id } \mathbf{V}_A) = \text{id } \mathbf{W}_B$.*

*Proof.* By definition the germs $\mathbf{V}_A$ and $\mathbf{W}_B$ are equivalent if and only if there exists a biholomorphic mapping $F$ from an open neighborhood of $A$ in $\mathbb{C}^n$ to an open neighborhood of $B$ in $\mathbb{C}^n$ such that $F(V) = W$, where $V$ and $W$ are representative subvarieties. The condition that $F(V) = W$ is equivalent to the condition that $F^*(\text{id } \mathbf{W}_A) = \text{id } \mathbf{W}_B$ in view of Theorem 4. The desired result is thus an immediate consequence of the preceding theorem.

This observation indicates that in a sense the local study of holomorphic subvarieties in $\mathbb{C}^n$, or more properly the study of equivalence classes of germs of holomorphic subvarieties in $\mathbb{C}^n$, can be reduced to the purely algebraic study of the local ring ${}_n\mathcal{O}$, for there is a natural one-to-one correspondence between equivalence classes of germs of holomorphic subvarieties in $\mathbb{C}^n$ and ideals $\mathfrak{A} \subseteq {}_n\mathcal{O}$ with the property that $\mathfrak{A} = \text{id loc } \mathfrak{A}$. To claim that this is a reduction to a purely algebraic matter requires a purely algebraic criterion for determining those ideals $\mathfrak{A} \subseteq {}_n\mathcal{O}$ with the property that $\mathfrak{A} = \text{id loc } \mathfrak{A}$; that will be obtained in section E. From the point of view of complex analysis, though, the interest lies not so much in the reduction of the local theory to a purely algebraic one as in the interplay between the algebra, geometry, and analysis involved.

There is, however, another slightly different matter that should be discussed first. Note that any two holomorphic subvarieties that are biholomorphic, in the sense that there is a biholomorphic mapping between them, are homeomorphic topological spaces but can be quite different as holomorphic subvarieties; they can, for instance, be holomorphic subvarieties of spaces $\mathbb{C}^n$ of quite different dimensions. On the other hand, it is apparent from Theorem 13 that such subvarieties have quite similar analytic properties, since their algebras of holomorphic functions are isomorphic and their local rings at corresponding points are also isomorphic. For example, two complex submanifolds are biholomorphic precisely when they are equivalent as complex manifolds. That suggests introducing biholomorphic equivalence classes of holomorphic subvarieties as new entities; more generally, it suggests introducing the analogue of a complex manifold as follows.

**16. DEFINITION.** *A* ***holomorphic variety*** *is a second-countable Hausdorff topological space $V$ for which there exist a covering by open subsets $V_\alpha$ and homeomorphisms $F_\alpha\colon V_\alpha \to W_\alpha$*

*between the subsets $V_\alpha \subseteq V$ and holomorphic subvarieties $W_\alpha$ of open sets $U_\alpha \subseteq \mathbb{C}^{n_\alpha}$, such that for each nonempty intersection $V_\alpha \cap V_\beta$ the composition*

$$F_{\alpha\beta} = F_\alpha \circ F_\beta^{-1} : F_\beta(V_\alpha \cap V_\beta) \to F_\alpha(V_\alpha \cap V_\beta)$$

*is a biholomorphic mapping.*

It is apparent from this definition that a complex manifold is a special case of a holomorphic variety, and moreover that the notion of a holomorphic variety is quite closely patterned upon that of a complex manifold. To emphasize this parallelism, with the notation as introduced in the preceding definition, $\{V_\alpha, F_\alpha\}$ will be called a **coordinate covering** of the holomorphic variety $V$, the subsets $V_\alpha \subset V$ being called the **coordinate neighborhoods** and the mappings $F_\alpha$ being called the **coordinate mappings**, and the mappings $F_{\alpha\beta}$ will be called the **coordinate transition mappings** for the given coordinate covering. A holomorphic variety can then be represented locally by a holomorphic subvariety of an open subset of some space $\mathbb{C}^n$, with the coordinate neighborhood $V_\alpha$ being represented by the holomorphic subvariety $W_\alpha = F_\alpha(V_\alpha) \subseteq U_\alpha \subseteq \mathbb{C}^{n_\alpha}$, for example. The local representative subvariety is of course determined only up to a biholomorphic mapping. Note particularly that it is not assumed that the entire holomorphic variety $V$ can be represented by a holomorphic subvariety; indeed, that is not always the case, as will later be established by simple examples. After having introduced the notion of a holomorphic variety in this manner, it is obvious how to introduce the notion of a **germ of a holomorphic variety**. Of course, any germ of a holomorphic variety can be represented by a germ of a holomorphic subvariety of some space $\mathbb{C}^n$. Indeed, two germs of holomorphic subvarieties can be considered **equivalent as germs of holomorphic varieties** if they have biholomorphic representative subvarieties; this is obviously an equivalence relation in the usual sense, and an equivalence class can be identified with a germ of a holomorphic variety. This is a weaker equivalence relation than that of equivalence as germs of holomorphic subvarieties, as considered earlier in this section. As usual, germs of holomorphic varieties will be denoted by boldface letters.

Since the rings of holomorphic functions on holomorphic subvarieties are preserved under biholomorphic mappings, it is possible to introduce the notion of a holomorphic function on a holomorphic variety. Formally a complex-valued function $f$ on a holomorphic variety $V$ is called a **holomorphic function** on $V$ if for each coordinate neighborhood $V_\alpha$ of the coordinate covering $\{V_\alpha, F_\alpha\}$, the function $(f \mid V_\alpha) \circ F_\alpha^{-1}$ is a holomorphic function on the holomorphic subvariety $F_\alpha(V_\alpha)$. The holomorphic functions are necessarily continuous functions and form a ring under pointwise addition and multiplication; this ring also will be denoted by ${}_V\mathcal{O}_V$. The ring ${}_V\mathcal{O}_V$ contains the complex numbers in a canonical way, by identifying $\mathbb{C}$ with the set of constant functions on $V$, and so ${}_V\mathcal{O}_V$ has the natural structure of an algebra over the complex numbers. Of course, the ring ${}_V\mathcal{O}_V$ may reduce to the field $\mathbb{C}$ of constants, as for instance when $V$ is a compact complex manifold. Similarly it is possible to introduce for any open subset $W \subseteq V$ the algebra ${}_V\mathcal{O}_W$ of holomorphic functions on that subset of $V$, and for any point $A \in V$ the algebra ${}_V\mathcal{O}_A$ of germs of holomorphic functions at the point $A$ on $V$. The ring ${}_V\mathcal{O}_A$ will be called the **local ring** of the holomorphic variety $V$ at the point $A$. It is apparent that this local ring really depends only on the germ of the holomorphic variety $V$ at the point $A$; so it can also be considered the local ring of the germ $\mathbf{V}_A$ and will correspondingly be denoted by ${}_{\mathbf{V}_A}\mathcal{O}$. The ring ${}_V\mathcal{O}_A = {}_{\mathbf{V}_A}\mathcal{O}$ can of course be identified with the local ring of any representative subvariety and so has the properties already established for these local rings. It is possible to speak of holomorphic subvarieties of holomorphic varieties, extending Definitions 1, 2, and 3 and all the results of Theorems 4, 6, and 10 in a straightforward manner. Nothing really new results, as in the case of the corresponding extension of holomorphic subvarieties in $\mathbb{C}^n$, but it is a very convenient terminology and way of thinking and will be used freely whenever convenient.

Similarly it is possible to introduce the notion of a holomorphic mapping between two holomorphic varieties. Formally a mapping $H: V \to W$ between two holomorphic varieties $V$ and $W$ with coordinate coverings $\{V_\alpha, F_\alpha\}$ and $\{W_\beta, G_\beta\}$ is a **holomorphic mapping** if whenever $A \in V_\alpha \subseteq V$ and $B = H(A) \in W_\beta \subseteq W$, the composition $G_\beta \circ H \circ F_\alpha^{-1}$ is a holomorphic mapping from an open neighborhood of $F_\alpha(A)$ in the holomorphic subvariety $F_\alpha(V_\alpha)$ into an open neighborhood of $G_\beta(B)$ in the holomorphic subvariety $G_\beta(W_\beta)$. The mapping $H$ is a biholomorphic mapping if $H$ is holomorphic and has a holomorphic inverse. It follows immediately from Theorem 13 that a continuous mapping $F: V \to W$ between two holomorphic varieties is holomorphic precisely when $F^*({}_W\mathcal{O}_B) \subseteq {}_V\mathcal{O}_A$ whenever $A$ is a point of $V$ with image $B = F(A) \in W$. Two holomorphic varieties that are biholomorphic, in the sense that there is a biholomorphic mapping between them, will be called **equivalent holomorphic varieties**; this is an equivalence relation in the usual sense, and it is clearly only these equivalence classes of holomorphic varieties that are really of interest. Again though in the general discussion it is probably less confusing just to ignore this equivalence relation, letting it remain understood implicitly rather than stated explicitly. However, it is worth pointing out explicitly in this context a further consequence of Theorem 14.

**17. COROLLARY.** *Two germs of holomorphic varieties are equivalent if and only if their local rings are isomorphic as complex algebras.*

*Proof.* By definition two germs of holomorphic varieties $\mathbf{V}_A$ and $\mathbf{W}_B$ are equivalent if and only if there exists a biholomorphic mapping $F: V \to W$ with $F(A) = B$, where $\mathbf{V}_A$ is the germ of the holomorphic variety $V$ at the point $A \in V$ and $\mathbf{W}_B$ is the germ of the holomorphic variety $W$ at the point $B \in W$. Any such biholomorphic mapping evidently induces an algebra isomorphism $F^*: {}_W\mathcal{O}_B \to {}_V\mathcal{O}_A$, and it follows directly from Theorem 14 that any algebra isomorphism $F^*: {}_W\mathcal{O}_B \to {}_V\mathcal{O}_A$ is induced by some biholomorphic mapping between an open neighborhood of $A$ in $V$ and an open neighborhood of $B$ in $W$. That suffices to conclude the proof.

Finally, one further construction involving holomorphic varieties should be introduced here. If $V$ and $W$ are any two holomorphic varieties, their Cartesian product $V \times W$ is a well-defined topological space that can be given the natural structure of a holomorphic variety. For any points $A \in V$ and $B \in W$, open neighborhoods $V_A$ and $W_B$ of these points can be represented by holomorphic subvarieties $V_A = \{Z \in \Delta(0; R) \subseteq \mathbb{C}^m : f_j(Z) = 0\}$ and $W_B = \{Z \in \Delta(0; S) \subseteq \mathbb{C}^n : g_k(Z) = 0\}$ for finitely many holomorphic functions $f_j \in {}_m\mathcal{O}_{\Delta(0;R)}$ and $g_k \in {}_n\mathcal{O}_{\Delta(0;S)}$. Then $V_A \times W_B \subseteq \Delta(0; R) \times \Delta(0; S) \subseteq \mathbb{C}^{m+n}$ can be described by $V_A \times W_B = \{(Z_1, Z_2) \in \Delta(0; R) \times \Delta(0; S) : f_j(Z_1) = g_k(Z_2) = 0\}$ and hence is a holomorphic subvariety of $\Delta(0; R) \times \Delta(0; S)$ and as such has the natural structure of a holomorphic variety. It is apparent that the resulting structure is really independent of the choice of representative subvarieties and hence leads to a well-defined structure of a holomorphic variety on $V \times W$.

# C
# Finite Branched Holomorphic Coverings

Holomorphic mappings in several variables are even locally considerably more diverse and complicated than the familiar mappings defined by holomorphic functions of one variable, but there is a special class of general holomorphic mappings having many of the same local geometric properties as the mappings defined by holomorphic functions of one variable. It is perhaps clearest and most convenient to begin the discussion of this class of mappings by considering their purely topological properties. Recall that a mapping between two topological spaces is called **finite** if the inverse image of each point is a finite set of points, and is called **proper** if the inverse image of each compact set is also compact.

1. DEFINITION. *A mapping $\pi$: $V \to W$ between two second-countable Hausdorff spaces is a* **finite branched covering** *if*:

(i) *$\pi$ is a continuous, finite, proper, surjective mapping.*

(ii) *There are dense open subsets $V_0 \subseteq V$, $W_0 \subseteq W$ such that $V_0 = \pi^{-1}(W_0)$ and the restriction $\pi | V_0 : V_0 \to W_0$ is a covering mapping.*

The space $W$ is called the **base space**, the space $V$ is called the **covering space**, and the restriction $\pi | V_0 : V_0 \to W_0$ is called a **regular part** of the finite branched covering $\pi$: $V \to W$. Obviously the regular part is not uniquely determined, for if $W_1$ is a dense open subset of $W_0$ and $V_1 = \pi^{-1}(W_1)$, then the restriction $\pi | V_1 : V_1 \to W_1$ is also a regular part of the branched covering $\pi$: $V \to W$. It should perhaps be recalled that the condition that $\pi | V_0 : V_0 \to W_0$ be a covering mapping is just that if $B$ is any point of $W_0$ and $\pi^{-1}(B) = \{A_j\} \subseteq V_0$, then there are an open neighborhood $W_B$ of the point $B$ in $W_0$ and disjoint open neighborhoods $V_j$ of the distinct points $A_j$ in $V_0$ such that $\pi^{-1}(W_B) = \bigcup_j V_j$ and each restriction $\pi | V_j : V_j \to W_B$ is a homeomorphism. The mapping $\pi$ may be rather more complicated than this over $W - W_0$ but still has some similar properties, particularly if some more topological regularity is assumed. Recall that a topological space $W$ is called **locally connected** at a point $B \in W$ is $B$ has arbitrarily small connected open neighborhoods, and is called merely locally connected if it is locally connected at each of its points. In extension of this definition a subset $W_0 \subseteq W$ will be called locally connected at a point $B \in W$ if $B$

has arbitrarily small open neighborhoods $W_B$ in $W$ such that $W_0 \cap W_B$ is connected, and $W_0$ will be called locally connected in $W$ if it is locally connected at each point of $W$. Note that if $W_0$ is a proper subset of $W$, the condition that $W_0$ be locally connected in $W$ is stronger than the condition that $W_0$ be merely locally connected. As a slightly weaker condition, a subset $W_0 \subseteq W$ will be called **locally finitely connected** at a point $B \in W$ if $B$ has arbitrarily small open neighborhoods $W_B$ in $W$ such that $W_0 \cap W_B$ has at most finitely many connected components, and $W_0$ will be called locally finitely connected in $W$ if it is locally finitely connected at each point of $W$.

2. LEMMA. *Suppose that $\pi: V \to W$ is a finite branched covering over a locally connected topological space $W$, and that this branched covering has a regular part $\pi | V_0 : V_0 \to W_0$ for which $W_0$ is locally finitely connected in $W$. If $B$ is any point of $W$ and $\pi^{-1}(B) = \{A_1, \ldots, A_n\}$ where $A_j$ are distinct points of $V$, then there are arbitrarily small connected open neighborhoods $W_B$ of the point $B$ such that $\pi^{-1}(W_B)$ consists of $n$ connected components $V_1, \ldots, V_n$ with $A_j \in V_j$ for each index $j$. As $W_B$ varies, the neighborhoods $V_j$ so arising form a basis for the open neighborhoods of the point $A_j$, and the restriction $\pi | V_j: V_j \to \pi(V_j)$ is a finite branched covering of its image $\pi(V_j) \subseteq W_B$. If the subset $W_0 \subseteq W$ is moreover locally connected at $B$, then $\pi(V_j) = W_B$; otherwise, $\pi(V_j)$ is the intersection of $W_B$ with the point set closure of some union of connected components of $W_0 \cap W_B$.*

*Proof.* Suppose that $V'_1, \ldots, V'_n$ are any disjoint open neighborhoods of the points $A_1, \ldots, A_n$, respectively. The first step in the proof is to show that whenever $W_B$ is a sufficiently small open neighborhood of the point $B$, then $\pi^{-1}(W_B) \subseteq \bigcup_j V'_j$. If that is not the case, there must exist a sequence of points $Y_\nu \in W$ converging to $B$ and a sequence of points $X_\nu \in V$ such that $\pi(X_\nu) = Y_\nu$ but $X_\nu \notin \bigcup_j V'_j$. The points $\{Y_\nu\}$ together with $B$ form a compact subset of $W$, so since $\pi$ is a proper mapping, the inverse image of this subset under $\pi$ is a compact subset of $V$. Hence, after passing to a subsequence if necessary, it can be assumed that the sequence $X_\nu$ converges to some point $X \in V$. From the continuity of the mapping $\pi$ it follows that $\pi(X) = B$, so that $X = A_j$ for some index $j$; but then $X_\nu$ converges to $A_j$, contradicting the assumption that $X_\nu \notin \bigcup_j V'_j$.

Now choose such a neighborhood $W_B$ with the additional property that $W_0 \cap W_B$ has finitely many connected components $\{W^\nu\}$. If $B \notin \overline{W^\nu}$, then $W_B$ can be replaced by the smaller open neighborhood $W_B - W_B \cap \overline{W^\nu}$ without harm; so it can be assumed that $B \in \overline{W^\nu}$ for each connected component $W^\nu$ of the intersection $W_0 \cap W_B$. Let $V_j = \pi^{-1}(W_B) \cap V'_j$, so that $\pi^{-1}(W_B) = \bigcup_j V_j$, and let $V_{0j} = V_0 \cap V_j = \pi^{-1}(W_0 \cap W_B) \cap V'_j$, so that $V_{0j}$ is a dense open subset of $V_j$. Each connected component $V^\nu_{0j}$ of the set $V_{0j}$ is a covering of some connected component $W^\nu$ of the intersection $W_0 \cap W_B$ under the mapping $\pi$, and it is easy to see that $\pi(\overline{V^\nu_{0j}}) = \overline{W^\nu}$. Indeed, since $\pi$ is continuous, obviously $\pi(\overline{V^\nu_{0j}}) \subseteq \overline{W^\nu}$. On the other hand, any point $Y \in \overline{W^\nu}$ is the limit of a sequence of points $Y_\mu \in W^\nu$, and $Y_\mu = \pi(X_\mu)$ for some point $X_\mu \in V^\nu_{0j}$. The points $\{Y_\mu\}$ together with $Y$ form a compact subset of $W$, and since $\pi$ is proper, the inverse image of this subset under $\pi$ is a compact subset of $V$. Hence, after passing to a subsequence if necessary, it can be assumed that the points $X_\mu$

converge to some point $X \in \bar{V}_{0j}^{\nu}$. From the continuity of $\pi$ it then follows that $Y = \pi(X) \in \pi(\bar{V}_{0j}^{\nu})$, so that $\bar{W}^{\nu} \subseteq \pi(\bar{V}_{0j}^{\nu})$, and therefore $\pi(\bar{V}_{0j}^{\nu}) = \bar{W}^{\nu}$ as asserted. In particular, then, $B \in \bar{W}^{\nu} = \pi(\bar{V}_{0j}^{\nu})$, so that $A_j \in \bar{V}_{0j}^{\nu}$; thus, $A_j$ must be contained in each connected component of $V_j$, so $V_j$ is necessarily connected. Since $V_j \subseteq V_j'$ and the neighborhoods $V_j'$ can be chosen arbitrarily small, it is clear that as $W_B$ varies, the sets $V_j$ form a basis for the open neighborhoods of the point $A_j$. It is also clear that the restriction $\pi | V_j : V_j \to \pi(V_j)$ is a finite branched covering with a regular part $\pi | V_{0j} : V_{0j} \to \pi(V_{0j})$; and the image $\pi(V_j) \subseteq W_B$ is evidently the subset $\bigcup_{\nu} \pi(\bar{V}_{0j}^{\nu}) \cap W_B = \bigcup_{\nu} \bar{W}^{\nu} \cap W_B$, where $V_{0j}^{\nu}$ are the finitely many connected components of $V_{0j}$ and their images $\pi(V_{0j}^{\nu}) = W^{\nu}$ are some of the finitely many connected components of $W_0 \cap W_B$, possibly with repetitions. Thus, if the subset $W_0 \subseteq W$ is locally connected at $B$, then actually $\bar{W}^{\nu} = W_0 \cap W_B$ for all indices $\nu$, so that $\pi(V_j) = W_B$. That suffices to conclude the proof.

It follows immediately from this lemma that if $\pi: V \to W$ is a finite branched covering over a locally connected topological space $W$ and if this branched covering has a regular part $\pi | V_0 : V_0 \to W_0$ for which $W_0$ is locally finitely connected in $W$, then each point $A \in V$ has arbitrarily small connected open neighborhoods $V_A$ such that the restriction $\pi | V_A : V_A \to \pi(V_A)$ is also a finite branched covering of its image $\pi(V_A) \subseteq W$. Thus under these circumstances the notion of the **germ** of a finite branched covering can be introduced in the obvious manner. It should be pointed out, though, that the germ of a finite branched covering $\pi: V \to W$ at a point $A \in V$ is not necessarily the germ of a finite branched covering over a full open neighborhood of the point $\pi(A) \in W$. That is the case, however, under the additional hypothesis that $W_0$ is locally connected in $W$, and then the mapping $\pi$ is an open mapping.

If $\pi: V \to W$ is a finite branched covering with a regular part $\pi | V_0 : V_0 \to W_0$, then over each connected component of $W_0$ this regular part is a covering space with a finite number of sheets, though not necessarily with the same number of sheets over different connected components. The maximum number of sheets of these coverings over all the connected components of $W_0$ will be called the **order** of the branched covering $\pi: V \to W$ and will be denoted by $o_{\pi}$. It may of course be the case what $W_0$ has infinitely many connected components and the number of sheets over the various components is unbounded, in which case $o_{\pi} = \infty$. Much of the interest in finite branched holomorphic coverings really lies in their local properties, though, so with rare exceptions when it makes a difference it will generally be supposed that $o_{\pi} < \infty$. If there are the same number of sheets over each connected component of $W_0$, the branched covering $\pi: V \to W$ will be said to be of **pure order** $o_{\pi}$. It is clear that the order is independent of the choice of a regular part. Note that if $W_0$ is connected, then the branched covering $\pi: V \to W$ is necessarily of pure order. Note also that if $o_{\pi} = 1$, then the branched covering $\pi: V \to W$ is necessarily of pure order one. In this case it should be observed that the mapping $\pi$ is not necessarily one-to-one everywhere; for example, if $V$ consists of two disjoint copies of a topological space, $W$ consists of two copies of the same space but with a point in one copy identified with the corresponding point in the other copy, and if $\pi: V \to W$ is the obvious mapping, then $o_{\pi} = 1$ but the two distinct points of $V$ identified in $W$ have the same image in $W$. However, if $W_0$ is locally connected in $W$, then it is clear from Lemma 2 that $o_\pi = 1$ implies that $\pi$ is a one-to-one mapping—indeed, that $\pi$ is a homeomorphism since it is also an open mapping.

If $\pi: V \to W$ is a finite branched covering over a locally connected topological space $W$ and has a regular part $\pi|V_0 : V_0 \to W_0$ for which $W_0$ is locally finitely connected in $W$, then as already observed each point $A \in V$ has arbitrarily small connected open neighborhoods $V_A$ such that the restriction $\pi|V_A : V_A \to \pi(V_A)$ is a finite branched covering. The order $o_{\pi|V_A}$ of this finite branched covering is a positive integer and can only decrease as the neighborhood $V_A$ shrinks; hence, this order must be the same for all sufficiently small such neighborhoods $V_A$. This common order will be called the **branching order** of the mapping $\pi$ at the point $A$ and will be denoted by $o_\pi(A)$. It can be viewed as the order of the germ $\pi_A$ of the finite branched covering $\pi$ at the point $A$. Note that for any point $A \in V_0$, clearly $o_\pi(A) = 1$. The complement of the set of points $A \in V$ at which $o_\pi(A) = 1$ will be called the **branch locus** of the mapping $\pi$ and will be denoted by $B_\pi$. Thus,

$$B_\pi = \{A \in V : o_\pi(A) > 1\} \tag{1}$$

and $B_\pi \subseteq V - V_0$ as already observed. If it is assumed moreover that $W_0$ is locally connected in $W$, then as in the preceding discussion $o_\pi(A) = 1$ implies that $\pi$ is a local homeomorphism at $A$. The set of points $A \in V$ at which $o_\pi(A) = 1$ is therefore an open subset of $V$—indeed, is a dense open subset since it contains $V_0$—or equivalently $B_\pi$ is a closed nowhere dense subset of $V$. If, even further, the set $W_0$ is connected, then the finite branched covering $\pi: V \to W$ is of pure order $o_\pi$. In this situation if $B$ is any point of $W$ and $\pi^{-1}(B) = \{A_1, \ldots, A_n\}$ where $A_j$ are distinct points of $V$, then it follows immediately from Lemma 2 that

$$\sum_{j=1}^{n} o_\pi(A_j) = o_\pi \tag{2}$$

In many applications it is convenient in these circumstances to write $\pi^{-1}(B) = \{A_1, A_1, \ldots, A_1, A_2, A_2, \ldots, A_2, \ldots, A_n, A_n, \ldots, A_n\}$ where the point $A_j$ is listed altogether $o_\pi(A_j)$ times, for when the points of $\pi^{-1}(B)$ are listed in this manner with repetitions according to the branching order, then there are precisely $o_\pi$ points in $\pi^{-1}(B)$ for each $B \in W$, although they are no longer necessarily distinct. This listing plays a useful role at several points in the subsequent discussion.

After having considered the purely topological properties of finite branched coverings in general, the particular finite branched coverings of analytic interest can be introduced as follows.

3. DEFINITION. *A holomorphic mapping $\pi: V \to W$ between two holomorphic varieties is a* **finite branched holomorphic covering** *if*:

(i) *$\pi$ is a finite branched covering.*

(ii) *There is a regular part $\pi|V_0 : V_0 \to W_0$ of this finite branched covering for which $W - W_0$ is a holomorphic subvariety of $W$ and $\pi|V_0$ is a locally biholomorphic mapping.*

It will be demonstrated later that any holomorphic variety $W$ is locally connected and that if $W_0 \subseteq W$ is the complement of any holomorphic subvariety of $W$, then $W_0$ is locally finitely connected in $W$. Thus some of the topological regularity conditions introduced in the discussion of general finite branched coverings are automatically fulfilled for all finite branched holomorphic coverings; so in particular it will always be possible to speak of the germ of a finite branched holomorphic covering and of its branching order. It is not always the case though that $W_0$ is locally connected in $W$; so in general some care must still be taken when considering arbitrary finite branched holomorphic coverings. However in the remainder of this section attention will be almost entirely limited to the consideration of finite branched holomorphic coverings $\pi: V \to W$ over open subsets $W \subseteq \mathbb{C}^n$. Such varieties $W$ are obviously locally connected, and it follows from Corollary I-D3 that the complement of a holomorphic subvariety of $W$ is always locally connected in $W$. These observations, together with the use of the extended Riemann removable singularities theorem, will be applied to derive a number of additional properties of finite branched holomorphic coverings over open subsets of $\mathbb{C}^n$. For example, such a covering $\pi$ is a holomorphic local homeomorphism at each point $A \in V$ at which $o_\pi(A) = 1$. Although it is only assumed that $\pi$ is locally biholomorphic at points $A \in V_0$, it is actually the case that $\pi$ is locally biholomorphic at each point $A \in V$ at which $o_\pi(A) = 1$.

**4. THEOREM.** *If $\pi: V \to W$ is a finite branched holomorphic covering over an open subset $W \subseteq \mathbb{C}^n$, then $\pi$ is locally biholomorphic at every point $A \in V$ for which $o_\pi(A) = 1$.*

*Proof.* Let $V_A$ be a sufficiently small open neighborhood of the point $A$ that the variety $V_A$ can be represented as a holomorphic subvariety of an open subset $U_A \subseteq \mathbb{C}^m$. If $o_\pi(A) = 1$ and $V_A$ is sufficiently small, then $\pi | V_A : V_A \to \pi(V_A)$ is a holomorphic mapping and a topological homeomorphism between $V_A$ and the open subset $\pi(V_A) \subseteq W \subseteq \mathbb{C}^n$. The inverse mapping $(\pi | V_A)^{-1}$ is a continuous mapping from the open subset $\pi(V_A) \subseteq \mathbb{C}^n$ to the holomorphic subvariety $V_A \subseteq U_A \subseteq \mathbb{C}^m$ and can be described by its $m$ coordinate functions. Since the regular part $\pi | V_0 : V_0 \to W_0$ is a locally biholomorphic mapping by Definition 3, it follows that the inverse mapping $(\pi | V_A)^{-1}$ is holomorphic in $\pi(V_A) \cap W_0$; hence, the coordinate functions are holomorphic in that subset. But it further follows from Definition 3 that $W - W_0$ is a holomorphic subvariety of $W$. Therefore, by the extended Riemann removable singularities theorem, Theorem I-D2, these $m$ coordinate functions must be holomorphic throughout $\pi(V_A)$, so $(\pi | V_A)^{-1}$ is a holomorphic mapping. That suffices to conclude the proof.

A finite branched holomorphic covering $\pi: V \to W$ over an open subset $W \subseteq \mathbb{C}^n$ induces a natural homomorphism $\pi^*: {}_n\mathcal{O}_W \to {}_V\mathcal{O}_V$ between the algebras of holomorphic functions on $V$ and $W$; since $\pi$ is a surjective mapping, this induced homomorphism $\pi^*$ is obviously injective. It is convenient in many instances to identify ${}_n\mathcal{O}_W$ with its image $\pi^*({}_n\mathcal{O}_W)$, and thus to view ${}_n\mathcal{O}_W$ as a subalgebra of ${}_V\mathcal{O}_V$ in a natural manner. Similarly for each point $A \in V$ the mapping $\pi: V \to W$ induces a natural homomorphism $\pi^*: {}_n\mathcal{O}_{\pi(A)} \to {}_V\mathcal{O}_A$ between the local rings of these two varieties; since $\pi$ is an open mapping, this homomorphism $\pi^*$ is also obviously injective. In the local case too it is often convenient to identify ${}_n\mathcal{O}_{\pi(A)}$ with its image $\pi^*({}_n\mathcal{O}_{\pi(A)})$ and thus to view ${}_n\mathcal{O}_{\pi(A)}$ as a subalgebra of ${}_V\mathcal{O}_A$ in a natural manner. There are various useful relations between these algebras, such as the following.

**5. THEOREM.** *If $\pi: V \to W$ is a finite branched holomorphic covering of pure order $v$ over an open subset $W \subseteq \mathbb{C}^n$, then to each holomorphic function $f \in {}_V\mathcal{O}_V$ there is canonically associated a monic polynomial $P_f(X) \in {}_n\mathcal{O}_W[X] \subseteq {}_V\mathcal{O}_V[X]$ of degree $v$ such that $P_f(f) = 0$ in ${}_V\mathcal{O}_V$.*

*Proof.* For any point $Z \in W$, let $\pi^{-1}(Z) = \{A_1(Z), \ldots, A_v(Z)\}$, where the $v$ points $A_j(Z)$ are listed with repetitions according to the branching order. No assumptions are made about the ordering of the points of $\pi^{-1}(Z)$ for different values of $Z$, so the functions $A_j(Z)$ do not necessarily depend continuously on $Z$. To any function $f \in {}_V\mathcal{O}_V$ associate the polynomial

$$\begin{aligned} P_f(X) &= \prod_{j=1}^{v} [X - f(A_j(Z))] \\ &= X^v + a_1(Z)X^{v-1} + \cdots + a_v(Z) \end{aligned}$$

This is the unique monic polynomial of degree $v$ having as roots the $v$ values $f(A_j(Z))$; the coefficients $a_i(Z)$ are the elementary symmetric functions of these $v$ roots and are well-defined functions on $W$, independent of the particular ordering chosen for the points $A_j(Z)$. Since any point $A \in V$ is of the form $A = A_j(\pi(A))$, for some index $j$ depending on $A$, it is evident that when the coefficients $a_i(Z)$ are viewed as functions $a_i(\pi(A))$ on $V$, then $P_f(f)(A) = \prod_j [f(A) - f(A_j(\pi(A)))] = 0$. To complete the proof it is only necessary to show that the coefficients $a_i$ are holomorphic functions on $W$. To see first that the functions $a_i$ are continuous on $W$, recall from elementary algebra that the elementary symmetric functions in the $v$ values $f(A_j(Z))$ are polynomials in the $v$ power sums $\sigma_\lambda(Z) = \sum_{j=1}^{v} f(A_j(Z))^\lambda$ for $1 \leq \lambda \leq v$; hence, it is enough to show that the functions $\sigma_\lambda$ are continuous on $W$. Given any point $B \in W$ and any constant $\varepsilon > 0$, let $A_1, \ldots, A_m$ be the distinct points among the $v$ points $A_j(B)$; thus, $o_\pi(A_i)$ of the points $A_j(B)$ coincide with $A_i$. Since $f$ is a continuous function on $V$, for any chosen constant $\delta > 0$ there will exist open neighborhoods $V_i'$ of the points $A_i$, respectively, such that $|f(A) - f(A_i)| < \delta$ whenever $A \in V_i'$; and if $\delta$ is chosen sufficiently small, then $|f(A)^\lambda - f(A_i)^\lambda| < \varepsilon/v$ whenever $A \in V_i'$ and $1 \leq \lambda \leq v$. It follows from Lemma 2 that there is an open neighborhood $W_B$ of the point $B$ in $W$ such that $\pi^{-1}(W_B)$ consists of $m$ connected components $V_1, \ldots, V_m$ with $A_i \in V_i \subseteq V_i'$; and whenever $Z \in W_B$, there will be $o_\pi(A_i)$ of the points $A_j(Z)$ in the open subset $V_i \subseteq V_i'$. Therefore whenever $Z \in W_B$ and $1 \leq \lambda \leq v$,

$$\begin{aligned} |\sigma_\lambda(Z) - \sigma_\lambda(B)| &= \left| \sum_{j=1}^{v} f(A_j(Z))^\lambda - \sum_{j=1}^{v} f(A_j(B))^\lambda \right| \\ &= \left| \sum_{i=1}^{m} \sum_{\{j: A_j(Z) \in V_i\}} [f(A_j(Z))^\lambda - f(A_i)^\lambda] \right| \\ &< \sum_{i=1}^{m} o_\pi(A_i) \frac{\varepsilon}{v} = \varepsilon \end{aligned}$$

so that the functions $\sigma_\lambda$ and hence also the functions $a_i$ are continuous on $W$. Next it is quite clear that the functions $a_i$ are holomorphic on the open subset $W_0 \subseteq W$, where $\pi | V_0 : V_0 + W_0$ is a regular part of the branched covering $\pi: V \to W$ for which $\pi | V_0$ is locally biholomorphic; for in some open neighborhood $W_B$ of any point $B \in W_0$ it is possible to relabel the points $A_j(Z)$ so that the mappings $Z \to A_j(Z)$ are holomorphic. Since $W - W_0$ is a holomorphic subvariety of $W$, it then follows from the extended Riemann removable singularities theorem again that the functions $a_i$ are actually holomorphic in all of $W$, and that concludes the proof.

It should be pointed out that the preceding theorem asserts that to any function $f \in {}_V\mathcal{O}_V$ there corresponds a canonical monic polynomial $P_f(X) \in {}_n\mathcal{O}_W[X]$ of degree $\nu$ for which $P_f(f) = 0$ in ${}_V\mathcal{O}_V$, but not that there is a unique such polynomial. For example, if $\pi: V \to W$ is a finite branched holomorphic covering of pure order $\nu > 1$ over an open subset $W \subseteq \mathbb{C}^n$ and if $f \in {}_n\mathcal{O}_W$, then for the function $\pi^*(f) \in {}_V\mathcal{O}_V$ it is evident that $P_{\pi^*(f)}(X) = (X - f)^\nu$; but if $Q(X) \in {}_n\mathcal{O}_W[X]$ is any monic polynomial of order $\nu - 1$, then $P(X) = (X - f)Q(X)$ is also a monic polynomial of order $\nu$ in ${}_n\mathcal{O}_W[X]$ such that $P(f) = 0$ in ${}_V\mathcal{O}_V$. On the other hand, there are many cases in which the polynomial $P_f(X)$ is actually uniquely determined. A holomorphic function $f \in {}_V\mathcal{O}_V$ is said to **separate the sheets** of a finite branched holomorphic covering $\pi: V \to W$ with a regular part $\pi | V_0 : V_0 + W_0$ provided that for each connected component of $W_0$ there is a point $B$ in that component such that if $\pi^{-1}(B) = \{A_1, \ldots, A_\nu\}$ where $A_j$ are distinct points, then $f(A_j) \neq f(A_k)$ whenever $j \neq k$. It is quite obvious that if the function $f \in {}_V\mathcal{O}_V$ in Theorem 5 separates the sheets of the finite branched holomorphic covering $\pi: V \to W$, then there is indeed a unique monic polynomial $P_f(X) \in {}_n\mathcal{O}_W[X]$ of degree $\nu$ such that $P_f(f) = 0$ in ${}_V\mathcal{O}_V$. Actually in this case something more can be said.

**6. THEOREM.** *If $\pi: V \to W$ is a finite branched holomorphic covering of pure order $\nu$ over an open subset $W \subseteq \mathbb{C}^n$ and if $f \in {}_V\mathcal{O}_V$ separates the sheets of this covering, then there is a unique monic polynomial $P_f(X) \in {}_n\mathcal{O}_W[X] \subseteq {}_V\mathcal{O}_V[X]$ of degree $\nu$ such that $P_f(f) = 0$ in ${}_V\mathcal{O}_V$, and the discriminant of this polynomial is a function $d_f \in {}_n\mathcal{O}_W$ that does not vanish identically on any connected component of $W$. Moreover, to any function $g \in {}_V\mathcal{O}_V$ there corresponds a unique polynomial $Q_{f,g}(X) \in {}_n\mathcal{O}_W[X] \subseteq {}_V\mathcal{O}_V[X]$ of degree $\nu - 1$ such that $d_f \cdot g = Q_{f,g}(f)$ in ${}_V\mathcal{O}_V$.*

*Proof.* As already mentioned, the uniqueness of the polynomial $P_f(X)$ is in this case quite obvious. With the same notation as in the proof of Theorem 5, the discriminant of the polynomial $P_f(X)$ is the function $d_f(Z) = \prod_{j \neq k} [f(A_j(Z)) - f(A_k(Z))]$ and is a well-defined complex-valued function on $W$. Since $d_f(Z)$ is evidently a symmetric polynomial in the values $f(A_j(Z))$, it is necessarily a polynomial in the elementary symmetric functions of the values $f(A_j(Z))$, the coefficients of the polynomial $P_f(X)$, and hence is actually a holomorphic function on $W$; and if $B$ is a point of $W$ for which $A_j(B)$ are distinct points and $f(A_j(B)) \neq f(A_k(B))$ whenever $j \neq k$, then clearly $d_f(B) \neq 0$. To complete the proof it is only necessary to find for any function $g \in {}_V\mathcal{O}_V$ a polynomial $Q_{f,g}(X) = a_1 X^{\nu-1} + \cdots + a_{\nu-2} X + a_{\nu-1} \in {}_n\mathcal{O}_W[X]$ such that

$$d_f(Z) g(A_j(Z)) = a_1(Z) f(A_j(Z))^{\nu-1} + \cdots + a_{\nu-2}(Z) f(A_j(Z)) + a_{\nu-1}(Z)$$

for all $Z \in W$ and all points $A_j(Z) \in \pi^{-1}(Z)$, and to see that it is uniquely determined. For any fixed point $Z \in W$ these conditions can be viewed as a set of $v$ linear equations in the $v$ unknown values of $a_k(Z)$, and by Cramer's rule the solutions $a_k(Z)$ are uniquely determined and must satisfy

$$a_k(Z)\cdot\det\{f(A_j(Z))^{v-1},\ldots,f(A_j(Z)),1\}$$
$$=\det\{f(A_j(Z))^{v-1},\ldots,f(A_j(Z))^{v-k+1},d_f(Z)g(A_j(Z)),f(A_j(Z))^{v-k-1},\ldots,1\}$$

where in each determinant only the $j$th row is written out. The determinant on the left-hand side of this equation is the van der Monde determinant $\Delta(Z)$, and it is a familiar result from elementary algebra that $\Delta(Z)^2 = d_f(Z)$. Hence, factoring $d_f(Z)$ out of the determinant on the right-hand side and dividing by $\Delta(Z)$ lead to

$$a_k(Z) = \Delta(Z)\cdot\det\{f(A_j(Z))^{v-1},\ldots,g(A_j(Z)),f(A_j(Z))^{v-k-1},\ldots,1\} \quad (3)$$

Both $\Delta(Z)$ and the determinant on the right-hand side of (3) change signs upon interchanging any two rows, so that the expression (3) for $a_k(Z)$ is independent of the order in which the points $A_j(Z)$ are listed. Now as in the proof of the preceding theorem the functions $A_j$ can be chosen so that they are holomorphic in an open neighborhood of any point of $W_0$; hence, the functions $a_k(Z)$ are holomorphic in $W_0$. Furthermore, also as in the proof of the preceding theorem, these functions $a_k$ are continuous on $W$. It then follows from the extended form of Riemann's removable singularities theorem that $a_k$ are holomorphic functions on all of $W$, so $Q_{f,g}(X) \in {}_n\mathcal{O}_W[X]$ as desired, and the proof is thereby concluded.

For later purposes it is worth noting here a simple consequence of a slight extension of the preceding theorem, as follows.

**7. COROLLARY.** *Let $\pi: V \to W$ be a finite branched holomorphic covering over an open subset $W \subseteq \mathbb{C}^n$, let $f \in {}_V\mathcal{O}_V$ be a holomorphic function separating the sheets of this covering, and let $d_f \in {}_n\mathcal{O}_W$ be the discriminant of the polynomial $P_f(X)$ canonically associated to $f$. If $D$ is a proper holomorphic subvariety of $W$ and $g$ is a bounded holomorphic function on $V - \pi^{-1}(D)$, then $d_f\cdot g$ extends to a holomorphic function on all of $V$.*

*Proof.* The preceding theorem applied to the restriction $\pi|V-\pi^{-1}(D): V - \pi^{-1}(D) \to W - D$ shows that there is a polynomial $Q_{f,g}(X)$ with coefficients holomorphic in $W - D$ such that $d_f\cdot g = Q_{f,g}(f)$ on $V - \pi^{-1}(D)$. The coefficients of the polynomial $Q_{f,g}(X)$ are polynomial functions in the values of $f$ and $g$ at various points of $V - \pi^{-1}(D)$, given explicitly in the proof of the preceding theorem, so these coefficients are clearly locally bounded functions in $W$. It then follows from the extended Riemann removable singularities theorem, Theorem I-D2, that these coefficients extend to holomorphic functions in all of $W$ and hence that $Q_{f,g}(X) \in {}_n\mathcal{O}_W[X] \subseteq {}_V\mathcal{O}_V[X]$. Then $Q_{f,g}(f) \in {}_V\mathcal{O}_V$ provides the desired extension of the function $d_f\cdot g$, thereby concluding the proof.

This observation can be viewed as providing an extension of the Riemann removable singularities theorem to finite branched holomorphic coverings over subsets of $\mathbb{C}^n$. However, the extension is somewhat weaker than the Riemann removable singularities theorem in $\mathbb{C}^n$, since it does not assert that a bounded holomorphic function $g$ in the complement of a holomorphic subvariety of the form $\pi^{-1}(D)$ in $V$ extends to a holomorphic function in $V$ but only that the product $d_f \cdot g$ so extends. The function $g$ at least extends to the meromorphic function $Q_{f,g}/d_f$ with a specified denominator independent of $g$, but that will be discussed in some detail later. At this point another consequence of the preceding theorem, with an interesting algebraic interpretation, should be mentioned.

**8. COROLLARY.** *If $\pi\colon V \to W$ is a finite branched holomorphic covering of pure order $\nu$ over an open subset $W \subseteq \mathbb{C}^n$, then the induced homomorphism $\pi^*$ exhibits ${}_V\mathcal{O}_V$ as an integral algebraic extension of ${}_n\mathcal{O}_W$ of degree at most $\nu$. If, moreover, there is a function in ${}_V\mathcal{O}_V$ that separates the sheets, then the degree of this extension is precisely $\nu$, and as an ${}_n\mathcal{O}_W$-module ${}_V\mathcal{O}_V$ is isomorphic to a submodule of a free ${}_n\mathcal{O}_W$-module of rank $\nu$.*

*Proof.* In these circumstances Theorem 5 shows that for any $f \in {}_V\mathcal{O}_V$ there is a monic polynomial $P_f(X) \in {}_n\mathcal{O}_W[X] \subseteq {}_V\mathcal{O}_V[X]$ such that $P_f(f) = 0$; that is just the condition that ${}_V\mathcal{O}_V$ be an integral algebraic extension of ${}_n\mathcal{O}_W$ of degree at most $\nu$. If there is a function $f \in {}_V\mathcal{O}_V$ that separates the sheets, then by Theorem 6 the polynomial $P_f(X)$ is of degree precisely $\nu$ and there is no nontrivial polynomial $P(X)$ of lower degree for which $P(f) = 0$, so the degree of this extension is precisely $\nu$. Moreover the discriminant $d_f \in {}_n\mathcal{O}_W$ of the polynomial $P_f(X)$ is not identically zero on any connected component; thus, multiplying by $d_f$ describes an isomorphism ${}_V\mathcal{O}_V \to d_f \cdot {}_V\mathcal{O}_V$ of ${}_n\mathcal{O}_W$-modules. Finally Theorem 6 also shows that whenever $g \in {}_V\mathcal{O}_V$, then the product $d_f \cdot g$ can be expressed as a linear combination of the elements $1, f, \ldots, f^{\nu-1}$ with coefficients from ${}_n\mathcal{O}_W$. Thus $d \cdot {}_V\mathcal{O}_V$ is a submodule of the submodule of ${}_n\mathcal{O}_W$ spanned by the elements $1, f, \ldots, f^{\nu-1}$. This latter module is actually a free module of rank $\nu$, since any nontrivial linear relation among the generators $1, f, \ldots, f^{\nu-1}$ would yield a nontrivial polynomial relation $P(f) = 0$ for $P(X) \in {}_n\mathcal{O}_W[X]$ of degree less than $\nu$, and as already noted there can be no such polynomial relation. That suffices to conclude the proof.

It should be noted that even when ${}_V\mathcal{O}_V$ is exhibited as a submodule of a finitely generated ${}_n\mathcal{O}_W$-module, that does not imply that ${}_V\mathcal{O}_V$ is itself a finitely generated ${}_n\mathcal{O}_W$-module, since the ring ${}_n\mathcal{O}_W$ is not Noetherian; but that difficulty does not arise in the local case, and something more can be said then.

**9. COROLLARY.** *If $\pi\colon \mathrm{V} \to \mathbb{C}^n$ is the germ of a finite branched holomorphic covering of order $\nu$, then the induced homomorphism $\pi^*$ exhibits ${}_\mathrm{V}\mathcal{O}$ as a finitely generated ${}_n\mathcal{O}$-module—indeed, as an integral algebraic extension of ${}_n\mathcal{O}$ of degree $\nu$. Moreover, if $V$ is irreducible, then the field ${}_\mathrm{V}\mathcal{M}$ is an algebraic extension of ${}_n\mathcal{M}$ of degree $\nu$ and is generated by any element $\mathrm{f} \in {}_\mathrm{V}\mathcal{O}$ that separates the sheets.*

*Proof.* Since the germ $\pi\colon \mathrm{V} \to \mathbb{C}^n$ can be represented by a finite branched holomorphic covering $\pi\colon V \to W$ of pure order $\nu$ over an open subset $W \subseteq \mathbb{C}^n$, and since there obviously exists a function $f \in {}_V\mathcal{O}_V$ that separates the sheets if the representative variety $V$ is sufficiently small, the first statement of this corollary follows immediately from Corollary 8. If the germ $\mathbf{V}$ is irreducible, then the inclusion ${}_n\mathcal{O} \subseteq {}_{\mathbf{V}}\mathcal{O}$ extends to an inclusion ${}_n\mathcal{M} \subseteq {}_{\mathbf{V}}\mathcal{M}$ of the quotient fields, and ${}_{\mathbf{V}}\mathcal{M}$ is an algebraic extension of ${}_n\mathcal{M}$ of degree $v$. The inclusion $\mathbf{d}_f \cdot {}_{\mathbf{V}}\mathcal{O} \subseteq {}_n\mathcal{O}[\mathbf{f}]$ implies immediately that ${}_{\mathbf{V}}\mathcal{O} \subseteq {}_n\mathcal{M}[f]$; since $\mathbf{f}$ is algebraic over ${}_n\mathcal{M}$, the extension ${}_n\mathcal{M}[\mathbf{f}]$ is also a field, and consequently ${}_{\mathbf{V}}\mathcal{M} \subseteq {}_n\mathcal{M}[\mathbf{f}]$. On the other hand, since ${}_n\mathcal{M} \subseteq {}_{\mathbf{V}}\mathcal{M}$ and $\mathbf{f} \in {}_{\mathbf{V}}\mathcal{O} \subseteq {}_{\mathbf{V}}\mathcal{M}$, then ${}_n\mathcal{M}[\mathbf{f}] \subseteq {}_{\mathbf{V}}\mathcal{M}$, so that actually ${}_{\mathbf{V}}\mathcal{M} = {}_n\mathcal{M}[\mathbf{f}]$. That suffices to conclude the proof.

It should be noted that although ${}_{\mathbf{V}}\mathcal{O}$ is a finitely generated ${}_n\mathcal{O}$-module and can be represented as a submodule of a free ${}_n\mathcal{O}$-module of rank $v$, the number of generators of ${}_{\mathbf{V}}\mathcal{O}$ as an ${}_n\mathcal{O}$-module may exceed $v$. A more detailed discussion of the minimal number of generators of ${}_{\mathbf{V}}\mathcal{O}$ as an ${}_n\mathcal{O}$-module will be given later. To turn now to other applications of the constructions in Theorems 5 and 6, the following simple result will be used later in examining the theory of functions on holomorphic varieties.

**10. THEOREM (maximum modulus theorem).** *If $\pi: V \to W$ is a finite branched holomorphic covering over an open subset $W \subseteq \mathbb{C}^n$, if $f \in {}_V\mathcal{O}_V$, and if there is a point $A \in V$ such that $|f(Z)| \leq |f(A)|$ for all points $Z$ in an open neighborhood of $A$, then $f$ is constant in an open neighborhood of $A$.*

*Proof.* Since the assertion is local, by applying Lemma 2 it can be assumed that $|f(Z)| \leq |f(A)|$ for all points $Z \in V$ and that if $\pi(A) = B \in W$, then $\pi^{-1}(B)$ consists of the point $A$ alone. If $\pi$ is a covering of pure order $v$ and $\pi^{-1}(Z) = \{A_1(Z), \ldots, A_v(Z)\}$ where the $v$ points $A_j(Z) \in V$ are listed with repetitions according to the branching order, then the power sums $\sigma_\lambda(Z) = \sum_{j=1}^{v} f(A_j(Z))^\lambda$ are holomorphic functions in $W$ as in the proof of Theorem 5. However, since $A_j(B) = A$, it follows from the hypotheses of the present theorem that $|\sigma_\lambda(Z)| \leq \sum_{j=1}^{v} |f(A_j(Z))|^\lambda \leq v|f(A)|^\lambda = |\sigma_\lambda(B)|$; but then the functions $\sigma_\lambda$ must be constant in $W$ as a consequence of the usual maximum modulus theorem, Theorem I-A4. The coefficients of the polynomial $P_f(X)$, being polynomials in the power sums $\sigma_\lambda$, must also be constant, and since $P_f(f) = 0$ by Theorem 5, it follows that the function $f$ takes only finitely many values in $V$ and hence by continuity must be constant near $A$. That suffices to conclude the proof.

**11. LEMMA.** *If $\pi: \mathbf{V} \to \mathbb{C}^n$ is any germ of a holomorphic mapping, then there exists a holomorphic subvariety $V \subseteq \Delta(0; R) \times \Delta(0; R') \subseteq \mathbb{C}^n \times \mathbb{C}^m$ representing the germ $\mathbf{V}$ such that the natural projection $\mathbb{C}^n \times \mathbb{C}^m \to \mathbb{C}^n$ induces a holomorphic mapping $\pi: V \to \Delta(0; R)$ representing $\pi$.*

*Proof.* Select a representative $\pi': V' \to \Delta(0; R)$ of the given germ $\pi$ of a holomorphic mapping, where $V'$ is a holomorphic subvariety of an open neighborhood $U$ of the origin in $\mathbb{C}^m$ such that $\mathbf{V}$ is the germ of $V'$ at the origin and $\pi'$ is a holomorphic mapping such that $\pi'(0) = 0$. The neighborhood $U$ can of course be taken to be as small as desired, so it can be assumed that there are holomorphic functions $f_j \in {}_m\mathcal{O}_U$

such that the restrictions $f_j|V'$ are the component functions of the mapping $\pi'$. Use these functions to introduce the holomorphic mapping $F: U \to \mathbb{C}^n \times \mathbb{C}^m$, taking a point $Z = (z_1, \ldots, z_m) \in U$ to the point $F(Z) = (f_1(Z), \ldots, f_n(Z), z_1, \ldots, z_m)$. Since the mapping $F$ is obviously nonsingular, it follows from Corollary I-C10 and Theorem I-C6 that if the neighborhood $U$ is sufficiently small, the image $F(U)$ is a complex submanifold of an open neighborhood $W$ of the origin in $\mathbb{C}^n \times \mathbb{C}^m$ and $F$ is a biholomorphic mapping from $U$ to $F(U)$. The image $V = F(V') \subseteq F(U)$ is therefore a holomorphic subvariety of $F(U)$ and hence of $W$, and is biholomorphic to $V'$. The composition $\pi = \pi' \circ (F|V')^{-1} : V \to \Delta(0; R)$ is therefore a holomorphic mapping that represents the same germ $\pi$, and by construction $\pi$ is the holomorphic mapping induced by the natural projection $\mathbb{C}^n \times \mathbb{C}^m \to \mathbb{C}^n$. That suffices to conclude the proof.

12. LEMMA. *For any space $\mathbb{C}^n$ and any integer $v \geqq 1$ there exist finitely many linear functions $f_1, \ldots, f_r$ that separate any $v$ points of $\mathbb{C}^n$, in the sense that if $Z_1, \ldots, Z_v$ are any distinct points of $\mathbb{C}^n$, there is an index $j$ such that the values $f_j(Z_1), \ldots, f_j(Z_v)$ are distinct.*

*Proof.* Choose any $r$ vectors $A_1, \ldots, A_r$ in $\mathbb{C}^n$ such that no $n$ of them are linearly dependent, where $r > \frac{1}{2}v(v-1)(n-1)$; and if $A_j = (a_{j1}, \ldots, a_{jn})$, set $f_j(Z) = \sum_{k=1}^{n} a_{jk} z_k$. To see that these functions have the desired property, suppose to the contrary that there are some distinct points $Z_1, \ldots, Z_v$ in $\mathbb{C}^n$ that cannot be separated by the functions $f_1, \ldots, f_r$. Thus, for any index $j$ in the interval $1 \leqq j \leqq r$ there are indices $p_j$ and $q_j$ in the interval $1 \leqq p_j \leqq q_j \leqq v$ such that $f_j(Z_{p_j}) = f_j(Z_{q_j})$ and hence such that $f_j(Z_{p_j} - Z_{q_j}) = 0$. This last equation can be viewed as a linear condition on the vector $A_j$. Now for any of the $\frac{1}{2}v(v-1)$ distinct pairs of integers $p, q$ satisfying $1 \leqq p < q \leqq v$ there are at most $n-1$ of the vectors $A_j$ such that $f_j(Z_p - Z_q) = 0$, since no $n$ of these vectors are linearly dependent. Thus, the preceding linear conditions on the vectors $A_j$ imply that there are at most $\frac{1}{2}v(v-1)(n-1)$ of these vectors, contradicting the assumption that $r > \frac{1}{2}v(v-1)(n-1)$. This contradiction suffices to conclude the proof.

13. THEOREM. *If $\pi: V \to W$ is a finite branched holomorphic covering over an open subset $W \subseteq \mathbb{C}^n$, then for any integer $\mu \geqq 1$ the subset*

$$B_\pi^\mu = \{A \in V : o_\pi(A) \geqq \mu\} \tag{4}$$

*is a holomorphic subvariety of $V$.*

*Proof.* Since this theorem is really a local assertion, it can be assumed as a consequence of Lemma 11 that $V$ is a holomorphic subvariety of an open polydisc $\Delta(0; R) \times \Delta(0; R') \subseteq \mathbb{C}^n \times \mathbb{C}^m$ and that $\pi: V \to W = \Delta(0; R)$ is induced by the natural projection mapping $\mathbb{C}^n \times \mathbb{C}^m \to \mathbb{C}^n$. Let $v$ be the order of this branched holomorphic covering, and choose linear functions $f_1, \ldots, f_r$ that separate any $v$ points of $\mathbb{C}^n \times \mathbb{C}^m$ in the sense of Lemma 12. To the holomorphic functions $f_j|V \in {}_V\mathcal{O}_V$ there are canonically associated monic polynomials $P_j(X) = P_{f_j}(X) \in {}_n\mathcal{O}_W[X]$ of degree $v$ as in Theorem 5, and these can be viewed as holomorphic functions $P_j(Z; X)$ of the variables $Z \in W$ and $X \in \mathbb{C}$. If $A \in V$ is a point such that $o_\pi(A) = \mu$ and if $B = \pi(A) \in W$, then by construction the polynomial $P_j(B; X)$ in $X$ has a root of order at least $\mu$ at the value $X = f_j(B)$. Hence, if $P_j^{(\lambda)}(B; X)$ denotes the derivative of order $\lambda$ of this polynomial, it follows that $P_j^{(\lambda)}(B; f_j(A)) = 0$ whenever $0 \leqq \lambda \leqq \mu - 1$. If $f_j$ separates the distinct points of $\pi^{-1}(B)$, then the polynomial $P_j(B; X)$ has a root of order precisely $\mu$ at the value $X = f_j(A)$, so in this case $P_j^{(\mu)}(B; f_j(A)) \neq 0$. Therefore clearly

$$B_\pi^\mu = \{A \in V : P_j^{(\lambda)}(\pi(A); f_j(A)) = 0 \text{ for } 0 \leqq \lambda < \mu \text{ and } 1 \leqq j \leqq r\}$$

and since $P_j^{(\lambda)}(\pi(A); f_j(A))$ is a holomorphic function of the point $A \in V$, that shows that $B_\pi^\mu$ is a holomorphic subvariety of $V$ and thereby concludes the proof.

In the preceding theorem clearly $B_\pi^1 = V$, so the assertion is rather trivial in the case $\mu = 1$; but $B_\pi^2 = B_\pi$ is the branch locus of the branched covering $\pi$ as defined in (1), so it follows that the branch locus itself is a holomorphic subvariety of $V$. The sets $B_\pi^\mu$ for $\mu \geqq 2$ are all holomorphic subvarieties of the branch locus. A simple consequence of the argument used to prove the above theorem is worth noting in passing.

**14. COROLLARY.** *If $\pi: V \to W$ is a finite branched holomorphic covering over an open subset $W \subseteq \mathbb{C}^n$, then the image under $\pi$ of the branch locus $B_\pi$ is a proper holomorphic subvariety $\pi(B_\pi) \subset W$.*

*Proof.* It is evident from Lemma 2 that this too is really a local assertion; so return to the situation in the proof of the preceding theorem, with the notation as introduced there. Since the functions $f_j | V$ separate the distinct points of $\pi^{-1}(B)$ for any $B \in W$, it is obvious that $B \in \pi(B_\pi)$ precisely when each polynomial $P_j(B; X)$ has a multiple root—hence, precisely when $d_j(B) = 0$ for $1 \leqq j \leqq r$ where $d_j \in {}_n\mathcal{O}_W$ is the discriminant of the polynomial $P_j(X)$. That shows that $\pi(B_\pi)$ is a holomorphic subvariety of $W$, necessarily a proper holomorphic subvariety of course, and concludes the proof.

It is obvious that if $\pi: V \to W$ is any holomorphic mapping between two holomorphic varieties and $Y$ is a holomorphic subvariety of $W$, then $\pi^{-1}(Y)$ is a holomorphic subvariety of $V$. However, it is not always the case that if $X$ is a holomorphic subvariety of $V$, then $\pi(X)$ is a holomorphic subvariety of $W$. For example, in the case of the mapping $\pi: \mathbb{C}^2 \to \mathbb{C}^2$ defined by $\pi(z_1, z_2) = (z_1, z_1 z_2)$, the image $\pi(\mathbb{C}^2)$ is the complement of the subset $\{(w_1, w_2) \in \mathbb{C}^2 : w_1 = 0, w_2 \neq 0\}$ and hence is not only not a holomorphic subvariety but also neither an open nor a closed subset of $\mathbb{C}^2$. Thus, there is some point to showing as in the preceding corollary that the image $\pi(B_\pi)$ is a holomorphic subvariety of $W$. Actually it will later be demonstrated that for a finite branched holomorphic covering $\pi: V \to W$, the image of any holomorphic subvariety $X \subseteq V$ is a holomorphic subvariety $\pi(X) \subseteq W$, subsuming the preceding corollary as a special case. But it is convenient to have the result of this corollary at hand for the following theorem.

**5. THEOREM.** *If $\pi: V \to W$ is a finite branched holomorphic covering of an open subset $W \subseteq \mathbb{C}^n$, if $D$ is a proper holomorphic subvariety of $W$, and if $V_1$ is a connected component of $V - \pi^{-1}(D) = \pi^{-1}(W - D)$, then the point set closure $\bar{V}_1$ of the subset $V_1$ in $V$ is a holomorphic subvariety of $V$.*

*Proof.* Since the image $\pi(B_\pi)$ of the branch locus is an analytic subvariety of $W$ by Corollary 14, it can be assumed without loss of generality that $B_\pi \subseteq D$; thus, the restriction $\pi: V - \pi^{-1}(D) \to W - D$ is a regular part of the finite branched covering $\pi$. Since, moreover, the theorem is really a local assertion, it can also be assumed as a consequence of Lemma 11 that $V$ is a holomorphic subvariety of an open polydisc $\Delta(0; R) \times \Delta(0; R') \subseteq \mathbb{C}^n \times \mathbb{C}^m$ and that $\pi: V \to W = \Delta(0; R)$ is induced by the natural projection $\mathbb{C}^n \times \mathbb{C}^m \to \mathbb{C}^n$. Recall from Corollary I-D3 that $W - D$ is connected; it is evident that $V - \pi^{-1}(D)$ has finitely many connected components, each of which is also a finite covering of $W - D$ under the mapping $\pi$. The set $V_1$ is of course one of these components, and the restriction $\pi_1 = \pi | V_1 : V_1 \to W - D$ is thus a finite unbranched holomorphic covering. The problem is just that of determining the behavior of $V_1$ at points of the branch locus $B_\pi$. If $f \in {}_V\mathcal{O}_V$ is any holomorphic function on all of $V$, then to the restriction $f | V_1 \in {}_{V_1}\mathcal{O}_{V_1}$ there is associated as in Theorem 5 a canonical polynomial $P_f(X) \in {}_n\mathcal{O}_{W-D}$ such that $P_f(f) = 0$ on $V_1$, by applying the construction of that theorem to the covering $\pi_1: V_1 \to W - D$. Since $f$ is holomorphic on all of $V$, though, it is clear that the coefficients of the polynomial $P_f(X)$ remain bounded in an open neighborhood of any point of $D$. Hence, from the extended Riemann removable singularities theorem, Theorem I-D2, it follows that these coefficients extend to holomorphic functions in all of $W$. Thus the polynomial $P_f(X)$ can also be viewed as an element of ${}_n\mathcal{O}_W$, and the composition $P_f(f)$ can consequently be viewed as an element of ${}_V\mathcal{O}_V$—that is, as a holomorphic function on all of $V$. This function vanishes on $V_1$ and hence by continuity also vanishes on the closure $\bar{V}_1$ of $V_1$ in $V$. Now choose a finite number of linear functions $f_1, \ldots, f_r$ in $\mathbb{C}^n \times \mathbb{C}^m$ such that these functions separate any $\nu$ points of $\mathbb{C}^n \times \mathbb{C}^m$ in the sense of Lemma 12, where $\nu$ is the order of the original branched covering $\pi: V \to W$. In terms of the associated holomorphic functions $p_j = P_{f_j}(f_j) \in {}_V\mathcal{O}_V$ introduce the holomorphic subvariety

$$V_* = \{Z \in V : p_1(Z) = \cdots = p_r(Z) = 0\}$$

As noted above, $\bar{V}_1 \subseteq V_*$; the theorem will be proved by showing conversely that $V_* \subseteq \bar{V}_1$ and hence that $\bar{V}_1$ is the holomorphic subvariety $V_*$. To see this suppose to the contrary that there is a point $A \in V_*$ such that $A \notin \bar{V}_1$. Let $B = \pi(A) \in W$ and let $A, A_1, A_2, \ldots$ be the distinct points of $\pi^{-1}(B)$. As in Lemma 2 choose an open neighborhood $W_B$ of the point $B$ so that the connected components of $\pi^{-1}(W_B)$ are disjoint open neighborhoods $V_A, V_{A_1}, V_{A_2}, \ldots$ of the points $A, A_1, A_2, \ldots$, respectively. Since $A \notin \bar{V}_1$, the neighborhood $W_B$ can be assumed chosen so small that $V_A \cap V_1 = \emptyset$. If the function $f_j$ takes distinct values at the points $A, A_1, A_2, \ldots$, the neighborhood $W_B$ can also be assumed chosen so small that the sets $f_j(V_A), f_j(V_{A_1}), f_j(V_{A_2}), \ldots$ are disjoint. Since $A \in V_*$, necessarily $0 = p_j(A) = P_{f_j}(B; f_j(A))$; thus, $f_j(A)$ is one of the roots of the polynomial $P_{f_j}(B; X)$. Whenever $B' \in W_B$ is near enough to $B$, then at least one of the roots of the polynomial $P_{f_j}(B'; X)$ must be contained in the ope[n] neighborhood $f_j(V_A)$ of the point $f_j(A)$, since it is either well known or eviden[t] from Lemma I-C3 that the roots of a polynomial are continuous functions of it[s] coefficients in this sense. However, if $B' \notin D$, the roots of the polynomial $P_{f_j}(B'; X$[)] are just the values of the function $f_j$ at the points of $\pi_1^{-1}(B')$. But since these point[s] $\pi_1^{-1}(B')$ lie in $V_1$ and $V_A \cap V_1 = \emptyset$, they must actually be contained in $V_{A_1} \cup V_{A_2} \cup \cdots$ and hence the values of the function $f_j$ at these points are in the set $f_j(V_{A_1}) \cup$ $f_j(V_{A_2}) \cup \cdots$ disjoint from $f_j(V_A)$. Thus no root of the polynomial $P_{f_j}(B'; X)$ can b[e] contained in the set $f_j(V_A)$ after all, which is a contradiction; that suffices to conclud[e] the proof.

# D

# Local Parametrization of Holomorphic Varieties

Although no examples of finite branched holomorphic coverings were given in the preceding section, there is in fact an abundance of examples. Perhaps the simplest are the following.

1. **THEOREM.** *If $U$ is an open subset of $\mathbb{C}^n$ and $P_j(X) \in {}_n\mathcal{O}_U[X]$ are monic polynomials with nonzero discriminants $d_j \in {}_n\mathcal{O}_U$ for $1 \leqq j \leqq m$, then the natural projection $\mathbb{C}^n \times \mathbb{C}^m \to \mathbb{C}^n$ exhibits the holomorphic subvariety*

$$V = \{Z \in U \times \mathbb{C}^m : P_j(z_1, \dots, z_n; z_{n+j}) = 0 \text{ for } 1 \leqq j \leqq m\} \tag{1}$$

*as a finite branched holomorphic covering of $U$.*

*Proof.* The restriction of the natural projection $\mathbb{C}^n \times \mathbb{C}^m \to \mathbb{C}^n$ to the subvariety $V \subseteq U \times \mathbb{C}^m$ is evidently a holomorphic mapping $\pi: V \to U$. For each fixed point $A \in U$ the polynomial $P_j(A; X)$ is a monic and hence nontrivial polynomial in $X$ and so has finitely many roots. Consequently, it is clear that the mapping $\pi$ is both finite and surjective. Furthermore, since $P_j(A; X)$ is monic and hence regular in the variable $X$, it is evident from Lemma I-C3 that the mapping $\pi$ is also proper, for that lemma essentially says that the roots of $P_j(A; X)$ vary continuously with $A$. Let $D = \{Z \in U : d_1(Z) = \cdots = d_m(Z) = 0\}$ be the analytic subvariety of $U$ defined by the discriminants of the polynomials $P_j(X)$, so the complement $U_0 = U - D$ is a connected dense open subset of $U$, and it follows immediately from Lemma I-C3 again that $V_0 = \pi^{-1}(U_0)$ is dense in $V$. Whenever $A \in U_0$, the polynomials $P_j(A; X)$ have only simple zeros, since $d_j(A) \neq 0$, and as a consequence of the implicit function theorem, Theorem I-C4, the mapping $\pi$ is a biholomorphic mapping in an open neighborhood of each point of $\pi^{-1}(A)$. That suffices to show that the mapping $\pi: V \to U$ is a finite branched holomorphic covering as desired.

Actually the assumption that the discriminants $d_j$ of the polynomials $P_j(X)$ are nonzero is not really essential, for if $d_j = 0$, then the polynomial $P_j(X)$ must have multiple factors and can be replaced by another polynomial defining the same subvariety but having a nonzero discriminant. There are some minor technical complications in this argument, though, and since the result will not be needed here, the details will be omitted; a more general assertion will be given later anyway. If the polynomial $P_j(X)$ is of degree $v_j$, then the finite branched holomorphic covering $\pi: V \to U$ is evidently of pure order $v = \prod_{j=1}^m v_j$. Note furthermore that the restriction to the subvariety $V$ of the coordinate function $z_j$ is a holomorphic function $f_j = z_j | V \in {}_V\mathcal{O}_V$, and to this function there is canonically associated a monic polynomial $P_{f_j}(X) \in {}_n\mathcal{O}_U[X]$ of degree $v$ as in Theorem C5. Indeed, it is clear that $P_{f_{n+j}}(X) = P_j(X)^{v/v_j}$ whenever $1 \leqq j \leqq m$, so in particular, $P_{f_{n+1}}(X) = P_1(X)$ in the special case that $m = 1$. In this special case something rather like a converse of the preceding theorem also holds.

**2. THEOREM.** *Let $U$ be an open subset of $\mathbb{C}^n$ and $V$ be a holomorphic subvariety of $U \times \mathbb{C} \subseteq \mathbb{C}^{n+1}$ such that the natural projection $U \times \mathbb{C} \to U$ exhibits $V$ as a finite branched holomorphic covering of $U$. If $P(X) \in {}_n\mathcal{O}_U[X]$ is the monic polynomial canonically associated to the function $z_{n+1} | V \in {}_V\mathcal{O}_V$ as in Theorem C5, then*

$$V = \{Z \in U \times \mathbb{C} : P(z_1, \ldots, z_n; z_{n+1}) = 0\}$$

*Furthermore at any point $A \in U \times \mathbb{C}$, the germ of the function $P(z_{n+1})$ generates the ideal* id $\mathbf{V}_A \subseteq {}_{n+1}\mathcal{O}_A$.

*Proof.* The set of zeros of the function $P(z_{n+1})$ is a holomorphic subvariety $W \subseteq U \times \mathbb{C}$, and by Theorem 1 the natural projection $U \times \mathbb{C} \to U$ also exhibits $W$ as a finite branched holomorphic covering of $U$. It is clear from the construction of the polynomial $P(X)$ that the branched coverings $V$ and $W$ have a common regular part, since whenever $Z' \in U$ is a point such that the polynomial $P(Z'; X)$ has distinct roots, then these roots are the distinct values of $z_{n+1}$ for which $(Z', z_{n+1}) \in V$; consequently $V = W$. If $\pi: V \to U$ is the given finite branched holomorphic covering, $A \in V$, and $\pi(A) = B \in U$, then as in Lemma C2 there is an open neighborhood $U_B$ of $B$ such that the connected components of $\pi^{-1}(U_B)$ are open neighborhoods $V_A$, $V_{A_1}, \ldots$ of the distinct points $A, A_1, \ldots$ of $\pi^{-1}(B)$, and the restrictions of $\pi$ to these components are finite branched holomorphic coverings of $U_B$. It is then obvious from the construction of the polynomial $P(X)$ that over $U_B$ there is a factorization $P(X) = P_A(X)P_{A_1}(X)\cdots$, where $P_A(X) \in {}_n\mathcal{O}_{U_B}[X]$ is the polynomial associated to the function $z_{n+1} | V_A$ as in Theorem C5 and similarly for the other factors. Note in particular therefore that the germs of the functions $P_{A_1}(z_{n+1})$, $P_{A_2}(z_{n+1})$, ... at the point $A$ are units in ${}_{n+1}\mathcal{O}_A$; hence, $\mathbf{P}(z_{n+1}) = \mathbf{P}_A(z_{n+1})\,\mathbf{u}$ where $\mathbf{u}$ is a unit in ${}_{n+1}\mathcal{O}_A$. The polynomial $\mathbf{P}_A(z_{n+1}) \in {}_n\mathcal{O}_B[z_{n+1}]$ is also clearly a Weierstrass polynomial of degree $o_\pi(A)$. Now if $\mathbf{f} \in \text{id}\,\mathbf{V}_A \in {}_{n+1}\mathcal{O}_A$, then apply the Weierstrass division theorem to write $\mathbf{f} = \mathbf{P}_A(z_{n+1})\mathbf{g} + \mathbf{r}$ where $\mathbf{g} \in {}_{n+1}\mathcal{O}_A$ and $\mathbf{r} \in {}_n\mathcal{O}_B[z_{n+1}]$ is a polynomial of degree strictly less than $o_\pi(A)$. However, since $\mathbf{r} \in \text{id}\,\mathbf{V}_A$ this polynomial generally vanishes at the $o_\pi(A)$ values corresponding to points on the finite branched holomorphic covering $\pi | V_A : V_A \to U_B$ of order $o_\pi(A)$. That can happen only when $r$ vanishes identically, though, so $\mathbf{f} = \mathbf{P}_A(z_{n+1})\mathbf{g} = \mathbf{P}(z_{n+1})\mathbf{u}^{-1}g$ is in the ideal in ${}_{n+1}\mathcal{O}_A$ generated by $\mathbf{P}(z_{n+1})$. That suffices to conclude the proof.

It is clear that there is no such simple converse to Theorem 1 in general, for if $V$ is a holomorphic subvariety of $U \times \mathbb{C}^m \subseteq \mathbb{C}^{n+m}$ and is exhibited as a finite branched holomorphic covering of $U$ by the natural projection $U \times \mathbb{C}^m \to U$, then the polynomials $P_j(X) \in {}_n\mathcal{O}_U[X]$ associated to the functions $z_{n+j}|V \in {}_V\mathcal{O}_V$ as in Theorem C5 for $1 \leqq j \leqq m$ describe a holomorphic subvariety properly larger than $V$. However, there actually is a set of polynomial equations that do serve indirectly to describe the subvariety $V$, and these equations can be used to show more generally that every irreducible germ of a holomorphic variety can be represented as the germ of a finite branched holomorphic covering of $\mathbb{C}^n$ for some index $n$. This in turn is really a very useful representation and is a basic tool in the subsequent more detailed study of the local properties of holomorphic varieties. This representation can be viewed as a normal form for an irreducible germ of a holomorphic variety, and it is in many ways analogous to the representation of a germ of a holomorphic function in terms of Weierstrass polynomials. The derivation of this representation indeed roughly parallels the earlier derivation of the Weierstrass preparation theorem. In this discussion it is convenient as before to identify ${}_k\mathcal{O}$ with the subring of ${}_l\mathcal{O}$ consisting of the germs of holomorphic functions that are independent of the variables $z_{k+1}, \ldots, z_l$ whenever $k \leqq l$—thus yielding canonical imbeddings ${}_k\mathcal{O} \subseteq {}_l\mathcal{O}$. Recall from Definition I-C1 that an element $\mathbf{f}_j \in {}_j\mathcal{O}$ is regular in the variable $z_j$ if it has a representative $f_j$ in some open neighborhood of the origin such that $f_j(0, \ldots, 0, z_j)$ is not identically zero.

**3. DEFINITION.** *An ideal $\mathfrak{A} \subseteq {}_m\mathcal{O}$ is* **regular in the variables $z_{n+1}, \ldots, z_m$** if ${}_n\mathcal{O} \cap \mathfrak{A} = 0$ *and if there is an element $\mathbf{f}_j \in {}_j\mathcal{O} \cap \mathfrak{A}$ that is regular in the variable $z_j$ for each index $j$ in the interval $n + 1 \leqq j \leqq m$.*

Note that the principal ideal ${}_m\mathcal{O}\mathbf{f}$ generated by an element $\mathbf{f} \in {}_m\mathcal{O}$ is regular in the variable $z_m$ precisely when $f$ is regular in the variable $z_m$. Thus the preceding definition can be viewed as an extension to more general ideals of Definition I-C1. Note further that this definition apparently depends on the order in which the variables are listed, not just on the set of variables in which the ideal is regular. It will shortly be shown that the order is really immaterial to regularity, but until that is done the order must be taken into account.

**4. LEMMA.** *Any nonzero ideal $\mathfrak{A} \subseteq {}_m\mathcal{O}$ can be made regular in some set of variables $z_{n+1}, \ldots, z_m$ by a suitable nonsingular linear change of coordinates in $\mathbb{C}^m$.*

*Proof.* Choose a nonzero element $\mathbf{f}_m \in \mathfrak{A}$. It follows from Lemma I-C2 that by a suitable nonsingular linear change of coordinates in $\mathbb{C}^m$ the element $\mathbf{f}_m$ can be made regular in the variable $z_m$, and if ${}_{m-1}\mathcal{O} \cap \mathfrak{A} = 0$, the ideal $\mathfrak{A}$ is itself regular in the variable $z_m$. If ${}_{m-1}\mathcal{O} \cap \mathfrak{A} \neq 0$, choose a nonzero element $\mathbf{f}_{m-1} \in {}_{m-1}\mathcal{O} \cap \mathfrak{A}$, and by applying Lemma I-C2 again note that there is a nonsingular linear change of coordinates in $\mathbb{C}^{m-1}$ after which $\mathbf{f}_{m-1}$ is regular in the variable $z_{m-1}$. Since this change of coordinates does not involve $z_m$, the element $\mathbf{f}_m$ remains regular in the variable $z_m$; so if ${}_{m-2}\mathcal{O} \cap \mathfrak{A} = 0$, the ideal $\mathfrak{A}$ is regular in the variables $z_{m-1}, z_m$. If ${}_{m-2}\mathcal{O} \cap \mathfrak{A} \neq$

0, the argument can be repeated, and repetition eventually leads to the proof of the desired result.

It should be noted incidentally that any finite collection of ideals $\mathfrak{A}_i \subseteq {}_m\mathcal{O}$ can simultaneously be made regular in some variables $z_{n_i+1}, \ldots, z_m$ by a suitable nonsingular linear change of coordinates in $\mathbf{C}^m$, since Lemma I-C2 holds for any finite collection of holomorphic functions as noted in the discussion of that lemma. As a useful notational convenience, for a fixed ideal $\mathfrak{A} \subseteq {}_m\mathcal{O}$ and any element $\mathbf{f} \in {}_m\mathcal{O}$ the image in the residue class ring ${}_m\mathcal{O}/\mathfrak{A}$ of the element $\mathbf{f}$ will be denoted by $\tilde{\mathbf{f}}$. Correspondingly the image in the residue class ring ${}_m\mathcal{O}/\mathfrak{A}$ of the subring ${}_j\mathcal{O} \subseteq {}_m\mathcal{O}$ will be denoted by ${}_j\tilde{\mathcal{O}}$, so in particular ${}_m\tilde{\mathcal{O}} = {}_m\mathcal{O}/\mathfrak{A}$.

**5. LEMMA.** *A nonzero ideal $\mathfrak{A} \subseteq {}_m\mathcal{O}$ is regular in the variables $z_{n+1}, \ldots, z_m$ precisely when ${}_n\mathcal{O}$ is isomorphic to its image ${}_n\tilde{\mathcal{O}} \subseteq {}_m\tilde{\mathcal{O}}$ and ${}_m\tilde{\mathcal{O}}$ is an integral algebraic extension of the subring ${}_n\tilde{\mathcal{O}}$ generated by the residue classes $\tilde{\mathbf{z}}_{n+1}, \ldots, \tilde{\mathbf{z}}_m$.*

*Proof.* If $\mathfrak{A}$ is regular in the variables $z_{n+1}, \ldots, z_m$, then ${}_n\mathcal{O} \cap \mathfrak{A} = 0$, so that ${}_n\tilde{\mathcal{O}} \cong {}_n\mathcal{O}/({}_n\mathcal{O} \cap \mathfrak{A}) \cong {}_n\mathcal{O}$. Furthermore, there are elements $\mathbf{f}_j \in {}_j\mathcal{O} \cap \mathfrak{A}$ that are regular in the variables $z_j$ for $n + 1 \leq j \leq m$, and by applying the Weierstrass preparation theorem it can even be assumed that these elements are Weierstrass polynomials and hence of the form

$$\mathbf{f}_j = z_j^{\nu_j} + \mathbf{a}_{j1} z_j^{\nu_j - 1} + \cdots + \mathbf{a}_{j\nu_j} \in {}_{j-1}\mathcal{O}[z_j] \cap \mathfrak{A}$$

Upon passing to the residue class ring ${}_m\tilde{\mathcal{O}}$, it follows that

$$0 = \tilde{\mathbf{f}}_j = \tilde{\mathbf{z}}_j^{\nu_j} + \tilde{\mathbf{a}}_{j1} \tilde{\mathbf{z}}_j^{\nu_j - 1} + \cdots + \tilde{\mathbf{a}}_{j\nu_j} \in {}_m\tilde{\mathcal{O}}$$

where $\tilde{\mathbf{a}}_{jk} \in {}_{j-1}\tilde{\mathcal{O}}$, and this is just the condition that $\tilde{\mathbf{z}}_j$ is integral over the subring ${}_{j-1}\tilde{\mathcal{O}} \subseteq {}_m\tilde{\mathcal{O}}$. In addition it follows from the Weierstrass division theorem that for any element $\mathbf{f} \in {}_j\mathcal{O}$ there are elements $\mathbf{g} \in {}_j\mathcal{O}$ and $\mathbf{r} \in {}_{j-1}\mathcal{O}[z_j]$ such that $\mathbf{f} = \mathbf{f}_j\mathbf{g} + \mathbf{r}$. Upon passing to the residue class ring ${}_m\tilde{\mathcal{O}}$ and recalling that $\tilde{\mathbf{f}}_j = 0$, it follows that $\tilde{\mathbf{f}} = \tilde{\mathbf{r}} \in {}_{j-1}\tilde{\mathcal{O}}[\tilde{z}_j]$; thus, ${}_j\tilde{\mathcal{O}}$ is actually the integral algebraic extension of the subring ${}_{j-1}\tilde{\mathcal{O}} \subseteq {}_j\tilde{\mathcal{O}}$ generated by the element $\tilde{\mathbf{z}}_j$. Altogether therefore the full residue class ring ${}_m\tilde{\mathcal{O}}$ arises from the subring ${}_n\tilde{\mathcal{O}}$ by successive integral algebraic extensions of the elements $\tilde{\mathbf{z}}_{n+1}, \ldots, \tilde{\mathbf{z}}_m$, and it follows from the well-known algebraic theorem of the transitivity of integral dependence that ${}_m\tilde{\mathcal{O}} = {}_n\tilde{\mathcal{O}}[\tilde{\mathbf{z}}_{n+1}, \ldots, \tilde{\mathbf{z}}_m]$ where all the elements $\tilde{\mathbf{z}}_{n+1}, \ldots, \tilde{\mathbf{z}}_m$ are integral over the subring ${}_n\tilde{\mathcal{O}} \subseteq {}_m\tilde{\mathcal{O}}$.

On the other hand, if ${}_n\mathcal{O} \cong {}_n\tilde{\mathcal{O}}$ and if ${}_m\tilde{\mathcal{O}} = {}_n\tilde{\mathcal{O}}[\tilde{\mathbf{z}}_{n+1}, \ldots, \tilde{\mathbf{z}}_m]$ where the elements $\tilde{\mathbf{z}}_{n+1}, \ldots, \tilde{\mathbf{z}}_m$ are integral over the subring ${}_n\tilde{\mathcal{O}} \subseteq {}_m\tilde{\mathcal{O}}$, then ${}_n\mathcal{O} \cap \mathfrak{A} = 0$ and there are monic polynomials $\mathbf{p}_j(X) \in {}_n\mathcal{O}[X] \cong {}_n\tilde{\mathcal{O}}[X]$ such that $\mathbf{p}_j(\tilde{\mathbf{z}}_j) = 0$ for $n + 1 \leq j \leq m$. The germs $\mathbf{p}_j(z_j) \in {}_n\mathcal{O}[z_j] \subseteq {}_m\mathcal{O}$ must then belong to the ideal $\mathfrak{A}$, and since $\mathbf{p}_j(z_j) \in {}_j\mathcal{O} \cap \mathfrak{A}$ is evidently regular in the variable $z_j$ for $n + 1 \leq j \leq m$, it follows that the ideal $\mathfrak{A}$ is regular in the variables $z_{n+1}, \ldots, z_m$. That concludes the proof.

It is clear from Lemma 5 that the condition that an ideal $\mathfrak{A} \subseteq {}_m\mathcal{O}$ be regular in the variables $z_{n+1}, \ldots, z_m$ is actually independent of the order in which these variables are taken. Indeed, it is even clear from this lemma that if an ideal $\mathfrak{A} \subseteq {}_m\mathcal{O}$ is regular in the variables $z_{n+1}, \ldots, z_m$ and if $z_j' = \sum_k c_{jk} z_k$ is a nonsingular linear change of variables such that $z_j' = \sum_{k=1}^n c_{jk} z_k$ for $1 \leq j \leq n$ and $z_j' = \sum_{k=n+1}^m c_{jk} z_k$ for $n+1 \leq j \leq m$, then the ideal $\mathfrak{A}$ is also regular in the variables $z_{n+1}', \ldots, z_m'$.

For prime ideals it is possible to say a good deal more. If the prime ideal $\mathfrak{p} \subseteq {}_m\mathcal{O}$ is regular in the variables $z_{n+1}, \ldots, z_m$, then by Lemma 5 there is an isomorphism ${}_m\mathcal{O}/\mathfrak{p} \cong {}_n\mathcal{O}[\mathbf{\tilde{z}}_{n+1}, \ldots, \mathbf{\tilde{z}}_m]$ where the elements $\mathbf{\tilde{z}}_j$ are integral over ${}_n\mathcal{O}$. Since ${}_m\mathcal{O}/\mathfrak{p}$ is an integral domain, it has a well-defined field of quotients ${}_m\tilde{\mathcal{M}}$, and ${}_m\tilde{\mathcal{M}} = {}_n\mathcal{M}[\mathbf{\tilde{z}}_{n+1}, \ldots, \mathbf{\tilde{z}}_m]$ where the elements $\mathbf{\tilde{z}}_{n+1}, \ldots, \mathbf{\tilde{z}}_m$ are algebraic over the subfield ${}_n\mathcal{M} \subseteq {}_m\tilde{\mathcal{M}}$. As is familiar from elementary algebra, each element $\mathbf{\tilde{z}}_j$ for $n+1 \leq j \leq m$ is the root of a unique irreducible monic polynomial $\mathbf{p}_j(X) \in {}_n\mathcal{M}[X]$, the **defining polynomial** for this element over the field ${}_n\mathcal{M}$. Since these elements $\mathbf{\tilde{z}}_j$ are integral over ${}_n\mathcal{O}$, the coefficients of the defining polynomial $\mathbf{p}_j(X)$ must be elements of ${}_n\mathcal{M}$ that are integral over ${}_n\mathcal{O}$ as well; but ${}_n\mathcal{O}$ is a unique factorization domain, and since it is well known that any unique factorization domain is integrally closed in its field of quotients, the coefficients of $\mathbf{p}_j(X)$ actually lie in ${}_n\mathcal{O}$. Thus, the defining polynomials are elements $\mathbf{p}_j(X) \in {}_n\mathcal{O}[X]$, and since they are irreducible over the field ${}_n\mathcal{M}$, their discriminants are nonzero elements $\mathbf{d}_j \in {}_n\mathcal{O}$. Note that since $\mathbf{p}_j(\mathbf{\tilde{z}}_j) = 0$, the elements $\mathbf{p}_j(z_j) \in {}_n\mathcal{O}[z_j]$ must be contained in the ideal $\mathfrak{p}$. These latter polynomials must indeed be Weierstrass polynomials, for otherwise an application of the Weierstrass preparation theorem would yield a monic polynomial $\mathbf{p}_j^*(z_j) \in \mathfrak{p} \cap {}_n\mathcal{O}[z_j]$ with degree $\mathbf{p}_j^*(X) <$ degree $\mathbf{p}_j(X)$, and since $\mathbf{p}_j^*(\mathbf{\tilde{z}}_j) = 0$ that contradicts the assumption that $\mathbf{p}_j(X)$ is the defining polynomial of the element $\mathbf{\tilde{z}}_j$ over the field ${}_n\mathcal{M}$. Finally extensions of fields, unlike extensions of rings, can in many circumstances be generated by a single appropriately chosen element, and that suggests considering the following extension of Definition 3.

**6. DEFINITION.** *A prime ideal $\mathfrak{p} \subseteq {}_m\mathcal{O}$ is* **strictly regular in the variables $\mathbf{z}_{n+1}, \ldots, \mathbf{z}_m$** *if it is regular in these variables and in addition the field ${}_m\tilde{\mathcal{M}}$ is generated over the subfield ${}_n\mathcal{M} \cong {}_n\tilde{\mathcal{M}}$ by the single element $\mathbf{\tilde{z}}_{n+1}$.*

It should be noted that the order of the variables is important in the definition of strict regularity, at least to the extent that the coordinate function $z_{n+1}$ plays a special role in this definition.

**7. LEMMA.** *A nonzero prime ideal $\mathfrak{p} \subseteq {}_m\mathcal{O}$ can be made strictly regular in some set of variables $z_{n+1}, \ldots, z_m$ by a suitable nonsingular linear change of coordinates in $\mathbb{C}^m$.*

*Proof.* It follows from Lemma 4 that the ideal $\mathfrak{p}$ can be made regular in some variables $z_{n+1}, \ldots, z_m$ by a suitable nonsingular linear change of coordinates in $\mathbb{C}^m$, so that ${}_m\tilde{\mathcal{M}} \cong {}_n\mathcal{M}[\mathbf{\tilde{z}}_{n+1}, \ldots, \mathbf{\tilde{z}}_m]$. As a consequence of the familiar algebraic theorem of the primitive element there are complex constants $c_j$ such that the extension ${}_m\tilde{\mathcal{M}}$ is generated over ${}_n\mathcal{M}$ by the single element $\mathbf{\tilde{z}}_{n+1}' = c_{n+1}\mathbf{\tilde{z}}_{n+1} + \cdots + c_m\mathbf{\tilde{z}}_m$, since $\mathbb{C}$ is an infinite subfield of ${}_n\mathcal{M}$. Then for any nonsingular linear change of coordinates in $\mathbb{C}^m$ of the form

$$z_j' = \begin{cases} z_j & \text{for } 1 \leqq j \leqq n, \\ c_{n+1} z_{n+1} + \cdots + c_m z_m & \text{for } j = n+1, \\ \sum_{k=n+1}^{m} c_{jk} z_k & \text{for } n+2 \leqq j \leqq m, \end{cases}$$

for some constants $c_{jk}$, the ideal $\mathfrak{p}$ remains regular in the variables $z'_{n+1}, \ldots, z'_m$, as noted before, while ${}_m\tilde{\mathcal{M}} \cong {}_n\mathcal{M}[\mathbf{\tilde{z}}'_{n+1}]$. That suffices to conclude the proof.

If the prime ideal $\mathfrak{p} \subseteq {}_m\mathcal{O}$ is strictly regular in the variables $z_{n+1}, \ldots, z_m$, then the field extension ${}_m\tilde{\mathcal{M}}$ over ${}_n\mathcal{M}$ is described completely by the defining polynomial $\mathbf{p}_{n+1}(X)$ alone. The elements $1, \mathbf{\tilde{z}}_{n+1}, \ldots, \mathbf{\tilde{z}}_{n+1}^{\nu-1}$ form a basis for ${}_m\tilde{\mathcal{M}}$ when it is viewed as a vector space over the field ${}_n\mathcal{M}$, where $\nu = \text{degree } \mathbf{p}_{n+1}(X)$, while multiplication in the field ${}_m\tilde{\mathcal{M}}$ is bilinear over ${}_n\mathcal{M}$ and the products of the basis elements can be read off easily from the defining polynomial. It is a familiar result from algebra that the ring extension ${}_m\tilde{\mathcal{O}}$ over ${}_n\mathcal{O}$ can be described in a roughly similar manner, although ring extensions are rather more complicated than field extensions and the defining polynomial $\mathbf{p}_{n+1}(X)$ does not quite serve to describe the ring ${}_m\tilde{\mathcal{O}}$ completely. Any element in the extension ring ${}_m\tilde{\mathcal{O}}$ over ${}_n\mathcal{O}$ can be written uniquely as a linear combination of the elements $1/\mathbf{d}, \mathbf{\tilde{z}}_{n+1}/\mathbf{d}, \ldots, \mathbf{\tilde{z}}_{n+1}^{\nu-1}/\mathbf{d}$ with coefficients from the ring ${}_n\mathcal{O}$, where $\mathbf{d} \in {}_n\mathcal{O}$ is the discriminant of the defining polynomial $\mathbf{p}_{n+1}(X)$, and the product of any two such elements is again determined by the defining polynomial in the obvious manner. The difficulty is that not all such linear expressions belong to the ring ${}_m\tilde{\mathcal{O}}$, so that

$${}_m\tilde{\mathcal{O}} \subseteq {}_n\mathcal{O} \cdot \frac{1}{\mathbf{d}} + {}_n\mathcal{O} \cdot \frac{\mathbf{\tilde{z}}_{n+1}}{\mathbf{d}} + \cdots + {}_n\mathcal{O} \cdot \frac{\mathbf{\tilde{z}}_{n+1}^{\nu-1}}{\mathbf{d}}$$

but this inclusion may well be a strict inclusion. This inclusion is already a useful result, however. In particular it implies that to each coordinate function $z_j$ for $n+2 \leqq j \leqq m$ there corresponds a unique polynomial $\mathbf{Q}_j(X) \in {}_n\mathcal{O}[X]$ of degree $\nu - 1$ such that $\mathbf{d} \cdot \mathbf{\tilde{z}}_j = \mathbf{Q}_j(\mathbf{\tilde{z}}_{n+1})$, or equivalently such that $\mathbf{q}_j = \mathbf{d} \cdot z_j - \mathbf{Q}_j(z_{n+1}) \in \mathfrak{p} \cap {}_n\mathcal{O}[z_{n+1}, z_j]$. The defining polynomials $\mathbf{p}_j(z_j) \in {}_n\mathcal{O}[z_j]$ for $n+1 \leqq j \leqq m$ together with the polynomials $\mathbf{q}_j(z_{n+1}, z_j) \in {}_n\mathcal{O}[z_{n+1}, z_j]$ for $n+2 \leqq j \leqq m$ form a useful canonical set of equations in the ideal $\mathfrak{p}$, and although they do not necessarily generate the ideal $\mathfrak{p}$, they do suffice to describe that ideal completely if indirectly. To analyze this situation further it is convenient to introduce the associated ideal $\mathfrak{A} \subseteq {}_m\mathcal{O}$ generated by the elements $\mathbf{p}_{n+1}, \ldots, \mathbf{p}_m, \mathbf{q}_{n+2}, \ldots, \mathbf{q}_m$, together with the smaller associated ideal $\mathfrak{A}' \subseteq {}_m\mathcal{O}$ generated by the elements $\mathbf{p}_{n+1}, \mathbf{q}_{n+2}, \ldots, \mathbf{q}_m$.

8. LEMMA. *If the prime ideal $\mathfrak{p} \subseteq {}_m\mathcal{O}$ is strictly regular in the variables $z_{n+1}, \ldots, z_m$, if $\mathbf{d} \in {}_n\mathcal{O}$ is the discriminant of the defining polynomial $\mathbf{p}_{n+1}(z_{n+1}) \in \mathfrak{p} \cap {}_n\mathcal{O}[z_{n+1}]$, and if $\mathfrak{A}$ and $\mathfrak{A}'$ are the associated ideals to $\mathfrak{p}$, then there is an integer $\nu \geqq 0$ such that:*

(i) *For any* $\mathbf{f} \in {}_m\mathcal{O}$ *there is a polynomial* $\mathbf{r} \in {}_n\mathcal{O}[z_{n+1}]$ *with degree* $\mathbf{r} <$ degree $\mathbf{p}_{n+1}$ *such that* $\mathbf{d}^\nu \cdot \mathbf{f} - \mathbf{r} \in \mathfrak{A}$.

(ii) $\mathbf{d}^\nu \cdot \mathfrak{p} \subseteq \mathfrak{A} \subseteq \mathfrak{p}$

(iii) $\mathbf{d}^\nu \cdot \mathfrak{A} \subseteq \mathfrak{A}' \subseteq \mathfrak{A}$

*Proof.* (i) For any element $\mathbf{f} \in {}_m\mathcal{O}$ it follows from the Weierstrass division theorem that $\mathbf{f} = \mathbf{p}_m\mathbf{g}_m + \mathbf{r}_m$ for some element $\mathbf{g}_m \in {}_m\mathcal{O}$ and some polynomial $\mathbf{r}_m \in {}_{m-1}\mathcal{O}[z_m]$ with degree $\mathbf{r}_m <$ degree $\mathbf{p}_m$, since the defining polynomial $\mathbf{p}_m \in {}_{m-1}\mathcal{O}[z_m]$ is a Weierstrass polynomial. Each coefficient of the polynomial $\mathbf{r}_m$ is an element of ${}_{m-1}\mathcal{O}$ and can be divided by the defining polynomial $\mathbf{p}_{m-1}$ to leave a remainder in ${}_{m-2}\mathcal{O}[z_{m-1}]$ by another application of the Weierstrass division theorem. By collecting these polynomials together it follows that $\mathbf{f} = \mathbf{p}_m\mathbf{g}_m + \mathbf{p}_{m-1}\mathbf{g}_{m-1} + \mathbf{r}_{m-1}$ where $\mathbf{g}_j \in {}_m\mathcal{O}$ and $\mathbf{r}_{m-1} \in {}_{m-2}\mathcal{O}[z_{m-1}, z_m]$. Repeating this argument leads eventually to the result that

$$\mathbf{f} = \mathbf{p}_m\mathbf{g}_m + \cdots + \mathbf{p}_{n+1}\mathbf{g}_{n+1} + \mathbf{r}_{n+1} \tag{2}$$

where $\mathbf{g}_j \in {}_m\mathcal{O}$ and $\mathbf{r}_{n+1} \in {}_n\mathcal{O}[z_{n+1}, \ldots, z_m]$ is a polynomial such that the degree of $\mathbf{r}_{n+1}$ in the variable $\mathbf{z}_j$ is strictly less than the degree of $\mathbf{p}_j$ for $n + 1 \leqq j \leqq m$. Next for any index $j$ in the interval $n + 2 \leqq j \leqq m$ and any positive integer $\lambda$, note that

$$\mathbf{d}^\lambda \cdot \mathbf{z}_j^\lambda = (\mathbf{q}_j + \mathbf{Q}_j)^\lambda = \mathbf{q}_j \cdot \mathbf{h}_{\lambda j} + \mathbf{Q}_j^\lambda$$

for some element $\mathbf{h}_{\lambda j} \in {}_m\mathcal{O}$, as is apparent upon using the binomial theorem and collecting together in the expression $\mathbf{q}_j \cdot \mathbf{h}_{\lambda j}$ all terms divisible by $\mathbf{q}_j$. So upon applying this observation to each term $\mathbf{z}_j^\lambda$ in the polynomial $\mathbf{r}_{n+1}$ and collecting together all expressions divisible by any of the polynomials $\mathbf{q}_j$ it follows that

$$\mathbf{d}^\nu \cdot \mathbf{r}_{n+1} = \mathbf{q}_m\mathbf{h}_m + \cdots + \mathbf{q}_{n+2}\mathbf{h}_{n+2} + \mathbf{r}' \tag{3}$$

where $\nu = \sum_{j=n+2}^{m} (\text{degree } \mathbf{p}_j - 1)$, $\mathbf{h}_j \in {}_m\mathcal{O}$, and $\mathbf{r}' \in {}_n\mathcal{O}[z_{n+1}]$. Apply the Weierstrass division theorem yet again to write $\mathbf{r}' = \mathbf{p}_{n+1} \cdot \mathbf{h}_{n+1} + \mathbf{r}$ where $\mathbf{r} \in {}_n\mathcal{O}[z_{n+1}]$ and degree $\mathbf{r} <$ degree $\mathbf{p}_{n+1}$. Substituting this in (3) and combining the result with (2) lead to

$$\mathbf{d}^\nu \cdot \mathbf{f} = \sum_{j=n+1}^{m} \mathbf{d}^\nu \mathbf{g}_j \mathbf{p}_j + \sum_{j=n+2}^{m} \mathbf{h}_j \mathbf{q}_j + \mathbf{h}_{n+1}\mathbf{p}_{n+1} + \mathbf{r}$$

so that $\mathbf{d}^\nu \cdot \mathbf{f} - \mathbf{r} \in \mathfrak{A}$ as desired.

(ii) It is of course clear from the definitions that $\mathfrak{A} \subseteq \mathfrak{p}$. If $\mathbf{f} \in \mathfrak{p}$, then by applying (i) write $\mathbf{d}^\nu \cdot \mathbf{f} = \mathbf{g} + \mathbf{r}$ where $\mathbf{g} \in \mathfrak{A} \subseteq \mathfrak{p}$ and $\mathbf{r} \in {}_n\mathcal{O}[z_{n+1}]$ is a polynomial with degree $\mathbf{r} <$ degree $\mathbf{p}_{n+1}$. But since both $\mathbf{f}$ and $\mathbf{g}$ belong to $\mathfrak{p}$, then necessarily $\mathbf{r} \in \mathfrak{p}$ also, and hence $\mathbf{r} = 0$ since $\mathbf{p}_{n+1}$ is the polynomial of lowest degree in $\mathfrak{p} \cap {}_n\mathcal{O}[z_{n+1}]$. Thus, $\mathbf{d}^\nu \cdot \mathbf{f} \in \mathfrak{A}$ as desired.

(iii) In order to prove the final assertion of the lemma, it is only necessary to show that $\mathbf{d}^\nu \cdot \mathbf{p}_j \in \mathfrak{A}'$ for $n + 2 \leqq j \leqq m$. Since $\mathbf{p}_j \in {}_n\mathcal{O}[z_j]$ and degree $\mathbf{p}_j \leqq \nu$, it follows as in the proof of part (i) that $\mathbf{d}^{\nu}\cdot\mathbf{p}_j = \mathbf{h}_j\mathbf{q}_j + \mathbf{r}'$ where $\mathbf{r}' \in {}_n\mathcal{O}[z_{n+1}]$. Then apply the Weierstrass division theorem to write $\mathbf{r}' = \mathbf{h}_{n+1}\mathbf{p}_{n+1} + \mathbf{r}$ where $\mathbf{r} \in {}_n\mathcal{O}[z_{n+1}]$ and degree $\mathbf{r} <$ degree $\mathbf{p}_{n+1}$. Since $\mathbf{r} \in \mathfrak{p}$ it follows as in the proof of part (ii) that $\mathbf{r} = 0$, so that $\mathbf{d}^{\nu}\cdot\mathbf{p}_j = \mathbf{h}_j\mathbf{q}_j + \mathbf{h}_{n+1}\mathbf{p}_{n+1} \in \mathfrak{A}'$, and that suffices to conclude the proof.

For later purposes it is worth noting here that part (i) of the preceding lemma is a consequence of a simple formal argument using only the explicit form of the polynomials $\mathbf{p}_j$ and $\mathbf{q}_j$. Indeed it is clear that part (i) of Lemma 8 holds more generally whenever $\mathfrak{A} \subseteq {}_m\mathcal{O}$ is an ideal generated by monic polynomials $\mathbf{p}_j \in {}_n\mathcal{O}[z_j]$ for $n + 1 \leq j \leq m$ and by polynomials $\mathbf{q}_j \in {}_n\mathcal{O}[z_{n+1}, z_j]$ for $n + 2 \leq j \leq m$ where $\mathbf{q}_j = \mathbf{d}\cdot z_j + \mathbf{Q}_j$ for some nonzero element $\mathbf{d} \in {}_n\mathcal{O}$ and some polynomials $\mathbf{Q}_j \in {}_n\mathcal{O}[z_{n+1}]$. If the polynomials $p_j$ are not themselves Weierstrass polynomials, the Weierstrass preparation theorem can be applied to replace them by Weierstrass polynomials of lower degrees. Parts (ii) and (iii) require some further properties of these polynomials, as derived in the preceding discussion. This lemma has an immediate geometric interpretation, which together with Lemma 8 itself can be viewed as in a sense a generalization of Theorem 2.

**9. LEMMA.** *If the prime ideal $\mathfrak{p} \subseteq {}_m\mathcal{O}$ is strictly regular in the variables $z_{n+1}, \ldots, z_m$, if $\mathfrak{A}$ and $\mathfrak{A}'$ are the associated ideals to $\mathfrak{p}$, and if $\mathbf{d} \in {}_n\mathcal{O}$ is the discriminant of the defining polynomial $\mathbf{p}_{n+1}(X) \in \mathfrak{p} \cap {}_n\mathcal{O}[z_{n+1}]$, then when $\mathbf{D} = \operatorname{loc} {}_m\mathcal{O}\mathbf{d}$ is the locus of the principal ideal in ${}_m\mathcal{O}$ generated by $\mathbf{d}$, it follows that*

(i) $\operatorname{loc}\mathfrak{p} \subseteq \operatorname{loc}\mathfrak{A} \subseteq \operatorname{loc}\mathfrak{A}'$
(ii) $\mathbf{D} \cup \operatorname{loc}\mathfrak{p} = \mathbf{D} \cup \operatorname{loc}\mathfrak{A} = \mathbf{D} \cup \operatorname{loc}\mathfrak{A}'$

*Proof.* (i) Since $\mathfrak{A}' \subseteq \mathfrak{A} \subseteq \mathfrak{p}$ as a consequence of the definitions of the associated ideals, the first assertion is quite trivial.

(ii) It follows from Lemma 8 that $\mathbf{d}^{2\nu}\cdot\mathfrak{p} \subseteq \mathfrak{A}'$ or equivalently that $({}_m\mathcal{O}\mathbf{d}^{2\nu})\cdot\mathfrak{p} \subseteq \mathfrak{A}'$. Hence, from Theorem B4 it follows that $\operatorname{loc}\mathfrak{A}' \subseteq (\operatorname{loc} {}_m\mathcal{O}\mathbf{d}^{2\nu}) \cup (\operatorname{loc}\mathfrak{p}) = \mathbf{D} \cup \operatorname{loc}\mathfrak{p}$. Assertion (ii) is an immediate consequence of this observation and assertion (i), and that concludes the proof.

The germs $\operatorname{loc}\mathfrak{A}$ and $\operatorname{loc}\mathfrak{A}'$ are thus very closely related to the germ $\operatorname{loc}\mathfrak{p}$. On the other hand, quite simple and explicit generators have been given for the ideals $\mathfrak{A}$ and $\mathfrak{A}'$, so their loci can be described rather completely. That leads to the following basic geometric property of the germ $\operatorname{loc}\mathfrak{p}$.

**10. THEOREM (local parametrization theorem).** *If the prime ideal $\mathfrak{p} \subseteq {}_m\mathcal{O}$ is strictly regular in the variables $z_{n+1}, \ldots, z_m$, and if $\mathbf{p}_{n+1}(z_{n+1}) \in {}_n\mathcal{O}[z_{n+1}] \cap \mathfrak{p}$ is the defining polynomial for the variable $z_{n+1}$ and is of degree r, then:*

(i) *The natural projection $\mathbb{C}^m = \mathbb{C}^n \times \mathbb{C}^{m-n} \to \mathbb{C}^n$ exhibits* $\operatorname{loc}\mathfrak{p}$ *as the germ of a finite branched holomorphic covering of pure order r over $\mathbb{C}^n$.*
(ii) *The natural projection $\mathbb{C}^m = \mathbb{C}^{n+1} \times \mathbb{C}^{m-n-1} \to \mathbb{C}^{n+1}$ exhibits* $\operatorname{loc}\mathfrak{p}$ *as the germ of a finite branched holomorphic covering of order* one *of* $\operatorname{loc} {}_{n+1}\mathcal{O}\mathbf{p}_{n+1}(z_{n+1})$.

*Proof.* As a notational convenience let $\mathbf{V} = \text{loc}\ \mathfrak{p} \subseteq \mathbb{C}^m$ and $\mathbf{W} = \text{loc}\ {}_{n+1}\mathcal{O}\mathbf{p}_{n+1}(z_{n+1}) \subseteq \mathbb{C}^{n+1}$, and if $\mathbf{d} \in {}_n\mathcal{O}$ is the discriminant of the polynomial $\mathbf{p}_{n+1}$, let $\mathbf{D} = \text{loc}\ {}_m\mathcal{O}\mathbf{d} \subseteq \mathbb{C}^m$ and note that $\mathbf{D} = \mathbf{D}_1 \times \mathbb{C}^{m-n}$ where $\mathbf{D}_1 = \text{loc}\ {}_n\mathcal{O}\mathbf{d} \subseteq \mathbb{C}^n$, since the germ $\mathbf{d}$ depends only on the variables $z_1, \ldots, z_n$. It follows from Theorem 1 that the natural projection $\mathbb{C}^{n+1} = \mathbb{C}^n \times \mathbb{C}^1 \to \mathbb{C}^n$ induces the germ $\pi''\colon W \to \mathbb{C}^n$ of a finite branched holomorphic covering. Indeed, as in the proof of that theorem, this covering is of pure order $r$, and if $\mathbf{W}_0 = \mathbf{W} - \mathbf{D} \cap \mathbf{W}$, then the restriction $\pi''|\mathbf{W}_0 : \mathbf{W}_0 \to \mathbb{C}^n - \mathbf{D}_1$ is the germ of a regular part of the finite branched covering $\pi''$. Furthermore, after introducing the germ

$$\mathbf{V}_1 = \{Z \in \mathbb{C}^m : \mathbf{p}_{n+1}(z_{n+1}) = \cdots = \mathbf{p}_m(z_m) = 0\} \subseteq \mathbb{C}^m$$

it also follows from Theorem 1 that the natural projection $\mathbb{C}^m = \mathbb{C}^n \times \mathbb{C}^{m-n} \to \mathbb{C}^n$ induces the germ $\pi_1\colon \mathbf{V}_1 \to \mathbb{C}^n$ of a finite branched holomorphic covering; so since $\mathbf{V} \subseteq \mathbf{V}_1$, the same projection induces the germ $\pi = \pi_1|\mathbf{V} : \mathbf{V} \to \mathbb{C}^n$ of a finite proper holomorphic mapping. The projection $\mathbb{C}^m \to \mathbb{C}^n$ can be factored as the composition of the projections $\mathbb{C}^m \to \mathbb{C}^{n+1}$ and $\mathbb{C}^{n+1} \to \mathbb{C}^n$, and correspondingly the germ $\pi$ can be factored as the composition $\pi = \pi'' \circ \pi'$ of the germ of a finite proper holomorphic mapping $\pi'\colon \mathbf{V} \to \mathbf{W}$ and the germ of the finite branched holomorphic covering $\pi''\colon \mathbf{W} \to \mathbb{C}^n$ already considered. The germ $\mathbf{V}_0 = \mathbf{V} - \mathbf{D} \cap \mathbf{V}$ is the germ of an open subset of $\mathbf{V}$, although it is not as yet clear that it is the germ of a dense open subset of $\mathbf{V}$. However, it follows immediately from Lemma 9 that $\mathbf{V}_0$ is nonempty—indeed, that $\mathbf{V}_0 = \text{loc}\ \mathfrak{A}' - \mathbf{D} \cap \text{loc}\ \mathfrak{A}'$. Thus after choosing representative subvarieties in a sufficiently small open polydisc $\Delta(0; R)$ about the origin in $\mathbb{C}^m$ and recalling the definition of the smaller associated ideal $\mathfrak{A}'$, it follows that a point $Z = (Z', z_{n+1}, \ldots, z_m) \in \Delta(0; R)$ belongs to the subset $V_0 \subseteq \Delta(0; R)$ precisely when $d(Z') \neq 0$, $p_{n+1}(Z'; z_{n+1}) = 0$, and $q_j(Z'; z_{n+1}, z_j) = d(Z')z_j - Q_j(Z'; z_{n+1}) = 0$ for $n + 2 \leqq j \leqq m$. It is obvious from these equations that the restriction $\pi'|V_0 : V_0 \to W_0$ is actually a biholomorphic mapping; hence, the restriction $\pi|V_0 : V_0 \to \Delta(0; R') - D_1$ is a locally biholomorphic covering mapping of order $r$ over the connected open set $\Delta(0; R') - D_1$. Since $V_0 \subseteq V \subseteq V_1$, the natural extensions of these mappings to the point set closure $\bar{V}_0$ of $V_0$ in $V_1$ are finite branched coverings $\pi'|\bar{V}_0 : \bar{V}_0 \to \bar{W}_0 = W$ and $\pi|\bar{V}_0 : \bar{V}_0 \to \Delta(0; R') \subseteq \mathbb{C}^n$, with regular parts $\pi'|V_0$ and $\pi|V_0$, respectively. So to complete the proof it is sufficient merely to show that $\bar{V}_0 = V$ near the origin. For this purpose note that it follows directly from Theorem C15 that $\bar{V}_0$ is a holomorphic subvariety of $V_1$, and hence of course it is also a holomorphic subvariety of $V$ since $V_0 \subset V$. However, since $\mathbf{V} = \bar{\mathbf{V}}_0 \cup (\mathbf{D} \cap \mathbf{V})$ by the definition of $V_0$ and since the germ $\mathbf{V} = \text{loc}\ \mathfrak{p}$ is irreducible and is not contained in $\mathbf{D}$, necessarily $\mathbf{V} = \bar{\mathbf{V}}_0$. That then suffices to conclude the proof.

In the proof of the preceding theorem the subset $V_0 \subset V$, the complement of the proper holomorphic subvariety $D \cap V \subset V$, is described very simply and explicitly by

$$V_0 = \{Z \in \Delta(0; R) : d(Z) \neq 0,\ p_{n+1}(Z) = q_{n+2}(Z) = \cdots = q_m(Z) = 0\}$$

It is clear that $V_0$ is actually an $n$-dimensional complex submanifold of $\Delta(0; R) - D$, and moreover that the functions $(p_{n+1}, q_{n+2}, \ldots, q_m)$ can be taken as part of a local system of coordinates in $\mathbb{C}^m$ near any point of $V_0$ exhibiting $V_0$ as such a submanifold. It is also quite evident that the point set closure $\bar{V}_0$ is a finite branched covering as desired; the subtle part of the proof is the demonstration that $\bar{V}_0 = V$. For that purpose it is necessary to show that there are no pieces of the subvariety $V$ contained in $D$ and not contained in the closure $\bar{V}_0$; Theorem C15 played a crucial role in that part of the proof.

Theorem 10 implies that for any prime ideal $\mathfrak{p} \subseteq {}_m\mathcal{O}$ the germ loc $\mathfrak{p}$ can be exhibited as the germ of a finite branched holomorphic covering of $\mathbb{C}^n$ for some dimension $n$, thus providing a vast number of examples of finite branched holomorphic coverings. This can also be viewed as establishing a standard local form or local parametrization for the germs loc $\mathfrak{p}$. These germs are not necessarily germs of complex manifolds, so there do not necessarily always exist germs of biholomorphic mappings $\pi$: loc $\mathfrak{p} \rightarrow \mathbb{C}^n$, but there do always exist germs of finite branched holomorphic coverings $\pi$: loc $\mathfrak{p} \rightarrow \mathbb{C}^n$. Actually the theorem asserts a bit more—namely, that the germ of the finite branched holomorphic covering $\pi$: loc $\mathfrak{p} \rightarrow \mathbb{C}^n$ can be factored as the composition $\pi = \pi' \circ \pi''$ of germs of finite branched holomorphic coverings $\pi'$: loc $\mathfrak{p} \rightarrow W$ and $\pi''$: $W \rightarrow \mathbb{C}^n$, where $W \subseteq \mathbb{C}^{n+1}$ is the germ of a holomorphic subvariety defined as the set of zeros of the Weierstrass polynomial $p_{n+1}(z_{n+1}) \in {}_n\mathcal{O}[z_{n+1}] \cap \mathfrak{p}$ and $\pi'$ is of branching order one. The finite branched holomorphic coverings $\pi$: loc $\mathfrak{p} \rightarrow \mathbb{C}^n$ and $\pi''$: $W \rightarrow \mathbb{C}^n$ have regular parts that are biholomorphic under the restriction of the mapping $\pi'$: loc $\mathfrak{p} \rightarrow W$; but $\pi'$ is not itself the germ of a biholomorphic mapping, only the germ of a finite branched holomorphic covering of branching order one. This finer assertion requires that the ideal $\mathfrak{p}$ be strictly regular in the variables $z_{n+1}, \ldots, z_m$, but the first assertion of the theorem does not really require strict regularity.

**11. COROLLARY.** *If the prime ideal* $\mathfrak{p} \subseteq {}_m\mathcal{O}$ *is regular in the variables* $z_{n+1}, \ldots, z_m$, *then the natural projection* $\mathbb{C}^m = \mathbb{C}^n \times \mathbb{C}^{m-n} \rightarrow \mathbb{C}^n$ *exhibits* loc $\mathfrak{p}$ *as the germ of a finite branched holomorphic covering of* $\mathbb{C}^n$.

*Proof.* It follows from Lemma 7 that the ideal $\mathfrak{p}$ can be made strictly regular in the last $m - n$ variables by a suitable nonsingular linear change of coordinates in $\mathbb{C}^m$. Indeed, as in the proof of that lemma, this linear change of coordinates can be taken to be a nonsingular linear change of coordinates in the space $\mathbb{C}^{m-n}$ of the last $m - n$ variables only. After this change of coordinates, it follows from Theorem 10 that the natural projection $\mathbb{C}^m \rightarrow \mathbb{C}^n$ exhibits loc $\mathfrak{p}$ as the germ of a finite branched holomorphic covering of $\mathbb{C}^n$; but this mapping is not affected by a change of coordinates in $\mathbb{C}^{m-n}$ and hence the desired result has been demonstrated.

A slightly weakened version of the second assertion of Theorem 10 also holds without the assumption of strict regularity.

**12. COROLLARY.** *If the prime ideal* $\mathfrak{p} \subseteq {}_m\mathcal{O}$ *is regular in the variables* $z_{n+1}, \ldots, z_m$ *and* $\pi$: loc $\mathfrak{p} \rightarrow \mathbb{C}^r$ *is the germ of a holomorphic mapping induced by the natural projection*

*$\mathbf{C}^m = \mathbf{C}^r \times \mathbf{C}^{m-r} \to \mathbf{C}^r$ where $n \leqq r \leqq m$, then the image of $\pi$ is the germ of a holomorphic subvariety of $\mathbf{C}^r$ and $\pi$ is the germ of a finite branched holomorphic covering of its image.*

*Proof.* As a notational convenience let $\mathbf{V} = \text{loc}\, \mathfrak{p} \subseteq \mathbf{C}^m$. The natural projection $\mathbf{C}^m \to \mathbf{C}^r$ induces the germ of a holomorphic mapping $\pi$: $\mathbf{V} \to \mathbf{C}^r$, and the natural projection $\mathbf{C}^m \to \mathbf{C}^n$ induces the germ of a holomorphic mapping $\pi'$: $\mathbf{V} \to \mathbf{C}^n$, which by Corollary 11 is the germ of a finite branched holomorphic covering of $\mathbf{C}^n$. In terms of the defining polynomials $\mathbf{p}_j(X) \in {}_n\mathcal{O}[X]$ for the residue classes of the coordinate functions $z_j$ in the residue class ring ${}_m\mathcal{O}/\mathfrak{p}$, introduce the germ

$$\mathbf{W} = \{Z \in \mathbf{C}^r : \mathbf{p}_{n+1}(z_{n+1}) = \cdots = \mathbf{p}_r(z_r) = 0\}$$

It follows from Theorem 1 that the natural mapping $\mathbf{C}^r \to \mathbf{C}^n$ induces the germ $\pi''$: $\mathbf{W} \to \mathbf{C}^n$ of a finite branched holomorphic covering of $\mathbf{C}^n$, and since $\mathbf{p}_j(z_j) \in \mathfrak{p}$, it is evident that $\pi(\mathbf{V}) \subseteq \mathbf{W}$, so that $\pi' = \pi'' \circ \pi$. For suitably chosen representatives of these germs the mappings $\pi'$: $V \to U$ and $\pi''$: $W \to U$ are finite branched holomorphic coverings of a connected open subset $U \subseteq \mathbf{C}^n$; and for a suitably chosen holomorphic subvariety $D \subseteq U$ the restrictions $\pi'|V_0 : V_0 \to U - D$ and $\pi''|W_0 : W_0 \to U - D$ are regular parts of these finite branched coverings, where $V_0 = V - (\pi')^{-1}(D)$ and $W_0 = W - (\pi'')^{-1}(D)$. Since $\pi' = \pi'' \circ \pi$, it is clear that $\pi$ is a finite proper holomorphic mapping and that $\pi| V_0 : V_0 \to \pi(V_0)$ is a locally biholomorphic covering mapping. The image $\pi(V_0)$ is a union of some of the connected components of $W_0$, so it follows from Theorem C15 that the point set closure of $\pi(V_0)$ in $W$ is a holomorphic subvariety of $W$; but since $V_0$ is a dense subset of $V$ and $\pi$ is a continuous and proper mapping, the closure of $\pi(V_0)$ in $W$ is just the image $\pi(V)$. Thus, $\pi(V)$ is a holomorphic subvariety of $W$, and $\pi$: $V \to \pi(V)$ is a finite branched holomorphic covering. That suffices to conclude the proof.

# E

# Some Applications of Local Parametrization

The local parametrization theorem is of fundamental importance in the study of holomorphic subvarieties and varieties, as will become evident in the subsequent discussion. Three different applications of that theorem will be discussed in this section: Hilbert's zero-theorem for holomorphic functions, the basic properties of finite holomorphic mappings, and some local topological properties of holomorphic varieties. The first of these applications settles one of the important questions raised but not answered in section B, by determining just which ideals in ${}_m\mathcal{O}$ are the ideals of germs of holomorphic subvarieties. The essential result in this direction is the following.

**1. THEOREM.** *If* $\mathfrak{p} \subseteq {}_m\mathcal{O}$ *is a prime ideal, then* $\mathfrak{p} = \text{id loc } \mathfrak{p}$.

*Proof.* It is of course clear that $\mathfrak{p} \subseteq \text{id loc } \mathfrak{p}$, as pointed out in Theorem B4, so it is only necessary to show conversely that $\text{id loc } \mathfrak{p} \subseteq \mathfrak{p}$. After a suitable nonsingular linear change of coordinates in $\mathbb{C}^m$, it can be assumed that the ideal $\mathfrak{p}$ is strictly regular in the variables $z_{n+1}, \ldots, z_m$, as in Lemma D7. The defining polynomial for the residue class of the variable $z_{n+1}$ in ${}_m\mathcal{O}/\mathfrak{p}$ is a polynomial $\mathbf{p}_{n+1}(z_{n+1}) \in \mathfrak{p} \cap {}_n\mathcal{O}[z_{n+1}]$ of some degree $r$, and it has a nonzero discriminant $\mathbf{d} \in {}_n\mathcal{O}$ for which $\mathbf{d} \notin \mathfrak{p}$. Now for any germ $\mathbf{f} \in \text{id loc } \mathfrak{p}$ it follows from Lemma D8 that $\mathbf{d}^\nu \cdot \mathbf{f} - \mathbf{q}(z_{n+1}) \in \mathfrak{p}$ for some integer $\nu \geqq 0$ and some polynomial $\mathbf{q}(z_{n+1}) \in {}_n\mathcal{O}[z_{n+1}]$ with degree $\mathbf{q} <$ degree $\mathbf{p}_{n+1} = r$, and it is obvious that $\mathbf{q}(z_{n+1}) \in \text{id loc } \mathfrak{p}$ as well. The local parametrization theorem, Theorem D10, implies that there is a holomorphic subvariety $V$ in an open neighborhood of the origin in $\mathbb{C}^m$ representing the germ $\text{loc } \mathfrak{p}$ and exhibited as a finite branched holomorphic covering of pure order $r$ over an open subset $U \subseteq \mathbb{C}^n$ by the natural projection. Moreover, there is a dense open subset $U_0 \subseteq U$ such that for each point $Z' = (z_1, \ldots, z_n) \in U_0$ there are $r$ distinct points of $V$ projecting to $Z'$, and these points have distinct coordinates $z_{n+1}$. If the subset $U$ is a sufficiently small open neighborhood of the origin, there will exist a holomorphic function $q \in {}_n\mathcal{O}_U[z_{n+1}]$ that represents the germ $\mathbf{q}(z_{n+1}) \in {}_n\mathcal{O}[z_{n+1}]$ and vanishes on the subvariety $V$. However for each fixed point $Z' \in U_0$ the polynomial $q(Z'; z_{n+1}) \in \mathbb{C}[z_{n+1}]$ is of degree strictly less than $r$ but vanishes for $r$ distinct values of $z_{n+1}$ and so must be the zero polynomial. Since $U_0$ is dense in $U$, the function $q$ actually vanishes identically, so $\mathbf{q} = 0$ and $\mathbf{d}^{\nu} \cdot \mathbf{f} \in \mathfrak{p}$. Finally, since $\mathfrak{p}$ is a prime ideal and $\mathbf{d} \notin \mathfrak{p}$ necessarily $\mathbf{f} \in \mathfrak{p}$, and that suffices to conclude the proof.

This theorem implies that every prime ideal in ${}_m\mathcal{O}$ is the ideal of some germ of a holomorphic subvariety in $\mathbb{C}^m$. Thus, in view of Theorem B6, the prime ideals in ${}_m\mathcal{O}$ are precisely the ideals of irreducible germs of holomorphic subvarieties in $\mathbb{C}^m$. The general result in this direction follows purely algebraically from Theorem 1. For this purpose recall the Lasker–Noether decomposition theorem, which asserts that any ideal in a Noetherian ring can be written as the intersection of finitely many primary ideals. Recall also that the **radical** of an ideal $\mathfrak{A} \subseteq {}_m\mathcal{O}$ is defined to be the ideal $\sqrt{\mathfrak{A}} = \{\mathbf{f} \in {}_m\mathcal{O} : \text{there is an integer } \nu > 0 \text{ for which } \mathbf{f}^{\nu} \in \mathfrak{A}\}$; in general, $\mathfrak{A} \subseteq \sqrt{\mathfrak{A}}$, and if equality holds the ideal $\mathfrak{A}$ is called a **radical ideal**. The radical of a primary ideal is a prime ideal, and if $\mathfrak{A} = \mathfrak{q}_1 \cap \cdots \cap \mathfrak{q}_r$ where $\mathfrak{q}_j$ are primary ideals, then $\sqrt{\mathfrak{A}} = \mathfrak{p}_1 \cap \cdots \cap \mathfrak{p}_r$ where $\mathfrak{p}_j = \sqrt{\mathfrak{q}_j}$ are prime ideals.

**2. THEOREM (Hilbert's zero-theorem).** *For any ideal* $\mathfrak{A} \subseteq {}_m\mathcal{O}$, *necessarily* $\operatorname{id} \operatorname{loc} \mathfrak{A} = \sqrt{\mathfrak{A}}$.

*Proof.* If $\mathfrak{A}$ is a prime ideal, then $\sqrt{\mathfrak{A}} = \mathfrak{A}$ and the desired result follows immediately from Theorem 1. If $\mathfrak{A}$ is a primary ideal, then $\sqrt{\mathfrak{A}}$ is a prime ideal, and since clearly $\operatorname{loc} \sqrt{\mathfrak{A}} = \operatorname{loc} \mathfrak{A}$, it follows from the result already established for prime ideals that $\operatorname{id} \operatorname{loc} \mathfrak{A} = \operatorname{id} \operatorname{loc} \sqrt{\mathfrak{A}} = \sqrt{\mathfrak{A}}$. If $\mathfrak{A}$ is an arbitrary ideal, then write $\mathfrak{A} = \mathfrak{q}_1 \cap \cdots \cap \mathfrak{q}_r$ where $\mathfrak{q}_j$ are primary ideals. By Theorem B4 and the result already established for primary ideals, note that $\operatorname{loc} \mathfrak{A} = \operatorname{loc} \mathfrak{q}_1 \cup \cdots \cup \operatorname{loc} \mathfrak{q}_r$ so that $\operatorname{id} \operatorname{loc} \mathfrak{A} = \operatorname{id} \operatorname{loc} \mathfrak{q}_1 \cap \cdots \cap \operatorname{id} \operatorname{loc} \mathfrak{q}_r = \mathfrak{p}_1 \cap \cdots \cap \mathfrak{p}_r = \sqrt{\mathfrak{A}}$ where $\mathfrak{p}_j = \sqrt{\mathfrak{q}_j}$. That suffices to conclude the proof.

Hilbert's zero-theorem can be restated in more elementary terms as follows.

**3. COROLLARY.** *If* $\mathbf{f} \in {}_m\mathcal{O}$ *vanishes on the germ* $\operatorname{loc} \mathfrak{A}$ *for an ideal* $\mathfrak{A} \subseteq {}_m\mathcal{O}$, *then there is an integer* $\nu > 0$ *such that* $\mathbf{f}^{\nu} \in \mathfrak{A}$.

*Proof.* If $\mathbf{f}$ vanishes on $\operatorname{loc} \mathfrak{A}$, then $\mathbf{f} \in \operatorname{id} \operatorname{loc} \mathfrak{A}$, so by the preceding theorem $\mathbf{f} \in \sqrt{\mathfrak{A}}$. The desired result is then a consequence of the definition of the radical of an ideal.

The zero-theorem can be used to extend in a quite algebraic form the discussion begun in section B. For example, Theorem B4(iv) asserts that $\mathfrak{A} \subseteq \operatorname{id} \operatorname{loc} \mathfrak{A}$ for any ideal $\mathfrak{A} \subseteq {}_m\mathcal{O}$, and the zero-theorem shows that this containment is an equality precisely when $\mathfrak{A}$ is a radical ideal. The correspondence introduced in Definition B3 is thus a one-to-one correspondence between germs of holomorphic subvarieties at the origin in $\mathbb{C}^m$ and radical ideals in ${}_m\mathcal{O}$. By Corollary B15 it induces a one-to-one correspondence between equivalence classes of germs of holomorphic subvarieties of $\mathbb{C}^m$ and equivalence classes of radical ideals in ${}_m\mathcal{O}$, where ideals $\mathfrak{A}$ and $\mathfrak{B}$ in ${}_m\mathcal{O}$ are equivalent if $\mathfrak{B} = \phi(\mathfrak{A})$ for some automorphism of algebras $\phi: {}_m\mathcal{O} \to {}_m\mathcal{O}$. By Corollary B17 it further induces a one-to-one correspondence between equivalence classes of germs of holomorphic varieties and isomorphism classes of complex algebras of the form ${}_m\mathcal{O}/\mathfrak{A}$ for all integers $m \geqq 1$ and all radical ideals $\mathfrak{A} \subseteq {}_m\mathcal{O}$.

The zero-theorem extends in a rather straightforward way to the corresponding assertion in the local ring of any germ of a holomorphic variety, as follows.

**4. COROLLARY.** *For any ideal* $\mathfrak{A} \subseteq {}_V\mathcal{O}$ *in the local ring of a germ* $V$ *of a holomorphic variety* id loc $\mathfrak{A} = \sqrt{\mathfrak{A}}$.

*Proof.* If $V$ is represented by a germ of a holomorphic subvariety at the origin in $\mathbb{C}^m$ and if $\mathfrak{B} = \text{id}\, V \subseteq {}_m\mathcal{O}$, then ${}_V\mathcal{O} = {}_m\mathcal{O}/\mathfrak{B}$. Let $\rho: {}_m\mathcal{O} \rightarrow {}_V\mathcal{O}$ be the natural ring homomorphism. For any ideal $\mathfrak{A} \subseteq {}_V\mathcal{O}$ it is clear that loc $\mathfrak{A} = \text{loc}\, \rho^{-1}(\mathfrak{A})$, and for any germ $W$ of a holomorphic subvariety of $V$ it is also clear that the ideals ${}_V\mathfrak{A} = \text{id}\, W \subseteq {}_V\mathcal{O}$ and ${}_m\mathfrak{A} = \text{id}\, W \subseteq {}_m\mathcal{O}$ are related by $\rho({}_m\mathfrak{A}) = {}_V\mathfrak{A}$. Therefore for any ideal $\mathfrak{A} \subseteq {}_V\mathcal{O}$, it follows from these observations and from Theorem 2 that id loc $\mathfrak{A} = \rho(\text{id loc}\, \rho^{-1}(\mathfrak{A})) = \rho(\sqrt{\rho^{-1}(\mathfrak{A})}) = \sqrt{\mathfrak{A}}$. That suffices to conclude the proof.

Hilbert's zero-theorem can be used in the other direction, in a sense, to obtain a useful geometric interpretation of the rather algebraic condition that an ideal $\mathfrak{A} \subseteq {}_m\mathcal{O}$ be regular, as follows.

**5. THEOREM.** *The following three conditions on an ideal* $\mathfrak{A} \subseteq {}_m\mathcal{O}$ *are equivalent*:

(i) *For each index* $j$ *in the interval* $n + 1 \leqq j \leqq m$ *there is an element* $f_j \in {}_j\mathcal{O} \cap \mathfrak{A}$ *that is regular in the variable* $z_j$.

(ii) *For each irreducible component* $V_i$ *of* loc $\mathfrak{A}$, *the natural projection* $\mathbb{C}^m = \mathbb{C}^n \times \mathbb{C}^{m-n} \rightarrow \mathbb{C}^n$ *induces the germ* $\pi_i: V_i \rightarrow \pi_i(V_i)$ *of a finite branched holomorphic covering, with the image* $\pi_i(V_i)$ *being the germ of a holomorphic subvariety of* $\mathbb{C}^n$.

(iii) *For the linear subspace* $L = \{Z \in \mathbb{C}^m : z_1 = \cdots = z_n = 0\}$ *the intersection* $L \cap \log \mathfrak{A} = 0$, *where* $0$ *is the germ of the holomorphic subvariety of* $\mathbb{C}^m$ *consisting of a single point, the origin in* $\mathbb{C}^m$.

*Proof.* The first step in the proof is to show that condition (i) implies condition (ii). Suppose therefore that the ideal $\mathfrak{A} \subseteq {}_m\mathcal{O}$ satisfies condition (i). This is part of the condition that the ideal $\mathfrak{A}$ be regular in some variables. Indeed, there exists an integer $k \leqq n$ such that after a suitable nonsingular linear change of coordinates in the space $\mathbb{C}^n$ of the variables $z_1, \ldots, z_n$, it can be assumed that there are also elements $f_j \in {}_j\mathcal{O} \cap \mathfrak{A}$ that are regular in the variable $z_j$ for $k + 1 \leqq j \leqq n$ and that ${}_k\mathcal{O} \cap \mathfrak{A} = 0$ and hence that the ideal $\mathfrak{A}$ is regular in the variables $z_{k+1}, \ldots, z_m$. Note that such a change of coordinates does not affect condition (ii). By the Lasker–Noether decomposition theorem the ideal $\mathfrak{A}$ can be written as an intersection $\mathfrak{A} = \mathfrak{q}_1 \cap \cdots \cap \mathfrak{q}_r$, where $\mathfrak{q}_i$ are primary ideals with prime radicals $\mathfrak{p}_i = \sqrt{\mathfrak{q}_i}$. Then loc $\mathfrak{A} = V_1 \cup \cdots \cup V_r$ where $V_i = \text{loc}\, \mathfrak{q}_i = \text{loc}\, \mathfrak{p}_i$ are irreducible germs of holomorphic varieties. It may happen that $V_i \subseteq V_j$ for some indices $i \neq j$, but after omitting any germs $V_i$ that are contained in one of the others, the remaining germs are just the irreducible components of loc $\mathfrak{A}$. Since $f_j \in {}_j\mathcal{O} \cap \mathfrak{A} \subseteq {}_j\mathcal{O} \cap \mathfrak{q}_i \subseteq {}_j\mathcal{O} \cap \mathfrak{p}_i$, clearly each ideal $\mathfrak{p}_i$ can be made regular in the variables $z_{k_i+1},\ldots,z_m$ for some $k_i \leq k$ by a suitable nonsingular linear change of coordinates in the space $\mathbb{C}^k$ of the variables $z_1,\ldots,z_k$. It follows from Corollary D12 though that even without this additional change of coordinates, the natural projection $\mathbb{C}^m \to \mathbb{C}^n$ induces the germ $\pi_i\colon \mathbf{V}_i \to \pi_i(\mathbf{V}_i)$ of a finite branched holomorphic covering of the germ $\pi_i(\mathbf{V}_i)$ of a holomorphic subvariety of $\mathbb{C}^n$. Thus condition (ii) holds as desired.

The next step is to show that condition (ii) implies condition (iii). Suppose therefore that the ideal $\mathfrak{A} \subseteq {}_m\mathcal{O}$ satisfies condition (ii). Let $V_i$ be holomorphic subvarieties of an open neighborhood of the origin in $\mathbb{C}^m$ such that the germs $\mathbf{V}_i$ are the irreducible components of loc $\mathfrak{A}$ and the natural projection $\mathbb{C}^m \to \mathbb{C}^n$ induces finite branched holomorphic coverings $\pi_i\colon V_i \to \pi_i(V_i)$ of holomorphic subvarieties $\pi_i(V_i)$ of an open neighborhood of the origin in $\mathbb{C}^n$. It is clear that $\pi_i^{-1}(0) = L \cap V_i$ where $L = \{Z \in \mathbb{C}^m : z_1 = \cdots = z_n = 0\}$; and since $\pi_i$ is a finite mapping, the subvariety $L \cap V_i$ must consist of finitely many points, one of which is of course the origin. Therefore $\mathbf{L} \cap \mathbf{V}_i = \mathbf{0}$, and consequently $\mathbf{L} \cap \text{loc}\,\mathfrak{A} = \bigcup_i(\mathbf{L} \cap \mathbf{V}_i) = \mathbf{0}$. Thus condition (iii) holds as desired.

The last step is to show that condition (iii) implies condition (i). First note that if $\mathfrak{A} \subseteq {}_m\mathcal{O}$ is any ideal that satisfies condition (iii) for some $n < m$, then there must exist an element $\mathbf{f}_m \in \mathfrak{A}$ that is regular in the variable $z_m$. Indeed by condition (iii) the ideal $\mathfrak{B} \subseteq {}_m\mathcal{O}$ generated by $\mathfrak{A}$ and the germs $\mathbf{z}_1,\ldots,\mathbf{z}_n$ evidently has the property that loc $\mathfrak{B} = \mathbf{0}$. Since the function $z_m$ vanishes on the germ loc $\mathfrak{B}$, it follows from Hilbert's zero-theorem that $\mathbf{z}_m^\nu \in \mathfrak{B}$ for some integer $\nu > 0$. Therefore $\mathbf{z}_m^\nu = \mathbf{f}_m + \mathbf{g}_1\mathbf{z}_1 + \cdots + \mathbf{g}_n\mathbf{z}_n$ for some elements $\mathbf{f}_m \in \mathfrak{A}$ and $\mathbf{g}_j \in {}_m\mathcal{O}$. The element $\mathbf{f}_m$ is nonzero since $n < m$, and it is clear from this identity that $\mathbf{f}_m$ is regular in $z_m$ as desired. Next note that if $\mathfrak{A} \subseteq {}_m\mathcal{O}$ is any ideal that satisfies condition (iii) for some $n < m$, then the ideal $\mathfrak{A}_{m-1} = {}_{m-1}\mathcal{O} \cap \mathfrak{A} \subseteq {}_{m-1}\mathcal{O}$ satisfies condition (iii) in $\mathbb{C}^{m-1}$ for the same value $n$. Indeed, since as just noted there must be an element $\mathbf{f}_m \in \mathfrak{A}$ that is regular in the variable $z_m$, and since it has already been demonstrated that condition (i) implies condition (ii), it follows that the natural projection $\mathbb{C}^m \to \mathbb{C}^{m-1}$ maps loc $\mathfrak{A}$ to some germ $\mathbf{V}_{m-1}$ of a holomorphic subvariety of $\mathbb{C}^{m-1}$. For this image $\mathbf{V}_{m-1}$ it is clear from condition (iii) that $\mathbf{L} \cap \mathbf{V}_{m-1} = \mathbf{0}$ where $L$ is now viewed as a subspace of $\mathbb{C}^{m-1}$. Now if $\mathfrak{A}_{m-1} = {}_{m-1}\mathcal{O} \cap \mathfrak{A}$, it is also clear that $\mathfrak{A}_{m-1} \subseteq \text{id}\,\mathbf{V}_{m-1}$. On the other hand, if $\mathbf{f} \in \text{id}\,\mathbf{V}_{m-1} \subseteq {}_{m-1}\mathcal{O}$, then when $\mathbf{f}$ is viewed as an element of ${}_m\mathcal{O}$ independent of $z_m$, that element vanishes on loc $\mathfrak{A}$, so by Hilbert's zero-theorem $\mathbf{f} \in {}_{m-1}\mathcal{O} \cap \sqrt{\mathfrak{A}} = \sqrt{\mathfrak{A}_{m-1}}$ and hence $\text{id}\,\mathbf{V}_{m-1} \subseteq \sqrt{\mathfrak{A}_{m-1}}$. From these two inclusions of ideals it follows that $\text{loc}\,\mathfrak{A}_{m-1} \supseteq \text{loc id}\,\mathbf{V}_{m-1} \supseteq \text{loc}\sqrt{\mathfrak{A}_{m-1}} = \text{loc}\,\mathfrak{A}_{m-1}$, so that $\text{loc}\,\mathfrak{A}_{m-1} = \text{id loc}\,\mathbf{V}_{m-1} = \mathbf{V}_{m-1}$ and hence $\mathfrak{A}_{m-1}$ also satisfies condition (iii) as desired. Finally if $\mathfrak{A} \subseteq {}_m\mathcal{O}$ is any ideal that satisfies condition (iii), then if $n = m$ the ideal $\mathfrak{A}$ trivially satisfies condition (i). If $n < m$, then there is an element $\mathbf{f}_m \in {}_m\mathcal{O} \cap \mathfrak{A}$ that is regular in the variable $z_m$ and the ideal ${}_{m-1}\mathcal{O} \cap \mathfrak{A}$ also satisfies condition (iii). If $n = m - 1$, that is all that is required; while if $n < m - 1$, then the same results hold for the ideal ${}_{m-1}\mathcal{O} \cap \mathfrak{A}$, namely, there is an element $\mathbf{f}_{m-1} \in {}_{m-1}\mathcal{O} \cap \mathfrak{A}$ that is regular in the variable $z_{m-1}$ and the intersection ${}_{m-2}\mathcal{O} \cap \mathfrak{A}$ also satisfies condition (iii). Repetition of this argument shows finally that the ideal $\mathfrak{A}$ satisfies condition (i), and that concludes the proof of the entire theorem.

As noted in the proof of the preceding theorem, condition (i) is equivalent to the condition that after a suitable nonsingular linear change of coordinates in the space $\mathbb{C}^n$ of the variables $z_1, \ldots, z_n$, the ideal $\mathfrak{A}$ will be regular in the variables $z_{k+1}, \ldots, z_m$ for some $k \leq n$. Thus condition (i) can be viewed as a partial regularity condition on the ideal $\mathfrak{A}$. It should also be noted that condition (ii) does not imply that the germ of a holomorphic mapping $\pi$: loc $\mathfrak{A} \to \pi$(loc $\mathfrak{A}$) is itself the germ of a finite branched holomorphic covering, for the images $\pi(V_i)$ of some components $V_i$ of loc $\mathfrak{A}$ may be contained in the images of other components. The equivalence between conditions (ii) and (iii) can be restated very conveniently in terms of a useful auxiliary concept. For this purpose note that if $\pi: V \to W$ is the germ of a holomorphic mapping between two germs of holomorphic varieties and $X$ is the germ of a holomorphic subvariety of $W$, then $\pi^{-1}(X)$ is the germ of a holomorphic subvariety of $V$. In particular, if $0$ is the base point of $W$, viewed as the germ of the holomorphic subvariety of $W$ consisting of that single point, then $\pi^{-1}(0)$ is the germ of a holomorphic subvariety of $V$. If $\pi: V \to W$ is the germ of a finite branched holomorphic covering, then of course $\pi^{-1}(0) = 0$.

**6. DEFINITION.** *A germ $\pi: V \to W$ of a holomorphic mapping between two germs of holomorphic varieties is called* **finite** *if $\pi^{-1}(0) = 0$.*

Any germ of a finite branched holomorphic covering is a finite germ of a holomorphic mapping, but not conversely; finite germs of holomorphic mappings need not be surjective, for instance. However finite germs of holomorphic mappings are quite closely related to germs of finite branched holomorphic coverings, as can readily be deduced from Theorem 5. Indeed, the precise statement of this relationship really amounts to the equivalence between conditions (ii) and (iii) of Theorem 5.

**7. THEOREM.** *A germ $\pi: V \to W$ of a holomorphic mapping between two germs of holomorphic varieties is finite if and only if for each irreducible component $V_i$ of $V$, the image $\pi(V_i)$ is the germ of a holomorphic subvariety of $W$ and the restriction $\pi: V_i \to \pi(V_i)$ is the germ of a finite branched holomorphic covering.*

*Proof.* First suppose that the germ $\pi: V \to W$ is finite. As in Lemma C11, the germs $V$ and $W$ can be represented by germs of holomorphic subvarieties at the origins in some spaces $\mathbb{C}^m$ and $\mathbb{C}^n$ in such a manner that the germ $\pi: V \to W$ is induced by the natural projection $\mathbb{C}^m = \mathbb{C}^n \times \mathbb{C}^{m-n} \to \mathbb{C}^n$. Then $\pi^{-1}(0) = L \cap V$ where $L$ is the germ of the linear subspace $L = \{Z \in \mathbb{C}^m : z_1 = \cdots = z_n = 0\}$; so since $\pi$ is finite, necessarily $0 = \pi^{-1}(0) = L \cap V$. It then follows immediately from Theorem 5 that for each irreducible component $V_i$ of $V$, the restriction $\pi: V_i \to \pi(V_i)$ is the germ of a finite branched holomorphic covering of the germ $\pi(V_i) \subseteq W$ of a holomorphic subvariety of $\mathbb{C}^n$. Since the converse assertion is trivial, that suffices to conclude the proof.

Note that if $\pi: V \to W$ is a finite germ of a holomorphic mapping and if $X \subseteq V$ is the germ of a holomorphic subvariety of $V$, then the restriction $\pi|X : X \to W$ is obviously also a finite germ of a holomorphic mapping; hence, by Theorem 7 the image $\pi(X)$ is the germ of a holomorphic subvariety of $W$. It was noted in the discussion after Corollary C14 that not all holomorphic mappings take subvarieties into subvarieties, so finite germs of holomorphic mappings do have quite special properties. This observation can be used to extend Corollary C14 as follows.

**8. COROLLARY.** *If $\pi: V \to W$ is a finite branched holomorphic covering over an open subset $W \subseteq \mathbb{C}^n$ and $X$ is a holomorphic subvariety of $V$, then $\pi(X)$ is a holomorphic subvariety of $W$.*

*Proof.* In view of Lemma C2 it is sufficient merely to show that the germ of a finite branched holomorphic covering $\pi: V \to W$ maps the germ of a holomorphic subvariety $X \subseteq V$ to the germ of a holomorphic subvariety $\pi(X) \subseteq W$; but that is an immediate consequence of Theorem 7.

**9. COROLLARY.** *A finite germ $\pi: V \to W$ of a holomorphic mapping between two germs of holomorphic varieties is surjective if and only if the induced homomorphism $\pi^*: {}_W\mathcal{O} \to {}_V\mathcal{O}$ is injective.*

*Proof.* If $\pi$ is surjective, it is obvious that the induced homomorphism $\pi^*$ is injective, and for that of course it is not necessary to require that the germ $\pi$ be finite. On the other hand, if $\pi$ is not surjective, then by Theorem 7 the image $\pi(V)$ will be a proper holomorphic subvariety of $W$. There must then be a nonzero element $f \in {}_W\mathcal{O}$ vanishing on $\pi(V)$, and for this element $\pi^*(f) = 0$, so $\pi$ is not injective. That suffices to complete the proof.

It should be noted that for an arbitrary germ $\pi: V \to W$ of a holomorphic mapping, the injectivity of $\pi^*$ does not imply the surjectivity of $\pi$. Indeed, in the example already considered of the holomorphic mapping $\pi: \mathbb{C}^2 \to \mathbb{C}^2$ defined by $\pi(z_1, z_2) = (z_1, z_1 z_2)$, the image of $\pi$ omits all points of the form $(0, \omega_2)$ where $w_2 \neq 0$, so the germ $\pi$ of $\pi$ at the origin is not surjective. But the image under $\pi$ of any connected open neighborhood of the origin is connected and contains an open set, so $\pi^*$ is injective.

**10. COROLLARY.** *If $\pi: U \to V$ is a one-to-one holomorphic mapping from an open subset $U \subseteq \mathbb{C}^n$ onto a subset $V \subseteq \mathbb{C}^n$, then $V$ is necessarily an open subset of $\mathbb{C}^n$ and $\pi$ is a biholomorphic mapping.*

*Proof.* The germ $\pi_A$ of the mapping $\pi$ at any point $A \in U$ is obviously a finite germ of a holomorphic mapping, so by Theorem 7 the image $\pi_A(U_A) = V_{\pi(A)}$ is the germ of a holomorphic subvariety of $\mathbb{C}^n$ at $\pi(A)$ and $\pi_A$ is the germ of a finite branched holomorphic covering. However, the germ $V_{\pi(A)}$ cannot be the germ of a proper subvariety of $\mathbb{C}^n$, for since $\pi_A$ is the germ of a finite branched holomorphic covering, a dense open subset of a representation of $V_{\pi(A)}$ must be an $n$-dimensional complex manifold, and that cannot be the case for a proper subvariety as is evident, for example, from the local parametrization theorem. Therefore $V$ is an open subset of $\mathbb{C}^n$, and since $\pi$ is one-to-one, the remainder of the desired result follows immediately from Theorem C4. That concludes the proof.

This observation settles a point left open in the discussion of biholomorphic mappings after Definition I-C7, for it shows that any one-to-one holomorphic mapping between open subsets of $\mathbb{C}^n$ is necessarily a biholomorphic mapping. Actually, of course, any one-to-one holomorphic mapping between arbitrary holomorphic varieties must be an open mapping, as a direct consequence of Theorem 7, but need not be a biholomorphic mapping. For example, the mapping $\pi: \mathbb{C} \to V$ defined by $\pi(t) = (t^2, t^3) \in V$ is a one-to-one holomorphic mapping from the complex manifold $\mathbb{C}$ onto the holomorphic variety $V = \{(z_1, z_2) \in \mathbb{C}^2 : z_2^2 - z_1^3 = 0\}$ discussed in the examples in section B; but $\pi$ is not a biholomorphic mapping, since the variety $V$ has a singularity at the origin. Even for the case of mappings between complex manifolds, though, this corollary is really a result in complex analysis and somewhat less trivial than it may at first appear to be. The mapping $\pi: \mathbb{R} \to \mathbb{R}$ defined by $\pi(x) = x^3$ is a one-to-one real analytic mapping but does not even have a differentiable inverse at the origin.

The definition of a finite germ of a holomorphic mapping and the characterization of such germs in Theorem 7 were of a rather geometric nature, so it may be of interest to see a complementary algebraic characterization of these germs. For this purpose note that any homomorphism $\pi^*: {}_{\mathbf{W}}\mathcal{O} \to {}_{\mathbf{V}}\mathcal{O}$ between two rings ${}_{\mathbf{W}}\mathcal{O}$ and ${}_{\mathbf{V}}\mathcal{O}$ naturally induces on the ring ${}_{\mathbf{V}}\mathcal{O}$ the structure of a module over the ring ${}_{\mathbf{W}}\mathcal{O}$ by defining the action of an element $\mathbf{g} \in {}_{\mathbf{W}}\mathcal{O}$ on an element $\mathbf{f} \in {}_{\mathbf{V}}\mathcal{O}$ to be the product $\pi^*(\mathbf{g}) \cdot \mathbf{f} \in {}_{\mathbf{V}}\mathcal{O}$. A ring homomorphism $\pi^*: {}_{\mathbf{W}}\mathcal{O} \to {}_{\mathbf{V}}\mathcal{O}$ is called **finite** if ${}_{\mathbf{V}}\mathcal{O}$, when viewed as an ${}_{\mathbf{W}}\mathcal{O}$-module in the natural way, is a finitely generated ${}_{\mathbf{W}}\mathcal{O}$-module.

**11. THEOREM.** *A germ $\pi: \mathbf{V} \to \mathbf{W}$ of a holomorphic mapping between two germs of holomorphic varieties is finite if and only if the induced homomorphism $\pi^*: {}_{\mathbf{W}}\mathcal{O} \to {}_{\mathbf{V}}\mathcal{O}$ is finite.*

*Proof.* Given a germ $\pi: \mathbf{V} \to \mathbf{W}$ of a holomorphic mapping between two germs of holomorphic varieties, choose representative germs of holomorphic subvarieties $V \subseteq \mathbb{C}^m$ and $W \subseteq \mathbb{C}^n$ such that the mapping $\pi$ is induced by the natural projection $\mathbb{C}^m = \mathbb{C}^n \times \mathbb{C}^{m-n} \to \mathbb{C}^n$, as in Lemma C11.

First suppose that the germ $\pi: \mathbf{V} \to \mathbf{W}$ is finite. If $\mathbf{V} = \mathbf{V}_1 \cup \cdots \cup \mathbf{V}_r$ where $\mathbf{V}_i$ are the irreducible components of $\mathbf{V}$, then the restrictions $\pi_i = \pi | \mathbf{V}_i : \mathbf{V}_i \to \mathbf{W}$ are also finite mappings. Associating to any germ $\mathbf{f} \in {}_{\mathbf{V}}\mathcal{O}$ the restrictions $\mathbf{f} | \mathbf{V}_i \in {}_{\mathbf{V}_i}\mathcal{O}$ clearly describes an injective homomorphism of ${}_{\mathbf{W}}\mathcal{O}$-modules of the form ${}_{\mathbf{V}}\mathcal{O} \to {}_{\mathbf{V}_1}\mathcal{O} + \cdots + {}_{\mathbf{V}_r}\mathcal{O}$; and since ${}_{\mathbf{W}}\mathcal{O}$ is a Noetherian ring, to show that ${}_{\mathbf{V}}\mathcal{O}$ is a finitely generated ${}_{\mathbf{W}}\mathcal{O}$-module it suffices to show that ${}_{\mathbf{V}_i}\mathcal{O}$ are finitely generated ${}_{\mathbf{W}}\mathcal{O}$-modules. Now since $\pi_i: \mathbf{V}_i \to \mathbb{C}^n$ is a finite mapping, it follows from Theorem 5 that for each index $j$ in the interval $n + 1 \leq j \leq m$ there is an element in ${}_j\mathcal{O} \cap \text{id}\, \mathbf{V}_i$ that is regular in the variable $z_j$. So after a suitable nonsingular linear change of coordinates in $\mathbb{C}^j$, the ideal id $\mathbf{V}_i$ will be regular in the variables $z_{n_i+1}, \ldots, z_m$ for some $n_i \leq n$. Then by Corollary D11 the natural projection $\mathbb{C}^m \to \mathbb{C}^{n_i}$ exhibits $\mathbf{V}_i$ as the germ of a finite branched holomorphic covering of $\mathbb{C}^{n_i}$. It follows from Corollary C9 that ${}_{\mathbf{V}_i}\mathcal{O}$ is a finitely generated ${}_{n_i}\mathcal{O}$-module and hence of course also a finitely generated ${}_n\mathcal{O}$-module, since $\mathbb{C}^{n_i} \subseteq \mathbb{C}^n$. Since $\pi(\mathbf{V}_i) \subseteq \mathbf{W} \subseteq \mathbb{C}^n$, it is also the case that ${}_{\mathbf{V}_i}\mathcal{O}$ is a finitely generated ${}_{\mathbf{W}}\mathcal{O}$-module as desired.

Next suppose that the induced homomorphism $\pi^*: {}_W\mathcal{O} \to {}_V\mathcal{O}$ is finite. The mapping $\pi: V \to W$ can be viewed as a mapping into $\mathbb{C}^n$, and the induced homomorphism $\pi^*: {}_n\mathcal{O} \to {}_V\mathcal{O}$ is also finite; so the natural projection $\mathbb{C}^m \to \mathbb{C}^n$ exhibits ${}_V\mathcal{O}$ as a finite ${}_n\mathcal{O}$-module. For each index $j$ in the interval $n + 1 \leqq j \leqq m$, the polynomial subring ${}_n\mathcal{O}[\mathbf{z}_j|\mathbf{V}] \subseteq {}_V\mathcal{O}$ is also a finite ${}_n\mathcal{O}$-module, so there must be a monic polynomial $\mathbf{P}_j(X) \in {}_n\mathcal{O}[X]$ of positive degree such that $\mathbf{P}_j(\mathbf{z}_j|\mathbf{V}) = 0$. That is an immediate consequence of the observation that for some integer $\nu > 0$ sufficiently large, $(\mathbf{z}_j|\mathbf{V})^\nu$ must be contained in the ${}_n\mathcal{O}$-module generated by $1, (\mathbf{z}_j|\mathbf{V}), \ldots, (\mathbf{z}_j|\mathbf{V})^{\nu-1}$. The germs $\mathbf{P}_j(z_j) \in {}_n\mathcal{O}[z_j] \subseteq {}_m\mathcal{O}$ are thus elements of ${}_m\mathcal{O} \cap \text{id } \mathbf{V}$ which are regular in the variable $z_j$, and it then follows immediately from Theorem 5 that the mapping $\pi: \mathbf{V} \to \mathbf{W} \subseteq \mathbb{C}^n$ is finite as desired. That suffices to conclude the proof.

The local parametrization theorem can also be used to demonstrate several important general topological properties of holomorphic varieties. The demonstrations are quite easy and straightforward, except that some care must be taken in piecing together the local representations of the irreducible components of a holomorphic variety as separate finite branched holomorphic coverings. The following rather technical and otherwise uninteresting lemma is convenient for this purpose.

**12. LEMMA.** *Each point $A$ on a holomorphic variety $V$ has arbitrarily small open neighborhoods of the form $V_A = V_1 \cup \cdots \cup V_r$, where $V_j$ are holomorphic varieties representing the irreducible components of the germ of the variety $V$ at the point $A$ and each $V_j$ can be represented as a finite branched holomorphic covering of a polydisc with the point $A \in V_j$ as the only point lying over the center of the polydisc.*

*Proof.* Since the desired result is really a local one, it can be assumed that the variety $V$ is represented as a holomorphic subvariety of an open neighborhood $U$ of the origin in $\mathbb{C}^m$, where the origin corresponds to the point $A \in V$. If the neighborhood $U$ is sufficiently small, it can further be assumed that $V = V_1 \cup \cdots \cup V_r$, where $V_j$ are holomorphic subvarieties of $U$ representing the irreducible components of the germ of the subvariety $V$ at the origin. After a suitable nonsingular linear change of coordinates in $\mathbb{C}^m$ and a renumbering of the subvarieties $V_j$ if necessary, it can still further be assumed that the prime ideal id $V_j \subseteq {}_m\mathcal{O}_0$ is regular in the variables $z_{n_j+1}, \ldots, z_m$ for $1 \leqq j \leqq r$, where $n_1 \leqq n_2 \leqq \cdots \leqq n_r$; for as noted Lemma D4 can be applied to any finite set of ideals simultaneously. The local parametrization theorem then implies that for any sufficiently small polyradius $R_j = (r_1, \ldots, r_{n_j})$ the intersection $V_j \cap (\Delta(0; R_j) \times \mathbb{C}^{m-n_j})$ is exhibited as a finite branched holomorphic covering of the polydisc $\Delta(0; R_j)$ under the natural projection $\mathbb{C}^m \to \mathbb{C}^{n_j}$; and by Lemma C2, if the polyradius $R_j$ is sufficiently small, the origin in $V_j$ will be the only point lying over the origin in $\Delta(0; R_j)$. Now if $U_0$ is any open neighborhood of the origin contained in $U$, then it follows from Lemma C2 that $V_r \cap (\Delta(0; R_r) \times \mathbb{C}^{m-n_r}) \subseteq U_0$ whenever the polyradius $R_r$ is sufficiently small. Furthermore after shrinking the polyradius $R_{r-1}$ even more if necessary, then

$$V_{r-1} \cap (\Delta(0; R_{r-1}) \times \mathbb{C}^{m-n_{r-1}}) \subseteq U_0 \cap (\Delta(0; R_r) \times \mathbb{C}^{m-n_r})$$

and this process can evidently be repeated. Finally then for the polyradius $R_r$ chosen in this manner, the intersections $V_j \cap (\Delta(0; R_r) \times \mathbb{C}^{m-n_r}) \subseteq U_0$ will have all the desired properties, thus concluding the proof.

**13. THEOREM.** *A holomorphic variety $V$ is a locally connected topological space. If $V_0$ is the complement of a holomorphic subvariety of $V$, then $V_0$ is locally finitely connected in $V$, and $V_0$ is locally connected at each point $A \in V$ at which the germ of $V$ is irreducible.*

*Proof.* Any point $A$ on a holomorphic variety $V$ has arbitrarily small open neighborhoods $V_A = V_1 \cup \cdots \cup V_r$ with the properties listed in Lemma 12. Note first that by Lemma C2 each variety $V_j$ is connected, and since $A \in V_j$ for each $V_j$, the union $V_A$ is also connected. Note next that if $X$ is a holomorphic subvariety of $V$ and $V_0 = V - X$, then $V_A \cap V_0 = (V_1 \cap V_0) \cup \cdots \cup (V_r \cap V_0)$; so the proof of the theorem can be concluded by showing that each set $V_j \cap V_0$ is connected. Now for the finite branched holomorphic covering $\pi_j: V_j \to \Delta(0; R_j) \subseteq \mathbb{C}^{n_j}$, it follows from Corollary 8 that the image $\pi_j(V_j \cap X) = D_j$ is a holomorphic subvariety of the polydisc $\Delta(0; R_j)$. The set $\pi_j^{-1}(\Delta(0; R_j) - D_j)$ has finitely many connected components, and the point set closure of each of these components is a holomorphic subvariety of $V_j$ by Theorem C15. However, since the variety $V_j$ can be taken to be irreducible, it follows that $\pi_j^{-1}(\Delta(0; R_j) - D_j)$ must actually be connected, and hence the larger set $V_j \cap V_0$ must also be connected. That suffices to conclude the proof.

In view of Theorem 13 finite branched holomorphic coverings, particularly finite branched holomorphic coverings over holomorphic varieties that are irreducible at each of their points, automatically satisfy many of the topological regularity conditions for finite branched coverings discussed in section C. It is convenient for some purposes to rephrase Theorem 13 in terms of the following auxiliary notion.

**14. DEFINITION.** *A* **thin set** *in a holomorphic variety $V$ is a subset $X \subseteq V$ with the property that for each point $A \in V$ there are an open neighborhood $U_A$ of the point $A$ in $V$ and a holomorphic function $f_A$ in $U_A$ such that the germ of $f_A$ in ${}_V\mathcal{O}_A$ does not vanish on any irreducible component of the germ of the variety $V$ at $A$ and $X \cap U_A \subseteq \{Z \in U_A: f_A(Z) = 0\}$.*

When $V$ is an open subset in $\mathbb{C}^m$, this definition reduces to the earlier Definition I-D1 of a thin set; in that case, a subset $X \subseteq V \subseteq \mathbb{C}^m$ is thin precisely when locally it is contained within a proper holomorphic subvariety of $V$. In general a subset $X \subseteq V$ is thin precisely when locally it is contained within a proper holomorphic subvariety of each irreducible component of $V$. Note that an irreducible component of a holomorphic variety $V$ is a proper holomorphic subvariety of $V$ but not a thin set in $V$. In these terms Theorem 13 can be rewritten as follows.

**15. THEOREM.** *A thin set in a holomorphic variety $V$ is nowhere dense in $V$. If $V_0$ is the complement of a thin set in $V$, then $V_0$ is locally finitely connected in $V$ and is locally connected at each point $A \in V$ at which the germ of $V$ is irreducible.*

*Proof.* If $\pi: V \to \Delta(0; R)$ is a finite branched holomorphic covering over a polydisc in $\mathbb{C}^n$ and $X \subseteq V$ is a proper holomorphic subvariety of $V$, then it follows from Corollary 8 that $\pi(X)$ is a holomorphic subvariety of $\Delta(0; R)$. If the variety $V$ is irreducible, $\pi(X)$ must indeed be a proper holomorphic subvariety of $\Delta(0; R)$. In this case it is obvious from Lemma C2 that $\pi^{-1}(\pi(X))$ is a nowhere dense subset of $V$, hence, $X$ itself is a nowhere dense subset of $V$. In view of Lemma 12 it is then evident that a thin set in any holomorphic variety $V$ is nowhere dense in $V$. The remainder of this theorem is an immediate consequence of Theorem 13, and that then suffices to conclude the proof.

To turn next to some more analytic applications of the local parametrization theorem, recall that as was already observed, complex manifolds are special cases of holomorphic varieties. It may happen that at some points a general holomorphic variety has the more regular structure of a complex manifold. A very simple but useful observation in this context is the following.

**16. THEOREM.** *If $V$ is a holomorphic subvariety of an open subset of $\mathbb{C}^m$, then $V$ is a complex manifold at a point $A \in V$ precisely when $V$ is a complex submanifold of $\mathbb{C}^m$ at the point $A$.*

*Proof.* If $V$ is a complex submanifold of $\mathbb{C}^n$ at a point $A \in V$, then obviously $V$ is a complex manifold at the point $A$. Conversely, if $V$ is a complex manifold at $A$, then for some open neighborhood $V_A$ of the point $A$ in $V$ there are holomorphic mappings $F: \Delta(0; R) \to V_A \subseteq \mathbb{C}^m$ and $G: V_A \to \Delta(0; R) \subseteq \mathbb{C}^n$ such that the composition $GF$ is a nonsingular holomorphic mapping. But then $F$ must be a nonsingular holomorphic mapping into $\mathbb{C}^n$, so by Corollary I-C10 the image $F(\Delta(0; R)) = V_A$ is a complex submanifold of $\mathbb{C}^m$ near $A$. That suffices to conclude the proof.

**17. THEOREM.** *A holomorphic variety is a complex manifold outside a thin subset.*

*Proof.* For any point $A$ in a holomorphic variety $V$ choose an open neighborhood $V_A = V_1 \cup \cdots \cup V_r$ of $A$ with the properties listed in Lemma 12. Each intersection $V_i \cap V_j$ for $i \neq j$ is necessarily a proper holomorphic subvariety of the component $V_j$ near $A$. Furthermore each variety $V_j$ can be represented as a finite branched holomorphic covering of a polydisc, so any regular part of that covering gives a subset of $V_j$ that is the complement of a proper holomorphic subvariety of $V_j$ and that is a complex manifold. Thus the union of the regular parts of these various finite branched coverings with the intersections removed is the complement of a proper holomorphic subvariety of each $V_j$ and is a complex manifold. That suffices to conclude the proof.

**18. DEFINITION.** *A point in a holomorphic variety $V$ at which $V$ is a complex manifold is called a* **regular point** *of $V$, and a point that is not a regular point is called a* **singular point** *of $V$. The set of all regular points comprise the* **regular locus** *$\mathfrak{R}(V) \subseteq V$, while the set of all singular points comprise the* **singular locus** *$\mathfrak{S}(V) = V - \mathfrak{R}(V) \subseteq V$.*

In a natural extension of this terminology, a complex manifold is sometimes called a **regular holomorphic variety**, since it is a holomorphic variety, all points of which are regular points. It is obvious that the regular locus $\mathfrak{R}(V)$ of a holomorphic variety $V$ is an open subset of $V$. Theorem 17 can be rephrased as the assertion that the singular locus $\mathfrak{S}(V)$ of $V$ is a thin set in $V$; thus by Theorem 15 the set $\mathfrak{S}(V)$ is nowhere dense in $V$, so the regular locus $\mathfrak{R}(V)$ is actually a dense open subset of $V$. It will be demonstrated later that $\mathfrak{S}(V)$ is actually a holomorphic subvariety of $V$ and hence is necessarily a proper holomorphic subvariety of each local irreducible component of $V$. Note incidentally that the set of points at which the germ of a holomorphic variety $V$ is reducible is evidently contained in the singular locus $\mathfrak{S}(V)$ and so is also a thin set in $V$. But this set is not necessarily a holomorphic subvariety of $V$—indeed, is not necessarily even a closed subset of $V$. For example, the subvariety $V = \{Z \in \mathbb{C}^3 : z_1^2 - z_2^2 z_3 = 0\}$ has as its singular locus the subvariety $\mathfrak{S}(V) = \{Z \in \mathbb{C}^3 : z_1 = z_2 = 0\}$ but is reducible precisely at the set of points $\{Z \in \mathbb{C}^3 : z_1 = z_2 = 0, z_3 \neq 0\}$ for whenever $z_3 \neq 0$, it is possible to choose a local branch of the holomorphic function $\sqrt{z_3}$ and to write $z_1^2 - z_2^2 z_3 = (z_1 + z_2\sqrt{z_3})(z_1 - z_2\sqrt{z_3})$, while the germ of the function $z_1^2 - z_2^2 z_3$ at the origin in $\mathbb{C}^3$ is easily seen to be irreducible.

**19. THEOREM.** *The point set closure of any connected component of the regular locus $\mathfrak{R}(V)$ of a holomorphic variety $V$ is a holomorphic subvariety of $V$. Thus, $V$ is irreducible precisely when $\mathfrak{R}(V)$ is connected.*

*Proof.* Let $U$ be a connected component of the regular locus $\mathfrak{R}(V)$ of a holomorphic variety $V$, and for any point $A \in V$ choose an open neighborhood $V_A = V_1 \cup \cdots \cup V_r$ of $A$ with the properties listed in Lemma 12. For each variety $V_j$ the regular locus $\mathfrak{R}(V_j)$ is the complement of a thin set in $V_j$ by Theorem 17 and hence is locally connected at $A$ by Theorem 15. The neighborhood $V_A$ can thus be so chosen that $\mathfrak{R}(V_j)$ is a connected dense open subset of $V_j$. The set $\mathfrak{R}(V_j) \cap (\bigcup_{k \neq j} V_k)$ is a proper holomorphic subvariety of the complex manifold $\mathfrak{R}(V_j)$, so by Corollary I-D3 its complement $V_j \cap \mathfrak{R}(V_A)$ is a connected dense open subset of $V_j$ as well. Now $U \cap V_A$ must be the union of some of the sets $V_j \cap \mathfrak{R}(V_A)$; hence, $\bar{U} \cap V_A$ must be the union of the corresponding sets $V_j$, so that $\bar{U}$ is a holomorphic subvariety of $V$. The last statement of the theorem follows immediately from this observation, and that concludes the proof.

**20. COROLLARY.** *If $W$ is a holomorphic subvariety of a holomorphic variety $V$, then the point set closure of any connected component of $V - W$ is a holomorphic subvariety of $V$.*

*Proof.* By enlarging the set $W$ if necessary it can be assumed that $W \supseteq \mathfrak{S}(V)$. Any connected component $U$ of $V - W$ is then the complement of a proper holomorphic subvariety of an irreducible component $V_j$ of $V$. Hence, $\bar{U} = V_j$ and the proof is thereby concluded.

# F
# Oka's Theorem

So far in the discussion of the local theory attention has been limited to the consideration of local rings at a single point. However it was recognized by K. Oka in his pioneering work on holomorphic functions of several variables that it is also essential to consider relations between local rings at nearby points; that leads to some subtle but very important concepts and results in the local theory. If $f$ is a holomorphic function in an open subset $U \subseteq \mathbb{C}^n$, it has a well-defined germ $\mathbf{f}_A \in {}_n\mathcal{O}_A$ at each point $A \in U$, and if $f_1, \ldots, f_\nu$ are holomorphic functions in $U$, their germs generate an ideal $\mathfrak{A}_A \subseteq {}_n\mathcal{O}_A$ at each point $A \in U$. This leads to a family of ideals $\{\mathfrak{A}_A\}$ which can be generated by the germs of a finite number of functions holomorphic in $U$. Of course, any single ideal $\mathfrak{A}_A$ is finitely generated, since ${}_n\mathcal{O}_A$ is Noetherian, but that is quite different from the assertion that there are finitely many holomorphic functions in $U$ such that their germs generate the given ideal $\mathfrak{A}_A$ at each point $A \in U$. If $V$ is a holomorphic subvariety of $U$ and $\mathbf{V}_A$ denotes the germ of the subvariety $V$ at a point $A \in U$, then there is an associated family of ideals $\mathfrak{A}_A = \text{id}\,\mathbf{V}_A \subseteq {}_n\mathcal{O}_A$ at all points $A \in U$, and it can be asked whether there are finitely many holomorphic functions in $U$ such that the germs of these functions generate the ideal $\mathfrak{A}_A$ at each point $A \in U$. This is just one instance of the general problem of determining whether various naturally arising families of ideals can be so generated. In practice it is essential to consider this question not just for families of ideals but even for families of modules; but only a special case will be considered now.

Recall that a **free module of rank** $\nu$ over the ring ${}_n\mathcal{O}_A$ is merely a direct sum

$${}_n\mathcal{O}_A^\nu = {}_n\mathcal{O}_A \oplus \cdots \oplus {}_n\mathcal{O}_A$$

of $\nu$ copies of the ring ${}_n\mathcal{O}_A$ with the obvious module operations. An element $\mathbf{F} \in {}_n\mathcal{O}_A^\nu$ is simply a $\nu$-tuple $\mathbf{F} = (\mathbf{f}_1, \ldots, \mathbf{f}_\nu)$ of germs $\mathbf{f}_j \in {}_n\mathcal{O}_A$; addition in ${}_n\mathcal{O}_A^\nu$ is defined componentwise, and the operation of an element $\mathbf{f} \in {}_n\mathcal{O}_A$ on the element $\mathbf{F} \in {}_n\mathcal{O}_A^\nu$ is defined by $\mathbf{f} \cdot \mathbf{F} = (\mathbf{f}\mathbf{f}_1, \ldots, \mathbf{f}\mathbf{f}_\nu)$. A **family of submodules** of the family of free modules $\{{}_n\mathcal{O}_A^\nu\}$ of rank $\nu$ over an open subset $U \subseteq \mathbb{C}^n$ is merely a collection $\{\mathscr{A}_A\}$, where $\mathscr{A}_A \subseteq {}_n\mathcal{O}_A^\nu$ is a submodule for each point $A \in U$. In particular a family of submodules of the family of free modules of rank one over $U$ is just a family of ideals in the family of local rings ${}_n\mathcal{O}_A$.

**1. DEFINITION.** *A family $\{\mathscr{A}_A\}$ of submodules of $\{{}_n\mathcal{O}_A^\nu\}$ over an open subset $U \subseteq \mathbb{C}^n$ is* **finitely generated over U** *if there are finitely many ν-tuples $F_j \in {}_n\mathcal{O}_U^\nu$ of holomorphic functions in U such that the germs $\mathbf{F}_{jA} \in {}_n\mathcal{O}_A^\nu$ generate $\mathscr{A}_A$ as an ${}_n\mathcal{O}_A$-module at each point $A \in U$. The ν-tuples $F_j$ are* **generators** *of the family $\{\mathscr{A}_A\}$ over U.*

If $F_1, \ldots, F_\mu \in {}_n\mathcal{O}_U^\nu$ are finitely many ν-tuples of holomorphic functions in $U$, the submodules $\mathscr{A}_A \subseteq {}_n\mathcal{O}_A^\nu$ they generate at various points $A \in U$ are not generally free modules themselves, since there may be some relations among the generators $\mathbf{F}_{1A}, \ldots, \mathbf{F}_{\mu A} \in {}_n\mathcal{O}_A^\nu$. A **relation** between these generators is a μ-tuple $\mathbf{G} = (\mathbf{g}_1, \ldots, \mathbf{g}_\mu) \in {}_n\mathcal{O}_A^\mu$ such that

$$\mathbf{g}_1\mathbf{F}_{1A} + \cdots + \mathbf{g}_\mu\mathbf{F}_{\mu A} = 0 \in {}_n\mathcal{O}_A^\nu$$

The set of all relations between these generators, which set is clearly a submodule of ${}_n\mathcal{O}_A^\mu$, is called the **module of relations** among $\mathbf{F}_{1A}, \ldots, \mathbf{F}_{\mu A}$ and is denoted by $\mathscr{R}(\mathbf{F}_{1A}, \ldots, \mathbf{F}_{\mu A})$. Alternatively, of course, to say that $\mathbf{F}_{1A}, \ldots, \mathbf{F}_{\mu A}$ generate the submodule $\mathscr{A}_A \subseteq {}_n\mathcal{O}_A^\nu$ is equivalent to the assertion that the homomorphism of ${}_n\mathcal{O}_A$-modules

$$\mathbf{F}: {}_n\mathcal{O}_A^\mu \to {}_n\mathcal{O}_A^\nu$$

defined by

$$\mathbf{F}(\mathbf{x}_1, \ldots, \mathbf{x}_\mu) = \mathbf{x}_1\mathbf{F}_{1A} + \cdots + \mathbf{x}_\mu\mathbf{F}_{\mu A}$$

for any $(\mathbf{x}_1, \ldots, \mathbf{x}_\mu) \in {}_n\mathcal{O}_A^\mu$ has as its image precisely the submodule $\mathscr{A}_A \subseteq {}_n\mathcal{O}_A^\nu$—that is, that $\mathbf{F}({}_n\mathcal{O}_A^\mu) = \mathscr{A}_A$. The kernel of this homomorphism is precisely the submodule $\mathscr{R}(\mathbf{F}_{1A}, \ldots, \mathbf{F}_{\mu A}) \subseteq {}_n\mathcal{O}_A^\mu$ so that, using the familiar terminology of exact sequences,

$$0 \longrightarrow \mathscr{R}(\mathbf{F}_{1A}, \ldots, \mathbf{F}_{\mu A}) \xrightarrow{\ \mathbf{i}\ } {}_n\mathcal{O}_A^\mu \xrightarrow{\ \mathbf{F}_A\ } \mathscr{A}_A \longrightarrow 0$$

is an exact sequence of ${}_n\mathcal{O}_A$-modules, where **i** denotes inclusion. The ${}_n\mathcal{O}_A$-module $\mathscr{A}_A$ is thus isomorphic to a quotient module of a free ${}_n\mathcal{O}_A$-module of the form $\mathscr{A}_A \cong {}_n\mathcal{O}_A^\mu/\mathscr{R}(\mathbf{F}_{1A}, \ldots, \mathbf{F}_{\mu A})$. It can then be asked whether the family of submodules $\{\mathscr{R}(\mathbf{F}_{1A}, \ldots, \mathbf{F}_{\mu A})\}$ of $\{{}_n\mathcal{O}_A^\mu\}$ is also finitely generated over $U$. One of the fundamental discoveries of Oka was that that is indeed the case, at least locally. It is not the case for arbitrary finitely generated families of modules, as distinct from finitely generated families of submodules of a family of free modules, that the module of relations is necessarily finitely generated, even locally. But the general discussion of this property, the property of coherence, will be postponed until the next volume, after the introduction of the machinery of sheaves, in terms of which the discussion is easier and more transparent. Some of the results of Oka do fit in quite naturally with the Weierstrass theorems and the local parametrization theorem and so will be discussed here.

**2. THEOREM (Oka's theorem).** *If $F_1, \ldots, F_\mu \in {}_n\mathcal{O}_U^\nu$ are finitely many ν-tuples of holomorphic functions in an open subset $U \subseteq \mathbb{C}^n$, then the family $\{\mathscr{R}(\mathbf{F}_{1A}, \ldots, \mathbf{F}_{\mu A})\}$ of submodules of $\{{}_n\mathcal{O}_A^\mu\}$ is finitely generated over an open neighborhood of each point of U.*

*Proof.* The proof of the theorem will be by induction on the dimension $n$. For the case $n = 0$, the functions involved are all merely complex constants, and the theorem as stated is trivially true. Assume then that the theorem holds for holomorphic functions of $n - 1$ variables; the proof that it holds for holomorphic functions of $n$ variables will be by induction on the index $\nu$. For the initial case $\nu = 1$, it is only necessary to show that if $f_1, \ldots, f_\mu \in {}_n\mathcal{O}_U$ are any holomorphic functions in an open neighborhood $U$ of the origin in $\mathbb{C}^n$, then the family $\{\mathcal{R}(\mathbf{f}_{1A}, \ldots, \mathbf{f}_{\mu A})\}$ of submodules of $\{{}_n\mathcal{O}_A^\mu\}$ is finitely generated over some subneighborhood $V \subseteq U$ of the origin. It can be assumed, after a nonsingular linear change of coordinates at the origin if necessary, that all the functions $f_j$ are regular in the variable $z_n$ at the origin. By the Weierstrass preparation theorem the germs at the origin of these functions $f_j$ can be written $\mathbf{f}_j = \mathbf{u}_j\mathbf{p}_j$, where $\mathbf{u}_j$ are units in ${}_n\mathcal{O}_0$ and $\mathbf{p}_j \in {}_{n-1}\mathcal{O}_0[z_n]$ are Weierstrass polynomials. The germs $\mathbf{u}_j$ and $\mathbf{p}_j$ can be represented by some holomorphic functions $u_j$ and $p_j$ in a polydisc $\Delta(0; R) \subseteq U$, and by choosing the polyradius $R$ sufficiently small, it can be assumed that $u_j(z) \neq 0$ for all $j$ and all points $z \in \Delta(0; R)$. It is then enough to show that the family $\{\mathcal{R}(\mathbf{p}_{1A}, \ldots, \mathbf{p}_{\mu A})\}$ of submodules of $\{{}_n\mathcal{O}_A^\mu\}$ is finitely generated over $\Delta(0; R)$ if $R$ is sufficiently small, since clearly whenever $G_j = (g_{j1}, \ldots, g_{j\mu})$ are generators of the family $\{\mathcal{R}(\mathbf{p}_{1A}, \ldots, \mathbf{p}_{\mu A})\}$, then $G_j' = (u_{j1}^{-1}g_{j1}, \ldots, u_{j\mu}^{-1}g_{j\mu})$ are generators of the family $\{\mathcal{R}(\mathbf{f}_{1A}, \ldots, \mathbf{f}_{\mu A})\}$. For this purpose let $d_j$ be the degree of the polynomial $\mathbf{p}_j$ and let $d = \max(d_1, \ldots, d_\mu)$, and as usual write $\Delta(0; R) = \Delta(0; R') \times \Delta(0; r'') \subseteq \mathbb{C}^{n-1} \times \mathbb{C}$ and correspondingly write $A = (A', a'') \in \mathbb{C}^{n-1} \times \mathbb{C}$ for any point $A \in \mathbb{C}^n$. Introduce now the subsets

$$\mathcal{R}^d(\mathbf{p}_{1A}, \ldots, \mathbf{p}_{\mu A}) = \{(\mathbf{x}_1, \ldots, \mathbf{x}_\mu) \in \mathcal{R}(\mathbf{p}_{1A}, \ldots, \mathbf{p}_{\mu A}) : \mathbf{x}_j \in {}_{n-1}\mathcal{O}_{A'}[z_n] \text{ and degree } \mathbf{x}_j \leqq d\}$$

Each polynomial $\mathbf{x}_j \in {}_{n-1}\mathcal{O}_{A'}[z_n]$ can be identified with its $d + 1$ coefficients and hence can be viewed also as an element of ${}_{n-1}\mathcal{O}_{A'}^{d+1}$; $\mathcal{R}^d(\mathbf{p}_{1A}, \ldots, \mathbf{p}_{\mu A})$ can be viewed correspondingly as a subset of ${}_{n-1}\mathcal{O}_{A'}^{\mu(d+1)}$ and is evidently a submodule of that free ${}_{n-1}\mathcal{O}_{A'}$-module. On the other hand, by considering the coefficients separately, the condition $\sum_j \mathbf{x}_j\mathbf{p}_{jA} = 0$ can be viewed as the condition that the coefficients of the polynomials $\mathbf{x}_j$ are relations between some combinations of the coefficients of the polynomials $\mathbf{p}_{jA}$. Thus, $\mathcal{R}^d(\mathbf{p}_{1A}, \ldots, \mathbf{p}_{\mu A})$ can be viewed as the module of relations between the germs of some holomorphic functions of $n - 1$ complex variables at each point $A' \in \Delta(0; R')$. It then follows from the induction hypothesis that if $R'$ is sufficiently small, the family $\{\mathcal{R}^d(\mathbf{p}_{1A}, \ldots, \mathbf{p}_{\mu A})\}$ of submodules of $\{{}_{n-1}\mathcal{O}_{A'}^{\mu(d+1)}\}$ is finitely generated over $\Delta(0; R')$. Let $G_j = (g_{j1}, \ldots, g_{j\mu}) \in {}_{n-1}\mathcal{O}_{\Delta(0;R')}[z_n]$ be the generators of this family of submodules, where $g_{jk}$ are polynomials in $z_n$ with coefficients holomorphic in $\Delta(0; R')$ and degree $g_{jk} \leqq d$, and note that when viewed as holomorphic functions of $n$ complex variables, the germs $\mathbf{G}_{jA}$ are contained in $\mathcal{R}(\mathbf{p}_{1A}, \ldots, \mathbf{p}_{\mu A})$ for each point $A \in \Delta(0; R)$. This part of the proof can then be concluded by showing that the germs $\mathbf{G}_{jA}$ actually generate $\mathcal{R}(\mathbf{p}_{1A}, \ldots, \mathbf{p}_{\mu A})$ as an ${}_n\mathcal{O}_A$-module at each point $A \in \Delta(0; R)$. For this purpose consider an arbitrary element $\mathbf{X} = (\mathbf{x}_1, \ldots, \mathbf{x}_\mu) \in \mathcal{R}(\mathbf{p}_{1A}, \ldots, \mathbf{p}_{\mu A})$ at some point $A \in \Delta(0; R)$. The function $p_1$ is of course regular in $z_n$ at the point $A$, and although it is a polynomial in $z_n$ and hence can be viewed as a polynomial in the local coordinate $z_n - a_n$ at $A$, it does not necessarily represent a Weierstrass polynomial in ${}_{n-1}\mathcal{O}_{A'}[z_n - a_n]$. However, from the Weierstrass preparation and division theorems it follows that $\mathbf{p}_{1A} =$

$\mathbf{p}'_{1A}\mathbf{p}''_{1A}$ where $\mathbf{p}'_{1A} \in {}_{n-1}\mathcal{O}_{A'}[z_n - a_n]$ is a Weierstrass polynomial of degree $d'_1$ and $\mathbf{p}''_{1A} \in {}_{n-1}\mathcal{O}_{A'}[z_n - a_n]$ is a polynomial of degree $d''_1$ which is a unit in ${}_n\mathcal{O}_A$, and of course $d_1 = d'_1 + d''_1$. Apply the Weierstrass division theorem to each component of $\mathbf{X} \in {}_n\mathcal{O}_A^\mu$ separately to write $\mathbf{X} = \mathbf{p}'_{1A}\mathbf{X}'' + \mathbf{R}'$ where $\mathbf{X}'' \in {}_n\mathcal{O}_A^\mu$ and $\mathbf{R}' \in {}_{n-1}\mathcal{O}_{A'}[z_n - a_n]^\mu$, with each component of $\mathbf{R}'$ having degree $< d'_1$. Then set $\mathbf{X}' = (\mathbf{p}''_{1A})^{-1}\mathbf{X}''$ and write

$$\mathbf{X} = \mathbf{p}_{1A}\mathbf{X}' + \mathbf{R}' \tag{1}$$

Note further that

$$\begin{aligned}\mathbf{p}_{1A}\mathbf{X}' &= \mathbf{p}_{1A}(\mathbf{x}'_1, \ldots, \mathbf{x}'_\mu)\\ &= (\mathbf{p}_{1A}\mathbf{x}'_1 + \mathbf{p}_{2A}\mathbf{x}'_2 + \cdots + \mathbf{p}_{\mu A}\mathbf{x}'_\mu, 0, \ldots, 0) + \sum_{j=2}^{\mu} \mathbf{x}'_j(-\mathbf{p}_{jA}, 0, \ldots, 0, \mathbf{p}_{1A}, 0, \ldots, 0)\end{aligned}$$

Now since $(-\mathbf{p}_{jA}, 0, \ldots, 0, \mathbf{p}_{1A}, 0, \ldots, 0) \in \mathscr{R}^d(\mathbf{p}_{1A}, \ldots, \mathbf{p}_{\mu A})$, it follows that $(-\mathbf{p}_{jA}, 0, \ldots, 0, \mathbf{p}_{1A}, 0, \ldots, 0) = \sum_k \mathbf{y}'_{jk}\mathbf{G}_{kA}$ for some germs $\mathbf{y}'_{jk} \in {}_{n-1}\mathcal{O}_{A'}$, and consequently $\mathbf{p}_{1A}\mathbf{X}' = \mathbf{H} + \sum_{j,k} \mathbf{x}'_j\mathbf{y}'_{jk}\mathbf{G}_{kA} = \mathbf{H} + \sum_j \mathbf{y}_j\mathbf{G}_{jA}$ for some elements $\mathbf{y}_j \in {}_n\mathcal{O}_A$ and $\mathbf{H} = (h_1, 0, \ldots, 0)$ where $\mathbf{h}_1 \in {}_n\mathcal{O}_A$. Substituting this into (1) yields

$$\mathbf{X} = \sum_j \mathbf{y}_j\mathbf{G}_{jA} + \mathbf{R} \tag{2}$$

where $\mathbf{R} = \mathbf{R}' + \mathbf{H} = (\mathbf{r}_1, \ldots, \mathbf{r}_\mu) \in {}_n\mathcal{O}_A^\mu$ and where the components $\mathbf{r}_2, \ldots, \mathbf{r}_\mu$ are actually polynomials in ${}_{n-1}\mathcal{O}_{A'}[z_n - a_n]$ of degree $< d'_1$. Since $\mathbf{X}$ and $\mathbf{G}_{jA}$ belong to $\mathscr{R}(\mathbf{p}_{1A}, \ldots, \mathbf{p}_{\mu A})$ necessarily $\mathbf{R} \in \mathscr{R}(\mathbf{p}_{1A}, \ldots, \mathbf{p}_{\mu A})$ as well. Thus, $\mathbf{p}'_{1A}\mathbf{p}''_{1A}\mathbf{r}_1 = \mathbf{p}_{1A}\mathbf{r}_1 = -\sum_{j=2}^{\mu} \mathbf{p}_{jA}\mathbf{r}_j$, so this element must be a polynomial in ${}_{n-1}\mathcal{O}_{A'}[z_n - a_n]$ of degree $< d + d'_1$. It follows from the Weierstrass division theorem that $\mathbf{p}''_{1A}\mathbf{r}_1$ is also a polynomial in ${}_{n-1}\mathcal{O}_{A'}[z_n - a_n]$ of degree $< d$; therefore, $\mathbf{p}''_{1A}\mathbf{R} \in \mathscr{R}^d(\mathbf{p}_{1A}, \ldots, \mathbf{p}_{\mu A})$, so that $\mathbf{p}''_{1A}\mathbf{R} = \sum_j \mathbf{y}''_j\mathbf{G}_{jA}$ for some elements $\mathbf{y}''_j \in {}_{n-1}\mathcal{O}_{A'}$. Substituting this into (2) again and recalling that $\mathbf{p}''_{1A}$ is a unit in ${}_n\mathcal{O}_A$ lead finally to $\mathbf{X} = \sum_j(\mathbf{y}_j + (\mathbf{p}''_{1A})^{-1}\mathbf{y}''_j)\mathbf{G}_{jA}$. That suffices to prove the induction step in the special case that $\nu = 1$.

Assume then that the theorem has been demonstrated for holomorphic functions of $n$ variables and for all indices $< \nu$ where $\nu > 1$, and consider a collection of $\nu$-tuples $F_1, \ldots, F_\mu \in {}_n\mathcal{O}_U^\nu$ over some open neighborhood $U$ of the origin in $\mathbb{C}^n$. Write $F_j = (F'_j, F''_j)$ where $F'_j \in {}_n\mathcal{O}_U^{\nu-1}$ and $F''_j \in {}_n\mathcal{O}_U^1$, and note that

$$\mathscr{R}(\mathbf{F}_{1A}, \ldots, \mathbf{F}_{\mu A}) = \mathscr{R}(\mathbf{F}'_{1A}, \ldots, \mathbf{F}'_{\mu A}) \cap \mathscr{R}(\mathbf{F}''_{1A}, \ldots, \mathbf{F}''_{\mu A})$$

It follows from the inductive hypothesis that the family $\{\mathscr{R}(\mathbf{F}'_{1A}, \ldots, \mathbf{F}'_{\mu A})\}$ of submodules of $\{{}_n\mathcal{O}_A^\mu\}$ is finitely generated over some polydisc $\Delta(0; R) \subseteq U$ about the origin. Let $G'_1, \ldots, G'_\lambda \in {}_n\mathcal{O}_{\Delta(0;R)}^\mu$ be a set of generators for this family of submodules, where $G'_k = (g'_{k1}, \ldots, g'_{k\mu})$ for some functions $g'_{kj} \in {}_n\mathcal{O}_{\Delta(0;R)}$. Then set $g''_k = \sum_j g'_{kj}F''_j \in {}_n\mathcal{O}_{\Delta(0;R)}$, and it follows from the inductive hypothesis that, after passing to a smaller polyradius $R$ if necessary, the family $\{\mathscr{R}(g''_{1A}, \ldots, g''_{\lambda A})\}$ of submodules of $\{{}_n\mathcal{O}_{\Delta(0;R)}^\lambda\}$ is also finitely generated over $\Delta(0; R)$. Let $H_\ell \in {}_n\mathcal{O}_{\Delta(0;R)}^\lambda$ be a set of generators of this family of submodules, where $H_\ell = (h_{\ell 1}, \ldots, h_{\ell\lambda})$ for some functions $h_{\ell k} \in {}_n\mathcal{O}_{\Delta(0;R)}$. Finally let $G_\ell = \sum_k h_{\ell k} G'_k \in {}_n\mathcal{O}^\mu_{\Delta(0;R)}$, and note that by construction $G_{\ell A} \in \mathcal{R}(\mathbf{F}_{1A}, \ldots, \mathbf{F}_{\mu A})$ at any point $A \in \Delta(0; R)$. If $\mathbf{X} = (\mathbf{x}_1, \ldots, \mathbf{x}_\mu) \in \mathcal{R}(\mathbf{F}_{1A}, \ldots, \mathbf{F}_{\mu A})$ for some point $A \in \Delta(0; R)$, then since $\mathbf{X} \in \mathcal{R}(\mathbf{F}'_{1A}, \ldots, \mathbf{F}'_{\mu A})$, it follows that $\mathbf{X} = \sum_k \mathbf{y}'_k \mathbf{G}'_{kA}$ for some elements $\mathbf{y}'_k \in {}_n\mathcal{O}_A$. But since $\mathbf{X} \in \mathcal{R}(\mathbf{F}''_{1A}, \ldots, \mathbf{F}''_{\mu A})$ as well, it follows that $0 = \sum_j \mathbf{x}_j \mathbf{F}''_{jA} = \sum_{jk} \mathbf{y}'_k \mathbf{g}'_{kjA} \mathbf{F}''_{jA} = \sum_k \mathbf{y}'_k \mathbf{g}''_{kA}$, so that $\mathbf{Y}' = (\mathbf{y}'_1, \ldots, \mathbf{y}'_\lambda) \in \mathcal{R}(g''_{1A}, \ldots, g''_{\lambda A})$ and hence $\mathbf{Y}' = \sum_\ell \mathbf{y}_\ell \mathbf{H}_{\ell A}$ for some elements $\mathbf{y}_\ell \in {}_n\mathcal{O}_A$. Combining these results yields $\mathbf{X} = \sum_k \mathbf{y}'_k \mathbf{G}'_{kA} = \sum_{k,\ell} \mathbf{y}_\ell \mathbf{h}_{\ell k} \mathbf{G}'_{kA} = \sum_\ell \mathbf{y}_\ell \mathbf{G}_{\ell A}$. Thus the $\mu$-tuples $G_\ell$ generate the family $\{\mathcal{R}(\mathbf{F}_{1A}, \ldots, \mathbf{F}_{\mu A})\}$ of submodules of $\{{}_n\mathcal{O}^\mu_A\}$ over $\Delta(0; R)$, and that suffices to conclude the proof of the theorem.

Oka's theorem seems perhaps somewhat technical and uninteresting; so to illustrate the power of this result, some useful corollaries will be deduced here.

**3. COROLLARY.** *If $\{\mathcal{A}_A\}$ and $\{\mathcal{B}_A\}$ are two finitely generated families of submodules of $\{{}_n\mathcal{O}^\nu_A\}$ over an open subset $U \subseteq \mathbb{C}^n$, then the intersections $\{\mathcal{A}_A \cap \mathcal{B}_A\}$ form a finitely generated family of submodules of $\{{}_n\mathcal{O}^\nu_A\}$ over an open neighborhood of any point of $U$.*

*Proof.* Suppose that $U$ is an open neighborhood of the origin in $\mathbb{C}^n$, and that $F_1, \ldots, F_\mu \in {}_n\mathcal{O}^\nu_U$ are generators of the family $\{\mathcal{A}_A\}$ over $U$ and $G_1, \ldots, G_\lambda \in {}_n\mathcal{O}^\nu_U$ are generators of the family $\{\mathcal{B}_A\}$ over $U$. It follows from Oka's theorem that the family $\{\mathcal{R}(\mathbf{F}_{1A}, \ldots, \mathbf{F}_{\mu A}, -\mathbf{G}_{1A}, \ldots, -\mathbf{G}_{\lambda A})\}$ of submodules of $\{{}_n\mathcal{O}^{\mu+\lambda}_A\}$ is finitely generated over some polydisc $\Delta(0; R) \subseteq U$. Let $H'_j \in {}_n\mathcal{O}^{\mu+\lambda}_{\Delta(0;R)}$ be generators of this family of submodules and write $H'_j = (h'_{j1}, \ldots, h'_{j\mu}, h''_{j1}, \ldots, h''_{j\lambda})$, noting that by construction $H_k = \sum_j h'_{kj} F_j \in {}_n\mathcal{O}^\nu_{\Delta(0;R)}$ has the property that $\mathbf{H}_{kA} \in \mathcal{A}_A \cap \mathcal{B}_A$ at each point $A \in \Delta(0; R)$. On the other hand, if $\mathbf{X} \in \mathcal{A}_A \cap \mathcal{B}_A \subseteq {}_n\mathcal{O}^\nu_A$ at some point $A \in \Delta(0; R)$, then $\mathbf{X} = \sum_j \mathbf{x}'_j \mathbf{F}_{jA} = \sum_j \mathbf{x}''_j \mathbf{G}_{jA}$ for some germs $\mathbf{x}'_j, \mathbf{x}''_j \in {}_n\mathcal{O}_A$. The element $(\mathbf{x}'_1, \ldots, \mathbf{x}'_\mu, \mathbf{x}''_1, \ldots, \mathbf{x}''_\mu) \in {}_n\mathcal{O}^{\mu+\lambda}_A$ evidently belongs to the submodule $\mathcal{R}(\mathbf{F}_{1A}, \ldots, \mathbf{F}_{\mu A}, -\mathbf{G}_{1A}, \ldots, -\mathbf{G}_{\lambda A})$ and hence can be written in the form $\sum_k \mathbf{y}_k \mathbf{H}'_{kA}$ for some germs $\mathbf{y}_k \in {}_n\mathcal{O}_A$. In particular therefore $\mathbf{x}'_j = \sum_k \mathbf{y}_k \mathbf{h}'_{kjA}$, so that $\mathbf{X} = \sum_{jk} \mathbf{y}_k \mathbf{h}'_{kjA} \mathbf{F}_{jA} = \sum_k \mathbf{y}_k \mathbf{H}_{kA}$. That shows that the $\nu$-tuples $H_k \in {}_n\mathcal{O}^\nu_{\Delta(0;R)}$ generate the family of submodules $\{\mathcal{A}_A \cap \mathcal{B}_A\}$ over $(0; R)$ and thereby completes the proof of the corollary.

A special case of the preceding corollary is the assertion that if $\{\mathfrak{A}_A\}$ and $\{\mathfrak{B}_A\}$ are two finitely generated families of ideals over an open subset $U \subseteq \mathbb{C}^n$, then their intersections $\{\mathfrak{A}_A \cap \mathfrak{B}_A\}$ form a finitely generated family of ideals over an open neighborhood of any point of $U$. It is of course obvious that the sums $\{\mathfrak{A}_A + \mathfrak{B}_A\}$ and products $\{\mathfrak{A}_A \cdot \mathfrak{B}_A\}$ are finitely generated families of ideals over $U$. It may be recalled that there is a fourth natural binary operation on ideals, the operation associating to two ideals $\mathfrak{A}$ and $\mathfrak{B}$ in a ring $\mathcal{O}$ the quotient ideal $\mathfrak{A} : \mathfrak{B}$ defined by $\mathfrak{A} : \mathfrak{B} = \{f \in \mathcal{O} : \mathbf{f}\mathfrak{B} \subseteq \mathfrak{A}\}$; the finite generation of families of ideals is also preserved at least locally by this operation.

**4. COROLLARY.** *If $\{\mathfrak{A}_A\}$ and $\{\mathfrak{B}_A\}$ are two finitely generated families of ideals in $\{{}_n\mathcal{O}_A\}$ over an open subset $U \subseteq \mathbb{C}^n$, then their quotients $\{\mathfrak{A}_A : \mathfrak{B}_A\}$ form a finitely generated family of ideals over an open neighborhood of any point of $U$.*

*Proof.* Suppose that $U$ is an open neighborhood of the origin in $\mathbb{C}^n$, that $f_1, \ldots, f_\mu \in {}_n\mathcal{O}_U$ are generators of the family $\{\mathfrak{A}_A\}$, and that $g_1, \ldots, g_\nu \in {}_n\mathcal{O}_U$ are generators of the family $\{\mathfrak{B}_A\}$ over $U$. For a fixed index $j$, $1 \leq j \leq \nu$, it follows from Oka's theorem that the family $\{\mathcal{R}(\mathbf{g}_{jA}, f_{1A}, \ldots, f_{\mu A})$ of submodules of $\{{}_n\mathcal{O}_A^{\mu+1}\}$ is finitely generated over some polydisc $\Delta(0; R) \subseteq U$. Let $H_k = (h_0^k, h_1^k, \ldots, h_\mu^k) \in {}_n\mathcal{O}_{\Delta(0;R)}^{\mu+1}$ be generators for this family of modules over $\Delta(0; R)$. It is easy to see that the functions $h_0^k \in {}_n\mathcal{O}_{\Delta(0;R)}$ are then generators for the family of ideals $\{\mathfrak{A}_A : {}_n\mathcal{O}_A \mathbf{g}_{jA}\}$ over $\Delta(0; R)$. Indeed, an element $\mathbf{x} \in {}_n\mathcal{O}_A$ at a point $A \in \Delta(0; R)$ belongs to the ideal $\mathfrak{A}_A : {}_n\mathcal{O}_A \mathbf{g}_{jA}$ precisely when there are elements $\mathbf{x}_k \in {}_n\mathcal{O}_A$ such that $\mathbf{x}\mathbf{g}_{jA} = \mathbf{x}_1 \mathbf{f}_{1A} + \cdots + \mathbf{x}_\mu \mathbf{f}_{\mu A}$—that is, such that $(\mathbf{x}, -\mathbf{x}_1, \ldots, -\mathbf{x}_\mu) \in \mathcal{R}(\mathbf{g}_{jA}, \mathbf{f}_{1A}, \ldots, \mathbf{f}_{\mu A})$—and that is the case precisely when $\mathbf{x}$ is in the ideal generated by the germs $\mathbf{h}_{0A}^k$. Since $\mathfrak{A}_A : \mathfrak{B}_A = \bigcap_j (\mathfrak{A}_A : {}_n\mathcal{O}_A \mathbf{g}_{jA})$, the family of ideals $\{\mathfrak{A}_A : \mathfrak{B}_A\}$ is an intersection of finitely generated families of ideals over $\Delta(0; R)$, so it follows from Corollary 3 that after shrinking the polyradius $R$ if necessary the family of ideals $\{\mathfrak{A}_A : \mathfrak{B}_A\}$ is also finitely generated over $\Delta(0; R)$, and that proof is thereby concluded.

To return now to the earlier question whether the family of ideals of a holomorphic subvariety of $U$ is finitely generated, the local answer is provided by an application of the local parametrization theorem and Oka's theorem, with a very simple proof due to K. Langmann resting on the following result.

**5. LEMMA.** *Let $f_1, \ldots, f_\nu$ be holomorphic functions in an open neighborhood $U$ of the origin $0 \in \mathbb{C}^n$, defining a holomorphic subvariety $V \subseteq U$ and a finitely generated family of ideals $\{\mathfrak{A}_A\}$ over $U$, and suppose that $\mathfrak{A}_0$ is a prime ideal in ${}_n\mathcal{O}_0$ and that $\mathfrak{A}_A = \operatorname{id} \mathbf{V}_A$ at all points $A \in V - W$ where $W$ is a holomorphic subvariety of $V$ with $\mathbf{W}_0 \neq \mathbf{V}_0$. Then there is an open polydisc $\Delta(0; R) \subseteq U$ such that $\mathfrak{A}_A = \operatorname{id} \mathbf{V}_A$ at all points $A \in \Delta(0; R)$.*

*Proof.* Since the functions $f_1, \ldots, f_\nu$ define the subvariety $V$, it follows from Hilbert's zero-theorem that $\mathfrak{A}_0 \subseteq \operatorname{id} V_0 \subseteq \sqrt{\mathfrak{A}_0}$; but since it was also assumed that $\mathfrak{A}_0$ is prime, necessarily $\mathfrak{A}_0 = \sqrt{\mathfrak{A}_0} = \operatorname{id} \mathbf{V}_0$. Choose a germ $\mathbf{d}_0 \in {}_n\mathcal{O}_0$ for which $\mathbf{d}_0 \in \operatorname{id} \mathbf{W}_0$ but $\mathbf{d}_0 \notin \operatorname{id} \mathbf{V}_0 = \mathfrak{A}_0$, as is evidently possible as a consequence of the hypothesis that $\mathbf{W}_0 \neq \mathbf{V}_0$, and choose a representative holomorphic function $d$ in some open polydisc $\Delta(0; R) \subseteq U$. The function $d$ can then be assumed to vanish identically on $W$ but not on $V$. It follows from Corollary 4 that the family of ideals $\{\mathfrak{A}_A : {}_n\mathcal{O}_A \mathbf{d}_A\}$ is finitely generated over $\Delta(0; R)$ if $R$ is sufficiently small. If $g_1, \ldots, g_\mu \in {}_n\mathcal{O}_{\Delta(0;R)}$ are generators for this family of ideals, then in particular $\mathbf{d}_0 \cdot \mathbf{g}_{j0} \in \mathfrak{A}_0$; but $\mathfrak{A}_0$ is prime and $\mathbf{d}_0 \notin \mathfrak{A}_0$, so that necessarily $\mathbf{g}_{j0} \in \mathfrak{A}_0$. It is thus possible to write $\mathbf{g}_{j0}$ as a linear combination of the generators $\mathbf{f}_{k0}$ of the ideal $\mathfrak{A}_0$ with coefficients from ${}_n\mathcal{O}_0$, and if $R$ is chosen sufficiently small that these coefficients have representatives in ${}_n\mathcal{O}_{\Delta(0;R)}$, it follows that $\mathbf{g}_{jA} \in \mathfrak{A}_A$ at all points $A \in \Delta(0; R)$. Thus if $R$ is sufficiently small, it can be assumed that $\mathfrak{A}_A : {}_n\mathcal{O}_A \mathbf{d}_A = \mathfrak{A}_A$ at all points $A \in \Delta(0; R)$.

Now consider any point $A \in V \cap \Delta(0; R)$ and any germ $\mathbf{f}_A \in \operatorname{id} \mathbf{V}_A \subseteq {}_n\mathcal{O}_A$. The germ $\mathbf{f}_A$ will have a representative $f \in {}_n\mathcal{O}_{U_A}$ in some open neighborhood $U_A$ of the point $A$, and it follows again from Corollary 4 that if this neighborhood is sufficiently small, the family of ideals $\{\mathfrak{A}_Z : {}_n\mathcal{O}_Z \mathbf{f}_Z\} = \{\mathfrak{B}_Z\}$ will be finitely generated over $U_A$.

Let $h_1, \ldots, h_\lambda \in {}_n\mathcal{O}_{U_A}$ be generators for this family of ideals. The hypothesis that $\mathfrak{A}_Z = \text{id } \mathbf{V}_Z$ whenever $Z \in V - W$ implies that $\mathfrak{B}_Z = {}_n\mathcal{O}_Z$ whenever $z \in U_A \cap (V - W)$, so the common zeros of the generators $h_1, \ldots, h_\lambda$ must lie in $W \cap U_A$. The function $d$ vanishes on $W$, so it follows from Hilbert's zero-theorem that $\mathbf{d}_A^r \in \mathfrak{B}_A$ for some integer $r > 0$; but that means that $\mathbf{d}_A^r \mathbf{f}_A \in \mathfrak{A}_A$, and since $\mathfrak{A}_A : {}_n\mathcal{O}_A \mathbf{d}_A = \mathfrak{A}_A$ necessarily $\mathbf{f}_A \in \mathfrak{A}_A$. Therefore $\mathfrak{A}_A = \text{id } \mathbf{V}_A$ at all points $A \in V \cap \Delta(0; R)$, and since that is also true trivially whenever $A \in \Delta(0; R) - V \cap \Delta(0; R)$, the proof is thereby completed.

**6. THEOREM (Cartan's theorem).** *If $V$ is a holomorphic subvariety of an open subset $U \subseteq \mathbb{C}^n$, then the family of ideals $\{\text{id } \mathbf{V}_A\}$ in $\{{}_n\mathcal{O}_A\}$ is finitely generated over an open neighborhood of each point of $U$.*

*Proof.* Fix a point of $V$, which to simplify notation can be assumed to be the origin $O \in \mathbb{C}^n$, and suppose at first that the germ $\mathbf{V}_O$ of $V$ at the origin is irreducible. After a nonsingular linear change of coordinates in $\mathbb{C}^n$, it can be assumed that the ideal $\text{id } \mathbf{V}_O \subseteq {}_n\mathcal{O}_O$ is strictly regular in the variables $z_{k+1}, \ldots, z_n$. Let $\mathbf{f}_1, \ldots, \mathbf{f}_\nu \in {}_n\mathcal{O}_O$ be generators of the ideal $\text{id } \mathbf{V}_O$, including among others the auxiliary polynomials $\mathbf{p}_j \in \text{id } \mathbf{V}_O \cap {}_k\mathcal{O}[z_j]$ for $k + 1 \leq j \leq n$ and $\mathbf{q}_j \in \text{id } \mathbf{V}_O \cap {}_k\mathcal{O}[z_{k+1}, z_j]$ for $k + 2 \leq j \leq n$, and choose an open neighborhood $U$ of the origin such that these germs can be represented by holomorphic functions $f_1, \ldots, f_\nu \in {}_n\mathcal{O}_U$. Now as observed after the proof of the local parametrization theorem, the subvariety $V$ is actually a complex submanifold outside a proper holomorphic subvariety $W \subset V$, and the functions $p_{k+1}, q_{k+2}, \ldots, q_n$ can be taken as part of a local system of coordinates in $\mathbb{C}^n$ near any point of $V - W$ exhibiting $V$ as such a submanifold. Thus these functions, and therefore the functions $f_1, \ldots, f_\nu$ as well, generate $\text{id } \mathbf{V}_A \subseteq {}_n\mathcal{O}_A$ at each point $A \in V - W$. It then follows immediately from Lemma 5 that the functions $f_1, \ldots, f_\nu$ generate $\text{id } \mathbf{V}_A \subseteq {}_n\mathcal{O}_A$ at all points $A$ in some open polydisc $\Delta(0; R) \subseteq U$ and hence that the family of ideals $\{\text{id } \mathbf{V}_A\}$ is finitely generated over $\Delta(0; R)$. If the germ $\mathbf{V}_O$ is reducible, then in some open neighborhood $U$ of the origin write $V \cap U = \bigcup V_i$ where $V_i$ are holomorphic subvarieties of $U$ and their germs at the origin are the irreducible components of $\mathbf{V}_O$. It follows from the first part of the proof that for each index $i$ the family of ideals $\{\text{id } \mathbf{V}_{iA}\}$ is finitely generated over some open polydisc $\Delta(0; R) \subseteq U$. But $\text{id } \mathbf{V}_A = \bigcap_i \text{id } \mathbf{V}_{iA}$, so by Corollary 3 the family of ideals $\{\text{id } \mathbf{V}_A\}$ is also finitely generated over $\Delta(0; R)$ after shrinking $R$ if necessary. That suffices to conclude the proof.

It is perhaps worth pointing out explicitly here that the lemma used in the proof of Cartan's theorem is in turn a simple consequence of the following corollary of that theorem.

**7. COROLLARY.** *If $f_1, \ldots, f_\nu$ are holomorphic functions in an open set $U \subseteq \mathbb{C}^n$ defining a holomorphic subvariety $V \subseteq U$ and a finitely generated family of ideals $\{\mathfrak{A}_A\}$ over $U$ and if $\mathfrak{A}_A = \text{id } \mathbf{V}_A$ at some point $A \in V$, then $\mathfrak{A}_Z = \text{id } \mathbf{V}_Z$ at all points $Z \in U - W$ where $W \subset V$ is a proper holomorphic subvariety of $V$.*

*Proof.* Since $\mathfrak{A}_Z \subseteq \text{id } \mathbf{V}_Z$ at every point $Z \in U$, the condition that $\mathfrak{A}_Z = \text{id } \mathbf{V}_Z$ is just that $\mathfrak{B}_Z = \mathfrak{A}_Z : \text{id } \mathbf{V}_Z = {}_n\mathcal{O}_Z$. Now by Cartan's theorem the family of ideals $\{\mathfrak{B}_Z\}$ is finitely generated over an open neighborhood $U_B$ of any point $B \in U$. if $g_j \in {}_n\mathcal{O}_{U_B}$ are generators of this family of ideals over $U_B$, then $\{Z \in U_B : \mathfrak{B}_Z \neq {}_n\mathcal{O}_Z\} = \{Z \in U_B : g_j(Z) = 0 \text{ for all } j\}$, and hence $\mathfrak{A}_Z = \text{id } \mathbf{V}_Z$ for all points $Z$ outside a holomorphic subvariety $W \subseteq U$. Since $\mathfrak{A}_Z = \text{id } \mathbf{V}_Z = {}_n\mathcal{O}_Z$ whenever $Z \in U - V$, it follows that $W \subseteq V$, while since $\mathfrak{A}_A = \text{id } \mathbf{V}_A$ at the point $A \in V$ by hypothesis it follows that $W$ is necessarily a proper holomorphic subvariety of $V$. That suffices for the proof.

Under the hypothesis of Lemma 5 the condition that the ideal $\mathfrak{A}_O$ be prime implies as noted in the proof of that lemma that $\mathfrak{A}_O = \text{id } \mathbf{V}_O$. So by Corollary 7 necessarily $\mathfrak{A}_Z = \text{id } \mathbf{V}_Z$ outside a proper holomorphic subvariety $W' \subseteq V$. Since $O \notin W'$, a full open neighborhood of $O$ is disjoint from $W'$; hence, $\mathfrak{A}_Z = \text{id } \mathbf{V}_Z$ at all points in an open neighborhood of $O$ as desired. The hypothesis in Lemma 5 that $\mathfrak{A}_O$ be prime, or at least that $\mathfrak{A}_O = \text{id } \mathbf{V}_O$, is clearly necessary. Indeed it is a triviality that there do exist holomorphic functions $f_j \in {}_n\mathcal{O}_U$, defining a holomorphic subvariety $V \subseteq U$ and a finitely generated family of ideals $\{\mathfrak{A}_A\}$ over $U$, such that $\mathfrak{A}_Z = \text{id } \mathbf{V}_Z$ for all points $Z \in V - W$, but $\mathfrak{A}_Z \neq \text{id } \mathbf{V}_Z$ for all points $Z \in W$, where $W$ is a nontrivial proper subvariety of $V$. If $O \in V$ and if $g_j \in {}_n\mathcal{O}_U$ generate the family of ideals $\{\text{id } \mathbf{V}_Z\}$ over $U$, the functions $f_{ij} = z_i g_j$ clearly have that property, where $W = 0$.

**8. COROLLARY.** *If $\{\mathfrak{A}_A\}$ is a finitely generated family of ideals in $\{{}_n\mathcal{O}_A\}$ over an open subset $U \subseteq \mathbb{C}^n$, then their radicals $\{\sqrt{\mathfrak{A}_A}\}$ form a finitely generated family of ideals over an open neighborhood of any point of $U$.*

*Proof.* If $\{\mathfrak{A}_A\}$ is a finitely generated family of ideals over $U$, then $V = \{A \in U : \mathfrak{A}_A \neq {}_n\mathcal{O}_A\}$ is clearly a holomorphic subvariety of $U$; for if $f_1, \ldots, f_\nu \in {}_n\mathcal{O}_U$ are generators of this family of ideals, then $V = \{Z \in U : f_1(Z) = \cdots = f_\nu(Z) = 0\}$. But it follows from Hilbert's zero-theorem that $\sqrt{\mathfrak{A}_A} = \text{id } \mathbf{V}_A$, and then $\{\sqrt{\mathfrak{A}_A}\}$ is locally a finitely generated family of ideals over $U$ as a consequence of Theorem 6.

It is of course also possible to speak of finitely generated families of ideals over open subsets of an arbitrary holomorphic variety $V$, and the results obtained for families of ideals over open subsets of $\mathbb{C}^n$ extend quite directly as a consequence of Cartan's theorem.

**9. COROLLARY (Oka's theorem on varieties).** *If $F_1, \ldots, F_\mu \in {}_V\mathcal{O}_V^\nu$ are finitely many $\nu$-tuples of holomorphic functions on a holomorphic variety $V$, then the family $\{\mathcal{R}(\mathbf{F}_{1A}, \ldots \mathbf{F}_{\mu A})\}$ of submodules of $\{{}_V\mathcal{O}_A^\mu\}$ is finitely generated over an open neighborhood of any point of $V$.*

*Proof.* For any point $O \in V$, an open neighborhood of that point can be represented by a holomorphic subvariety of an open neighborhood $U$ of the origin in $\mathbb{C}^n$, where the origin represents the point $O$. To simplify the notation this representative subvariety will also be denoted by $V$. If the neighborhood $U$ is sufficiently small, there will be $\nu$-tuples of holomorphic functions $\tilde{F}_1, \ldots, \tilde{F}_\mu \in {}_n\mathcal{O}_U^\nu$ such that $\tilde{F}_j | V = F_j$, and moreover by Cartan's theorem, Theorem 6, there will exist holomorphic functions $g_1, \ldots, g_\lambda \in {}_n\mathcal{O}_U$ that generate the ideal of the subvariety $V$ at each point of $U$.

If $G_{k\ell} \in {}_n\mathcal{O}_U^\nu$ is the $\nu$-tuple of holomorphic functions consisting of the zero function except for the function $g_\ell$ is the $k$th entry, it follows from Oka's theorem, Theorem 2, that after shrinking the neighborhood $U$ if necessary, the family $\{\mathcal{R}(\mathbf{\tilde{F}}_{jA}, \mathbf{G}_{k\ell A})\}$ of submodules of $\{{}_U\mathcal{O}_A^{\mu+\nu\lambda}\}$ will be finitely generated over $U$. If $H_i = \{h_{ij}, h_{ik\ell}\}$ are generators of this family over $U$, the proof of the corollary will be completed by showing that the functions $H_j' = \{h_{ij}|V\} \in {}_V\mathcal{O}_V^\mu$ are generators of the family $\{\mathcal{R}(F_{1A}, \ldots, F_{\mu A})\}$ over $V$. Whenever $A \in V$ and $(\mathbf{x}_1, \ldots, \mathbf{x}_\mu) \in \mathcal{R}(\mathbf{F}_{1A}, \ldots, \mathbf{F}_{\mu A})$ for some germs $\mathbf{x}_j \in {}_V\mathcal{O}_A$, then $\sum_j \mathbf{x}_j F_{jA} = 0$. Choose germs $\tilde{\mathbf{x}}_j \in {}_n\mathcal{O}_A$ such that $\tilde{\mathbf{x}}_j | V = \mathbf{x}_j$. Then each entry in the $\nu$-tuple $\sum_j \tilde{\mathbf{x}}_j \mathbf{\tilde{F}}_{jA} \in {}_n\mathcal{O}_A^\nu$ must be contained in the ideal of the subvariety $V$ at the point $A$; hence, there are germs $\tilde{\mathbf{x}}_{k\ell} \in {}_n\mathcal{O}_A$ such that

$$\sum_j \tilde{\mathbf{x}}_j \mathbf{\tilde{F}}_{jA} + \sum_{k\ell} \tilde{\mathbf{x}}_{k\ell} \mathbf{G}_{k\ell A} = 0$$

That means that $(\tilde{\mathbf{x}}_j, \tilde{\mathbf{x}}_{k\ell}) \in \mathcal{R}(\mathbf{\tilde{F}}_{jA}, \mathbf{G}_{k\ell A})$, and consequently $(\tilde{\mathbf{x}}_j, \tilde{\mathbf{x}}_{k\ell}) = \sum \tilde{\mathbf{y}}_i \mathbf{H}_{iA}$ for some germs $\tilde{\mathbf{y}}_i \in {}_n\mathcal{O}_A$. But then

$$\mathbf{x}_j = \tilde{\mathbf{x}}_j | V = \sum_i \mathbf{y}_i (\mathbf{h}_{ij} | V)$$

where $\mathbf{y}_i = \tilde{\mathbf{y}}_i | V$, showing that the functions $H_j' = \{h_{ij}|V\}$ are generators of the family $\mathcal{R}(\mathbf{F}_{1A}, \ldots, \mathbf{F}_{\mu A})$ over $V$ and thereby concluding the proof.

**10. COROLLARY.** *If $V$ is a holomorphic variety and $\{\mathfrak{A}_A\}$ and $\{\mathfrak{B}_A\}$ are finitely generated families of ideals in $\{{}_V\mathcal{O}_A\}$ over $V$, then $\{\mathfrak{A}_A + \mathfrak{B}_A\}$, $\{\mathfrak{A}_A \cdot \mathfrak{B}_A\}$, $\{\mathfrak{A}_A \cap \mathfrak{B}_A\}$, and $\{\mathfrak{A}_A : \mathfrak{B}_A\}$ are finitely generated families of ideals over an open neighborhood of any point of $V$.*

*Proof.* It is obvious that $\{\mathfrak{A}_A + \mathfrak{B}_A\}$ and $\{\mathfrak{A}_A \cdot \mathfrak{B}_A\}$ are finitely generated families of ideals over $V$, but the remaining two cases are not really trivial. For any point $O \in V$, an open neighborhood of that point in $V$ can be represented by a holomorphic subvariety of an open neighborhood $U$ of the origin in $\mathbb{C}^n$, where the origin represents the point $O$. To simplify the notation this representative subvariety will also be denoted by $V$. If the neighborhood $U$ is sufficiently small, the generators of the families of ideals $\{\mathfrak{A}_A\}$ and $\{\mathfrak{B}_A\}$ can be extended to holomorphic functions in all of $U$. They are then generators of finitely generated families of ideals $\{\tilde{\mathfrak{A}}_A\}$ and $\{\tilde{\mathfrak{B}}_A\}$ in $\{{}_n\mathcal{O}_A\}$ over $U$ such that $\tilde{\mathfrak{A}}_A | V = \mathfrak{A}_A$ and $\tilde{\mathfrak{B}}_A | V = \mathfrak{B}_A$ whenever $A \in V$. Furthermore, it follows from Cartan's theorem, Theorem 6, that if the neighborhood $U$ is sufficiently small, the family of ideals $\{\text{id } \mathbf{V}_A\}$ in $\{{}_n\mathcal{O}_A\}$ is also finitely generated over $U$. Now by Corollary 3 the family of ideals $\{(\tilde{\mathfrak{A}}_A + \text{id } \mathbf{V}_A) \cap (\tilde{\mathfrak{B}}_A + \text{id } \mathbf{V}_A)\}$ will also be finitely generated over $U$ if $U$ is sufficiently small, and since it is clear that

$$\mathfrak{A}_A \cap \mathfrak{B}_A = (\tilde{\mathfrak{A}}_A + \text{id } \mathbf{V}_A) \cap (\tilde{\mathfrak{B}}_A + \text{id } \mathbf{V}_A) | V$$

whenever $A \in V$, it follows immediately that the family of ideals $\{\mathfrak{A}_A \cap \mathfrak{B}_A\}$ is finitely generated over $U \cap V$. Similarly by Corollary 4 the family of ideals $(\tilde{\mathfrak{A}}_A + \text{id } \mathbf{V}_A) : \tilde{\mathfrak{B}}_A$ will be finitely generated over $U$ if $U$ is sufficiently small, and since it is clear that

$$\mathfrak{A}_A : \mathfrak{B}_A = (\tilde{\mathfrak{A}}_A + \text{id } \mathbf{V}_A) : \tilde{\mathfrak{B}}_A | V$$

whenever $A \in V$, it follows immediately that the family of ideals $\{\mathfrak{A}_A : \mathfrak{B}_A\}$ is finitely generated over $U \cap V$. That suffices to conclude the proof.

**11. COROLLARY (Cartan's theorem on varieties).** *If $W$ is a holomorphic subvariety of a holomorphic variety $V$, then the family of ideals $\{\text{id } \mathbf{W}_A\}$ in $\{{}_V\mathcal{O}_A\}$ is finitely generated over an open neighborhood of any point of $V$.*

*Proof.* Again for any point $O \in V$ an open neighborhood of $O$ in $V$ can be represented by a holomorphic subvariety of an open neighborhood $U$ of the origin in $\mathbf{C}^n$, with the origin representing $O$ and the subvariety also being denoted by $V$; $W$ will then be represented by a subvariety in $U$ as well. If the neighborhood $U$ is sufficiently small, then by Cartan's theorem, Theorem 6, there will exist holomorphic functions $\tilde{f}_1, \ldots, \tilde{f}_r \in {}_n\mathcal{O}_U$ generating the ideal id $\mathbf{W}_A \subseteq {}_n\mathcal{O}_A$ at each point $A \in U$. But the restrictions $f_j = \tilde{f}_j | V \in {}_V\mathcal{O}_{V \cap U}$ clearly generate the ideal id $\mathbf{W}_A \subseteq {}_V\mathcal{O}_A$ at each point $A \in U \cap V$, and the proof is thereby concluded.

It should be noted in conclusion that as in the proof of Corollary 8, it follows from the preceding result that if $\{\mathfrak{A}_A\}$ is a finitely generated family of ideals over a holomorphic variety $V$, then $\{\sqrt{\mathfrak{A}_A}\}$ form a finitely generated family of ideals over an open neighborhood of any point of $V$.

# G
# Dimension

A connected complex manifold has a well-defined dimension, as discussed in section I-C. By dimension here will be meant the complex dimension, so an $n$-dimensional complex manifold is locally biholomorphic to $\mathbb{C}^n$ but as a topological space has dimension $2n$. It is customary to define the dimension of a not necessarily connected complex manifold to be the supremum of the dimensions of its connected components. The regular locus $\mathfrak{R}(V)$ of any holomorphic variety $V$ is a dense open subset of $V$ by Theorem E17, and has the natural structure of a complex manifold; that suggests introducing the following.

1. **DEFINITION.** *The* **dimension** *of a holomorphic variety $V$ is the dimension of the complex manifold $\mathfrak{R}(V)$ and is denoted by* dim $V$.

It is possible that the regular locus $\mathfrak{R}(V)$ of a holomorphic variety $V$ has infinitely many connected components and that the dimensions of these components are unbounded; in that case, the preceding definition requires that $\dim V = \infty$. That case will generally be excluded from consideration henceforth, so that the dimension will normally be assumed to be a well-defined finite integer. If the complex manifold $\mathfrak{R}(V)$ is pure dimensional—that is, if all the connected components of $\mathfrak{R}(V)$ have the same dimension—the variety $V$ will also be called **pure dimensional**. Any irreducible holomorphic variety $V$ is of course pure dimensional, since $\mathfrak{R}(V)$ is then connected by Theorem E19. Moreover, if $V = \bigcup_j V_j$ where $V_j$ are the irreducible components of $V$, then clearly $\dim V = \sup_j \dim V_j$. Note that $\dim(V_1 \times V_2) = \dim V_1 + \dim V_2$ for any two holomorphic varieties, since $\mathfrak{R}(V_1) \times \mathfrak{R}(V_2)$ is clearly a dense open subset of $V_1 \times V_2$ and a complex manifold of dimension equal to $\dim V_1 + \dim V_2$.

For any point $A$ of a holomorphic variety $V$, the dimension of the union of those connected components of $\mathfrak{R}(V)$ containing $A$ in their closures will be called the **dimension of the variety $V$ at the point $A$**, and will be denoted by $\dim_A V$. In this context it is worth recalling from Theorems E15 and E17 that $\mathfrak{R}(V)$ is locally finitely connected at any point $A \in V$, so that this local dimension is always a finite integer. It is clear that all sufficiently small open neighborhoods of $A$ in $V$ will be holomorphic varieties of dimensions equal to $\dim_A V$. Thus this local dimension really depends only on the germ of the variety $V$ at the point $A$ and so will also be denoted by dim $\mathbf{V}_A$. Conversely if $\mathbf{V}$ is any germ of a holomorphic variety and is represented by a variety $V$ at a point $A \in V$, then the dimension of the germ $\mathbf{V}$ is defined by $\dim \mathbf{V} = \dim_A V$ and is well defined. If all the connected components of $\mathfrak{R}(V)$ containing $A$ in their closures have the same dimension, the variety $V$ will be called **pure dimensional at** $A$. That is, of course, equivalent to the condition that all sufficiently small open neighborhoods of $A$ in $V$ are pure dimensional holomorphic varieties. Correspondingly a germ of a holomorphic variety will be called **pure dimensional** if it has a pure dimensional representative. Any irreducible germ of a holomorphic variety is pure dimensional by Theorem E15; moreover, if $\mathbf{V}_j$ are the irreducible components of a germ $\mathbf{V}$, then $\dim \mathbf{V} = \max_j \dim \mathbf{V}_j$.

Dimension is really the only local invariant of a complex manifold, since clearly two germs of regular holomorphic varieties are biholomorphic precisely when they have the same dimension. That is not the case for arbitrary holomorphic varieties, though, as is quite apparent from the examples that have already been discussed, but dimension is nonetheless the basic and most useful of the invariants associated to holomorphic varieties. To begin the discussion, there is the following algebraic interpretation of dimension for varieties represented as holomorphic subvarieties of a space $\mathbb{C}^m$, relating the rather geometrical definition of dimension given here to the algebraic properties underlying the proof of the local parametrization theorem.

**2. THEOREM.** *If an ideal* $\mathfrak{A} \subseteq {}_m\mathcal{O}$ *is regular in the variables* $z_{n+1}, \ldots, z_m$, *then* $\dim \operatorname{loc} \mathfrak{A} = n$.

*Proof.* First if $\mathfrak{p} \subseteq {}_m\mathcal{O}$ is a prime ideal that is regular in the variables $z_{n+1}, \ldots, z_m$, then by Corollary D11 the natural projection $\mathbb{C}^m \to \mathbb{C}^n$ exhibits loc $\mathfrak{p}$ as the germ of a finite branched holomorphic covering of $\mathbb{C}^n$. The regular part of a representative covering is then a dense open subset that has the structure of an $n$-dimensional complex manifold, so that $\dim \operatorname{loc} \mathfrak{p} = n$. Next if $\mathfrak{A} \subseteq {}_m\mathcal{O}$ is an arbitrary ideal, write $\operatorname{loc} \mathfrak{A} = \bigcup_i \mathbf{V}_i$ where $\mathbf{V}_i$ are the irreducible components of loc $\mathfrak{A}$. The ideals $\mathfrak{p}_i = \operatorname{id} \mathbf{V}_i \subseteq {}_n\mathcal{O}$ are prime, and by Hilbert's zero-theorem $\mathfrak{A} \subseteq \sqrt{\mathfrak{A}} = \operatorname{id} \operatorname{loc} \mathfrak{A} = \bigcap_i \mathfrak{p}_i$. If, further, the ideal $\mathfrak{A}$ is regular in the variables $z_{n+1}, \ldots, z_m$, there are elements $\mathbf{f}_j \in {}_j\mathcal{O} \cap \mathfrak{A} \subseteq \bigcap_i ({}_j\mathcal{O} \cap \mathfrak{p}_i)$ regular in the variables $z_j$ for $n + 1 \leq j \leq m$. So after a suitable nonsingular change of coordinates in $\mathbb{C}^n$, it can be supposed that each ideal $\mathfrak{p}_i$ is regular in the variables $z_{n_i+1}, \ldots, z_m$ for some index $n_i \leq n$. It follows from the first part of the proof that $\dim \mathbf{V}_i = n_i$; so in order to conclude the proof it is only necessary to show that $n = \max_i n_i$, since $\dim \mathbf{V} = \max_i \dim \mathbf{V}_i$. If, to the contrary, $n_i < n$ for each $i$, then there are elements $\mathbf{g}_i \in {}_n\mathcal{O} \cap \mathfrak{p}_i$ that are regular in the variable $z_n$. The product $\mathbf{g} = \prod_i \mathbf{g}_i \in {}_n\mathcal{O} \cap \sqrt{\mathfrak{A}}$ is then also regular in the variable $z_n$, as is any positive integral power $\mathbf{g}^\nu$, but if $\nu$ is sufficiently large, then $\mathbf{g}^\nu \in {}_n\mathcal{O} \cap \mathfrak{A}$ and that contradicts the hypothesis that $\mathfrak{A}$ is regular in the variables $z_{n+1}, \ldots, z_m$. With this contradiction the proof is completed.

This algebraic interpretation of the dimension of a holomorphic subvariety can be used to derive some other simple properties and characterizations of dimension.

3. THEOREM. *If* $\mathbf{V}$ *and* $\mathbf{W}$ *are germs of holomorphic varieties with* $\mathbf{V} \subseteq \mathbf{W}$, *then* $\dim \mathbf{V} \leqq \dim \mathbf{W}$, *and equality holds precisely when* $\mathbf{V}$ *and* $\mathbf{W}$ *have a common irreducible component* $\mathbf{U}$ *with* $\dim \mathbf{U} = \dim \mathbf{V} = \dim \mathbf{W}$.

*Proof.* It can of course be assumed that $\mathbf{V}$ and $\mathbf{W}$ are germs of holomorphic subvarieties at the origin in $\mathbb{C}^k$ with $\mathbf{V} \subseteq \mathbf{W}$, and after a suitable nonsingular linear change of coordinates in $\mathbb{C}^k$, that id $\mathbf{W}$ is regular in the coordinates $z_{n+1}, \ldots, z_k$ and id $\mathbf{V} \supseteq$ id $\mathbf{W}$ is regular in the coordinates $z_{m+1}, \ldots, z_k$, where $m \leqq n$. Since $n = \dim \mathbf{W}$ and $m = \dim \mathbf{V}$ by the preceding theorem, the first assertion of the present theorem is evidently true. Next suppose that $\dim \mathbf{V} = \dim \mathbf{W}$, and let $\mathbf{U}$ be an irreducible component of $\mathbf{V}$ for which $\dim \mathbf{U} = \dim \mathbf{V} = \dim \mathbf{W}$. Since $\mathbf{U} \subseteq \mathbf{W}$ and $\mathbf{U}$ is irreducible, clearly $\mathbf{U}$ must be contained in some irreducible component $\mathbf{W}_1$ of the germ $\mathbf{W}$. Then by the first part of the proof $\dim \mathbf{U} \leqq \dim \mathbf{W}_1$, so actually $\dim \mathbf{U} = \dim \mathbf{W}_1$. After a nonsingular linear change of coordinates in $\mathbb{C}^k$, it can be assumed that id $\mathbf{U}$ and id $\mathbf{W}_1$ are both regular in the coordinates $z_{n+1}, \ldots, z_k$ where $n = \dim \mathbf{U} = \dim \mathbf{W}_1$, by the preceding theorem. The natural projection $\mathbb{C}^k \rightarrow \mathbb{C}^n$ then exhibits both $\mathbf{U}$ and $\mathbf{W}_1$ as germs of finite branched holomorphic coverings. Choose representatives $U$ and $W_1$ that are finite branched holomorphic coverings of a common polydisc $\Delta \subseteq \mathbb{C}^n$. Outside a proper subvariety $D \subseteq \Delta$, both $U$ and $W_1$ will be unbranched coverings. Both can be assumed connected since $\mathbf{U}$ and $\mathbf{W}_1$ are irreducible, by Theorem E13, and since $U \subseteq W_1$ it follows that these unbranched coverings must coincide and hence that $\mathbf{U} = \mathbf{W}_1$ as desired. The converse implication is trivial, and the proof is thereby concluded.

When considering a holomorphic subvariety $V$ of an open subset $U \subseteq \mathbb{C}^m$ it is sometimes convenient to introduce the difference $m - \dim V$, which is called the **codimension** of $V$ in $\mathbb{C}^m$ and is denoted by codim $V$. Similarly when considering the germ $\mathbf{V}$ of a holomorphic subvariety at a point in $\mathbb{C}^m$, the difference $m - \dim \mathbf{V}$ is called the **codimension** of the germ $\mathbf{V}$ and is denoted by codim $\mathbf{V}$. More generally if $V$ is a holomorphic subvariety of a holomorphic variety $W$, the difference $\dim W - \dim V$ is called the codimension of $V$ in $W$ and is denoted by **codim** $V$; there is a corresponding local definition. Although this extension is also useful and natural, particularly when the ambient variety $W$ is pure dimensional, some care should be taken to avoid confusion and to ensure that the ambient variety is clearly specified.

4. THEOREM. *If* $\mathbf{V}$ *is the germ of a holomorphic subvariety at the origin in* $\mathbb{C}^m$, *then* codim $\mathbf{V}$ *is the largest integer* $k$ *for which there is a* $k$*-dimensional linear subspace* $\mathbf{L}$ *through the origin in* $\mathbb{C}^m$ *with* $\mathbf{V} \cap \mathbf{L} = 0$.

*Proof.* First, if there is a $k$-dimensional linear subspace $L$ through the origin in $\mathbb{C}^m$ with $\mathbf{V} \cap \mathbf{L} = 0$, then with coordinates in $\mathbb{C}^m$ for which $L = \{Z \in \mathbb{C}^m : z_1 = \cdots = z_{m-k} = 0\}$, it follows from Theorem E5 that there are elements $\mathbf{f}_j \in {}_m\mathcal{O} \cap \text{id}\,\mathbf{V}$ regular in the variables $z_j$ for $m - k + 1 \leqq j \leqq m$. But after a nonsingular linear change of coordinates in the space $\mathbb{C}^{m-k}$ of the first $m - k$ variables, the ideal id $\mathbf{V}$ will be regular in the variables $z_{n+1}, \ldots, z_m$ for some $n \leqq m - k$, so by Theorem 2 necessarily $\dim \mathbf{V} = n \leqq m - k$ or equivalently codim $\mathbf{V} \geqq k$. On the other hand, suppose that $\dim \mathbf{V} = m - k$, and write $\mathbf{V} = \bigcup_i \mathbf{V}_i$ where $\mathbf{V}_i$ are the irreducible components of $\mathbf{V}$

and $n_i = \dim \mathbf{V}_i$. After a nonsingular linear change of coordinates in $\mathbb{C}^m$, it can be assumed that id $\mathbf{V}_i$ is regular in the variables $z_{n_i+1}, \ldots, z_m$, by Theorem 2 again, and it then follows from Theorem E5 that $\mathbf{L}_i \cap \mathbf{V}_i = \mathbf{0}$ where $L_i = \{Z \in \mathbb{C}^m : z_1 = \cdots = z_{n_i} = 0\}$. However, since $\max_i n_i = m - k$, necessarily $\mathbf{L} \cap \mathbf{V}_i = 0$ for each $i$, where $L = \{Z \in \mathbb{C}^m : z_1 = \cdots z_{m-k} = 0\}$ is a $k$-dimensional linear subvariety of $\mathbb{C}^m$, and therefore $\mathbf{L} \cap \mathbf{V} = \mathbf{0}$. That suffices to conclude the proof.

The relations between the dimension of a holomorphic subvariety of $\mathbb{C}^m$ and defining equations for that subvariety are particularly interesting and important. The first basic result in that direction is the following.

**5. THEOREM.** *If* $\mathbf{f} \in {}_m\mathcal{O}$ *is a nonunit and* $\mathbf{f} \neq \mathbf{0}$, *then* loc ${}_m\mathcal{O}\mathbf{f}$ *has pure dimension* $m - 1$. *A germ* $\mathbf{V}$ *of a holomorphic subvariety at the origin in* $\mathbb{C}^m$ *has pure dimension* $m - 1$ precisely when id $\mathbf{V}$ *is a principal ideal in* ${}_m\mathcal{O}$.

*Proof.* Since ${}_m\mathcal{O}$ is a unique factorization domain by Theorem A7 any nonzero nonunit $\mathbf{f} \in {}_m\mathcal{O}$ can be written $\mathbf{f} = \prod_j \mathbf{f}_j$, where $\mathbf{f}_j \in {}_m\mathcal{O}$ are irreducible elements generating prime ideals ${}_m\mathcal{O}\mathbf{f}_j \subseteq {}_m\mathcal{O}$. Then loc ${}_m\mathcal{O}\mathbf{f} = \bigcup_j$ loc ${}_m\mathcal{O}\mathbf{f}_j$ where loc ${}_m\mathcal{O}\mathbf{f}_j$ are the irreducible components of loc ${}_m\mathcal{O}\mathbf{f}$. In order to prove the first assertion of the theorem, it is thus only necessary to show that dim loc ${}_m\mathcal{O}\mathbf{f} = m - 1$ whenever $\mathbf{f} \in {}_m\mathcal{O}$ is irreducible. Now after a suitable nonsingular linear change of coordinates in $\mathbb{C}^m$, it can be assumed that $\mathbf{f}$ is regular in $z_m$, and after multiplying $\mathbf{f}$ by a unit, it can also be assumed that $\mathbf{f}$ is a nontrivial Weierstrass polynomial in ${}_{m-1}\mathcal{O}[z_m]$. It is clear that ${}_{m-1}\mathcal{O} \cap {}_m\mathcal{O}\mathbf{f} = \mathbf{0}$. Indeed, if $\mathbf{g} \in {}_{m-1}\mathcal{O} \cap {}_m\mathcal{O}\mathbf{f}$, then $\mathbf{g} = \mathbf{f}\mathbf{h} \in {}_{m-1}\mathcal{O}$ for some $\mathbf{h} \in {}_m\mathcal{O}$; but since for each point $A' \in \mathbb{C}^{m-1}$ sufficiently near the origin the polynomial $\mathbf{f}(A', z_n)$ will have a root near the origin, it follows that $\mathbf{g}(A') = 0$ and hence that $\mathbf{g} = 0$. The ideal ${}_m\mathcal{O}\mathbf{f}$ is consequently regular in the variable $z_m$, so by Theorem 2 necessarily dim loc ${}_m\mathcal{O}\mathbf{f} = m - 1$ as desired.

To prove the remainder of the theorem note first that by what has already been proved, whenever id $\mathbf{V} \subseteq {}_m\mathcal{O}$ is principal, then $\mathbf{V} =$ loc id $\mathbf{V}$ has pure dimension $m - 1$. On the other hand, suppose that $\mathbf{V}$ is the germ of a holomorphic subvariety of pure dimension $m - 1$ at the origin in $\mathbb{C}^m$, and write $\mathbf{V} = \bigcup_j \mathbf{V}_j$ where $\mathbf{V}_j$ are the irreducible components of $\mathbf{V}$ and $\dim \mathbf{V}_j = m - 1$. After a suitable nonsingular linear change of coordinates in $\mathbb{C}^m$, it can be assumed that each prime ideal id $\mathbf{V}_j$ is strictly regular in $z_m$, by Theorem 2 again. The local parametrization theorem shows that $\mathbf{V}_j$ is exhibited as the germ of a finite branched holomorphic covering by the natural projection $\mathbb{C}^m \to \mathbb{C}^{m-1}$, and it then follows from Theorem D2 that id $\mathbf{V}_j$ is generated by a polynomial $\mathbf{p}_j \in {}_{m-1}\mathcal{O}[z_m]$. The element $\mathbf{p}_j \in {}_m\mathcal{O}$ must be irreducible, since id $\mathbf{V}_j = {}_m\mathcal{O}\mathbf{p}_j$ is prime. Finally id $\mathbf{V} = \bigcap_j$ id $\mathbf{V}_j = \bigcap_j({}_m\mathcal{O}\mathbf{p}_j)$, so since ${}_m\mathcal{O}$ is a unique factorization domain, id $\mathbf{V} = {}_m\mathcal{O}\mathbf{p}$ where $\mathbf{p} = \prod_j \mathbf{p}_j$. That then concludes the proof.

In view of the preceding theorem it is convenient to define a **local divisor** at a point $A \in \mathbb{C}^m$ to be a finite formal sum $\mathfrak{d}_A = \sum_j \nu_j \mathbf{V}_j$, where $\nu_j \in \mathbb{Z}$ and $\mathbf{V}_j$ is the germ of an irreducible holomorphic subvariety of codimension one at the point $A$. Such a divisor is called **positive**, or sometimes **effective**, if $\nu_j \geqq 0$ for all indices $j$. The set of all local divisors at a point $A$ has the natural structure of an abelian group, with the group operations defined by $(\sum_j \mu_j \mathbf{V}_j) \pm (\sum_j \nu_j \mathbf{V}_j) = \sum_j (\mu_j \pm \nu_j)\mathbf{V}_j$. This group $\mathscr{D}_A$, called the **group of local divisors** at the point $A$, is thus just the free abelian group generated by the irreducible germs of holomorphic subvarieties of codimension one at the point $A$. The subset $\mathscr{D}_A^+$ of positive local divisors is a semigroup contained in $\mathscr{D}_A$. Since ${}_m\mathcal{O}_A$ is a unique factorization domain, any germ $\mathbf{f} \in {}_m\mathcal{O}_A$ can be written uniquely up to units and the order of the terms as a finite product $\mathbf{f} = \mathbf{u} \prod_j \mathbf{f}_j^{\nu_j}$, where $\mathbf{u} \in {}_m\mathcal{O}_A$ is a unit, $\mathbf{f}_j \in {}_m\mathcal{O}_A$ are irreducible germs no two of which differ by a unit factor, and $\nu_j \in \mathbb{Z}$ are positive integers. The **divisor of the germ f** is then defined to be the local divisor $\mathfrak{d}_A(\mathbf{f}) = \sum_j \nu_j \mathbf{V}_j$ where $\mathbf{V}_j = \text{loc}\ {}_m\mathcal{O}_A \mathbf{f}_j$, and by Theorem 5 this is a well-defined local divisor. It is clear that $\mathfrak{d}_A(\mathbf{f})$ is a positive divisor, that two germs have the same local divisor precisely when they differ by a unit, and that $\mathfrak{d}_A(\mathbf{fg}) = \mathfrak{d}_A(\mathbf{f}) + \mathfrak{d}_A(\mathbf{g})$. The mapping $\mathbf{f} \in {}_m\mathcal{O}_A \to \mathfrak{d}_A(\mathbf{f}) \in \mathscr{D}_A$ is thus a homomorphism from the multiplicative semigroup of germs of holomorphic functions at the point $A$ to the semigroup of positive local divisors at the point $A$, the kernel of which is the multiplicative group of units in ${}_m\mathcal{O}_A$. The homomorphism is surjective, since it follows immediately from Theorem 5 that any local divisor at $A$ is the divisor of some germ of a holomorphic function at $A$.

As a natural global analogue a **divisor** in an open subset $D \subseteq \mathbb{C}^m$ is a formal sum $\mathfrak{d} = \sum_j \nu_j V_j$, where $\nu_j \in \mathbb{Z}$ and $V_j$ is an irreducible holomorphic subvariety of codimension one in $D$. In this case, though, the sum is not required to be finite, but it is required that whenever $U \subseteq D$ is an open subset for which $\bar{U}$ is a compact subset of $D$, then $U$ meets only finitely many of the subvarieties $V_j$, so that the sum is at least locally finite. Again such a divisor is called **positive** or **effective** if $\nu_j \geqq 0$ for all indices $j$. The set of all divisors in $D$ has the natural structure of an abelian group, which is denoted by $\mathscr{D}_D$ and called the **group of divisors** in $D$. The subset of positive divisors form a semigroup $\mathscr{D}_D^+ \subseteq \mathscr{D}_D$. For the special case $m = 1$, a divisor has the form $\mathfrak{d} = \sum_j \nu_j p_j$ where $\{p_j\}$ is a countable discrete sequence of points in $D$ so is a very familiar entity in complex analysis. Clearly each divisor $\mathfrak{d} = \sum_j \nu_j V_j \in \mathscr{D}_D$ in an open subset $D \subseteq \mathbb{C}^m$ has a well-defined germ $\mathfrak{d}_A \in \mathscr{D}_A$ at each $A \in D$, where $\mathfrak{d}_A$ is the local divisor $\mathfrak{d}_A = \sum_j \nu_j(\mathbf{V}_j' + \mathbf{V}_j'' + \cdots)$ and $\mathbf{V}_j', \mathbf{V}_j'', \ldots$ are the irreducible components of the germ of the holomorphic subvariety $V_j$ at the point $A$. Conversely, of course, any local divisor $\mathfrak{d}_A \in \mathscr{D}_A$ at a point $A \in \mathbb{C}^m$ is the germ at that point of a divisor in some open neighborhood of $A$ in $\mathbb{C}^m$.

If a germ $\mathbf{f} \in {}_m\mathcal{O}_A$ is represented by a holomorphic function $f \in {}_m\mathcal{O}_U$ in some open neighborhood $U$ of the point $A$ and if the local divisor $\mathfrak{d}_A(\mathbf{f})$ is represented correspondingly by a divisor $\mathfrak{d} \in \mathscr{D}_U$, then it is easy to see but perhaps not outrightly obvious that *the germ $\mathfrak{d}_B$ of the divisor $\mathfrak{d}$ at any point $B \in U$ is the local divisor at $B$ of the germ $\mathbf{f}_B \in {}_m\mathcal{O}_B$ provided that $U$ is sufficiently small.* It is clearly sufficient just to demonstrate this assertion in the special case that the germ $\mathbf{f}$ is irreducible in ${}_m\mathcal{O}_A$. The divisor of interest can be taken to be $\mathfrak{d} = 1 \cdot V$ where $V = \{Z \in U : f(z) = 0\}$, since the germ $\mathbf{V}_A$ is irreducible and the subvariety $V$ will also be irreducible if $U$ is sufficiently small. After shrinking the neighborhood $U$ further if necessary, it can be assumed that the germ $\mathbf{f}_B \in {}_m\mathcal{O}_B$ generates the ideal $\text{id}\ \mathbf{V}_B \subseteq {}_m\mathcal{O}_B$ at each point $B \in U$, by Cartan's theorem, Theorem F6. Now at any point $B \in U$, the germ of the divisor $\mathfrak{d}$ has the form $\mathfrak{d}_B = 1 \cdot \mathbf{V}_B' + 1 \cdot \mathbf{V}_B'' + \cdots$, where $\mathbf{V}_B', \mathbf{V}_B'', \ldots$ are the distinct irreducible components of the germ $\mathbf{V}_B$, and the divisor of the germ $\mathbf{f}_B$ has the form $\mathfrak{d}(\mathbf{f}_B) = 1 \cdot \tilde{\mathbf{V}}_B' + 1 \cdot \tilde{\mathbf{V}}_B'' + \cdots$, where $\mathbf{f}_B = \mathbf{u} \cdot \mathbf{f}_B' \cdot \mathbf{f}_B'' \cdots$ is a decomposition of $\mathbf{f}_B$ into irreducible factors and $\tilde{\mathbf{V}}_B' = \mathrm{loc}\,_m\mathcal{O}_B\mathbf{f}_B'$ and so on. Each germ $\tilde{\mathbf{V}}_B'$, $\tilde{\mathbf{V}}_B''$, ... is an irreducible component of $\mathbf{V}_B$ and all the irreducible components of $\mathbf{V}_B$ occur among these germs, so to show the desired result, it is finally enough just to show that all the germs $\tilde{\mathbf{V}}_B'$, $\tilde{\mathbf{V}}_B''$, ... are distinct. If that is not the case, it is possible to write $\mathbf{f}_B = \mathbf{u} \cdot (\mathbf{f}_B')^\nu \cdot \mathbf{f}_B'' \cdots$ for some integer $\nu > 1$, where the germ $\tilde{\mathbf{V}}_B' = \mathrm{loc}\,_m\mathcal{O}_B\mathbf{f}_B'$ is distinct from the germs $\tilde{\mathbf{V}}_B'' = \mathrm{loc}\,_m\mathcal{O}_B\mathbf{f}_B''$ and so on. In some open neighborhood $U_B$ of $B$ there will be a corresponding decomposition of holomorphic functions $f = u(f_B')^\nu f_B'' \cdots$, and if $U_B$ is sufficiently small, the germ of the function $f_B'$ at any point $Z \in U_B$ will generate the ideal of the germ of the subvariety $\tilde{V}_B' = \{Z \in U_B : f_B'(Z) = 0\}$ at that point. However, at any point $Z \in \tilde{V}_B'$ that does not lie in any other component $\tilde{V}_B'', \ldots$, the germ of $\tilde{V}_B'$ is the same as the germ of $V$. But then the germ of $f$ at that point is the $\nu$th power of a generator of the ideal of $V$, contradicting the condition that the germ of $f$ itself generates the ideal of $V$ at each point. That contradiction concludes the proof of the assertion.

To turn then to the global situation, if $f \in {}_m\mathcal{O}_D$ is any holomorphic function in an open subset $D \subseteq \mathbb{C}^m$ and if $V_j$ are the irreducible components of the subvariety $V = \{Z \in D : f(Z) = 0\}$, then there is a divisor $\mathfrak{d}(f) = \sum_j \nu_j V_j \in \mathcal{D}_D$ such that the germ of this divisor at any point $A \in D$ is the divisor of the germ $\mathbf{f}_A \in {}_m\mathcal{O}_A$. This divisor $\mathfrak{d}(f)$ will be called the **divisor** of the function $f$ in $D$. To demonstrate this assertion note first that as a consequence of the observation made in the preceding paragraph, each point $A \in D$ has an open neighborhood $U$ in which there is a divisor $\mathfrak{d}_U(f) = \sum_j (\nu_j' V_j' + \nu_j'' V_j'' + \cdots)$ having the desired property, where $V_j'$, $V_j''$, ... are the irreducible components of $V_j \cap U$; so it is just necessary to show that $\nu_j' = \nu_j'' = \cdots$. If $A \in \mathfrak{R}(V)$, then $A$ belongs to a single component $V_j$, and this divisor has the simpler form $\mathfrak{d}_U(f) = \nu_j \cdot V_j$. If $A' \in \mathfrak{R}(V) \cap V_j$ is another such point, there will correspondingly be another divisor $\mathfrak{d}_{U'}(f) = \nu_j' \cdot V_j$, and if $U \cap U' \cap V_j \neq \varnothing$, then since these divisors have the same germ at any point of $U \cap U' \cap V_j$, it is evident that $\nu_j = \nu_j'$. The set $\mathfrak{R}(V) \cap V_j$ is connected by Theorem E19, and therefore the integer $\nu_j$ is the same for all points $A \in \mathfrak{R}(V) \cap V_j$. Then at a general point $A \in V_j$ at which the divisor has as above the form $\mathfrak{d}_V(f) = \sum_j (\nu_j' V_j' + \nu_j'' V_j'' + \cdots)$, it follows immediately upon considering points in $\mathfrak{R}(V) \cap V_j$ in this neighborhood that $\nu_j' = \nu_j'' = \cdots = \nu_j$ as desired, to conclude the proof.

Not every positive divisor in an open subset $D \subseteq \mathbb{C}^m$ is the divisor of a holomorphic function in $D$. The examples discussed in section I-P already demonstrated that for divisors of the particularly simple form $\mathfrak{d} = 1V$ where $V$ is a connected complex submanifold of dimension $m - 1$ in $D$. The argument of the proof of Corollary I-P5 does show, however, that if $D \subseteq \mathbb{C}^m$ is a domain of holomorphy, then a positive divisor in $D$ is the divisor of a holomorphic function if there is no topological obstruction. Since this will be discussed in more detail in Volume III, nothing more will be added here.

The preceding considerations have been limited to holomorphic functions and their divisors on the space $\mathbb{C}^m$, and it is natural to seek to extend them to the case of holomorphic functions and their divisors on an arbitrary holomorphic variety. Unfortunately only the first part of Theorem 5 holds on arbitrary holomorphic varieties, so the discussion of divisors on general holomorphic varieties is a somewhat different and considerably more difficult undertaking even locally. For the most part, the discussion of these matters must be left to a later point, after the development of other local properties of holomorphic varieties. In order to establish at least part of Theorem 5 here in general, it is useful to have available the following auxiliary result.

6. **LEMMA (semicontinuity lemma).** *Suppose that $f_1, \ldots, f_v$ are continuous functions in an open subset $D \times E \subseteq \mathbb{C}^m \times \mathbb{C}^n$ where $O \in D$, and that $f_j(Z, T)$ is holomorphic in $Z \in D$ for each fixed point $T \in E$. For any value $T \in E$ let $V(T)$ be the holomorphic subvariety of $D$ defined by*

$$V(T) = \{Z \in D : f_1(Z, T) = \cdots = f_v(Z, T) = 0\}$$

*Then if $\mathbf{V}(T)$ is the germ of the subvariety $V(T)$ at the origin $O \in D$, it follows that* $\dim \mathbf{V}(T)$ *is an upper semicontinuous function of $T$ in $E$.*

*Proof.* Since $\dim \mathbf{V}(T)$ takes only integral values, it is upper semicontinuous precisely when for any point $T_0 \in E$ there is an open neighborhood $U$ of $T_0$ in $D$ such that $\dim \mathbf{V}(T) \leqq \dim \mathbf{V}(T_0)$ for all $T \in U$. To demonstrate that, consider a point $T_0 \in D$ and let $k_0 = \operatorname{codim} \mathbf{V}(T_0)$. It follows from Theorem 4 that there is a $k_0$-dimensional linear subspace $L$ through the origin in $\mathbb{C}^m$ such that $\mathbf{V}(T_0) \cap \mathbf{L} = \mathbf{0}$. It is then possible to find an open neighborhood $W$ of $T_0$ in $D$ such that $\overline{W}$ is a compact subset of $D$ and $\partial W \cap V(T_0) \cap L = \emptyset$. Since $\partial W \cap L$ is a compact subset of $D$ disjoint from $V(T_0)$, there is clearly a constant $\varepsilon > 0$ such that $\sum_j |f_j(T_0, Z)| \geqq \varepsilon$ for all points $Z \in \partial W \cap L$. It then follows from the continuity of the functions $f_j$ and the compactness of $\partial W \cap L$ that there is an open neighborhood $U$ of $T_0$ in $D$ such that $\sum_j |f_j(T, Z)| \geqq \varepsilon/2$ for all points $T \in U$ and $Z \in \partial W \cap L$. However, that in turn implies that $\partial W \cap V(T) \cap L = \emptyset$ whenever $T \in U$, so that $W \cap V(T) \cap L$ is necessarily a compact holomorphic subvariety of the open subset $W \cap L \subseteq L \cong \mathbb{C}^{k_0}$. It is easy to see, and will be demonstrated explicitly in Corollary H4, that $V(T) \cap W \cap L$ is then just a finite set of points in $W$, so that either $\mathbf{V}(T) \cap \mathbf{L} = \emptyset$ or $\mathbf{V}(T) \cap \mathbf{L} = \mathbf{0}$. In the first case $\dim \mathbf{V}(T) = -1 \leqq \dim \mathbf{V}(T_0)$, while in the second case it follows from Theorem 4 again that $\operatorname{codim} \mathbf{V}(T) \geqq k_0 = \operatorname{codim} \mathbf{V}(T_0)$ and hence $\dim \mathbf{V}(T) \leqq \dim \mathbf{V}(T_0)$. That suffices to conclude the proof.

An immediate consequence of this lemma is the observation that $\dim_Z V$ is an upper semicontinuous function of the point $Z \in D$ whenever $V$ is a holomorphic subvariety of an open subset $D \subseteq \mathbb{C}^m$. Although useful, this result will very shortly be succeeded by a stronger result. Another easy consequence is the following, extending at least a part of Theorem 5 to general holomorphic varieties.

7. **THEOREM.** *If $\mathbf{V}$ is the germ of a holomorphic variety of pure dimension $n$ and $\mathbf{f} \in {}_{\mathbf{V}}\mathcal{O}$ is a nonunit and not a zero divisor in the local ring ${}_{\mathbf{V}}\mathcal{O}$, then* $\operatorname{loc} {}_{\mathbf{V}}\mathcal{O}\mathbf{f}$ *is the germ of a holomorphic subvariety of $\mathbf{V}$ of pure dimension $n - 1$.*

*Proof.* Let $\mathbf{W} = \operatorname{loc} {}_{\mathbf{V}}\mathcal{O}\mathbf{f} \subseteq \mathbf{V}$ and represent $\mathbf{V}$ and $\mathbf{W}$ as the germs at the origin of holomorphic subvarieties $V$ and $W$ of an open polydisc $\Delta = \Delta(0; R)$ about the origin in some space $\mathbb{C}^m$. If $R$ is sufficiently small, there will exist a holomorphic function $f \in {}_m\mathcal{O}_\Delta$ such that $f|V$ represents the germ $\mathbf{f}$, and there will exist holomorphic functions $f_1, \ldots, f_\nu$ in ${}_m\mathcal{O}_\Delta$ such that

$$V = \{Z \in \Delta : f_1(Z) = \cdots = f_\nu(Z) = 0\}$$

It can of course also be assumed that

$$W = \{Z \in V : f(Z) = 0\}$$

That $\mathbf{f}$ is not a zero divisor in ${}_{\mathbf{V}}\mathcal{O}$ means as noted in section B that $\mathbf{f}$ does not vanish identically on any irreducible component of $\mathbf{V}$. By choosing $R$ sufficiently small, it can thus be assumed that $f$ does not vanish identically on any irreducible component of $V$. The theorem will then be proved by showing that $\dim_A W = n - 1$ at any point $A \in W \cap \Delta(0; R/2)$. For this purpose consider the translate of the subvariety $V$ by the vector $-T$ for any point $T \in V \cap \Delta(0; R/2)$. This translate can be viewed as the subvariety

$$V(T) = \left\{Z \in \Delta\left(0; \frac{R}{2}\right) : Z + T \in V\right\}$$

$$= \left\{Z \in \Delta\left(0; \frac{R}{2}\right) : f_1(Z + T) = \cdots = f_\nu(Z + T) = 0\right\}$$

and an open neighborhood of the origin on $V(T)$ is biholomorphic to an open neighborhood of the point $T$ on $V$. Correspondingly consider the subvariety

$$W(T) = \{Z \in V(T) : f(Z + A) = 0\}$$

$$= \left\{Z \in \Delta\left(0; \frac{R}{2}\right) : f(Z + A) = f_1(Z + T) = \cdots = f_\nu(Z + T) = 0\right\}$$

note that an open neighborhood of the origin on $W(A)$ is biholomorphic to an open neighborhood of the point $A$ on $W$, and let $\mathbf{W}(T)$ be the germ of the subvariety $W(T)$ at the origin. If $T \in \mathfrak{R}(V)$, then $V(T)$ is an $n$-dimensional complex submanifold in an open neighborhood of the origin in $\mathbb{C}^m$, and since it is nonempty by construction it follows from Theorem 5 that $\dim \mathbf{W}(T) = n - 1$. However, $\dim \mathbf{W}(T)$ is an upper semicontinuous function of $T$ by Lemma 6, so by considering a sequence of points $T_j$ in the dense subset $\mathfrak{R}(V)$ of $V$ for which $T_j$ approaches $A$, it follows that $n - 1 = \lim_j \dim \mathbf{W}(T_j) \leq \dim \mathbf{W}(A) = \dim_A(W)$. On the other hand, if $\dim_A(W) = n$, then $V$ and $W$ must have a common irreducible component at $A$ by Theorem 3, and that is impossible since $f$ does not vanish identically on any irreducible component of $V$. Therefore $\dim_A W = n - 1$, and the proof is thereby concluded.

The global version of the preceding theorem is an almost immediate consequence, and asserts that if $V$ is an irreducible holomorphic variety of dimension $n$

and if $f \in {}_V\mathcal{O}_V$ is a nonunit, then the holomorphic subvariety $W = \{Z \in V : f(Z) = 0\} \subseteq V$ has pure dimension $n - 1$. More generally a holomorphic function on a holomorphic variety either is identically zero on an irreducible component, is nowhere vanishing on an irreducible component, or defines a holomorphic subvariety of pure codimension one in an irreducible component. More succinctly if less precisely, a single holomorphic function can decrease the dimension by at most one.

**8. COROLLARY.** *If* $\mathbf{V}$ *is the germ of a holomorphic variety of pure dimension* $n$ *and* $\mathfrak{A} \subseteq {}_{\mathbf{V}}\mathcal{O}$ *is an ideal generated by* $k$ *elements of* ${}_{\mathbf{V}}\mathcal{O}$, *then* $\dim \mathbf{W} \geqq n - k$ *for any irreducible component* $\mathbf{W}$ *of* loc $\mathfrak{A}$.

*Proof.* This is an immediate consequence of the preceding theorem.

It was already mentioned that the last part of Theorem 5 does not hold on an arbitrary holomorphic variety, but there is at least the following general result.

**9. THEOREM.** *If* $\mathbf{V}$ *is the germ of a holomorphic variety of dimension* $n$ *and* $\mathbf{W} \subseteq \mathbf{V}$ *is the germ of a subvariety of dimension* $m$, *then there exists an ideal* $\mathfrak{A} \subseteq {}_{\mathbf{V}}\mathcal{O}$ *generated by* $n$ *elements of* ${}_{\mathbf{V}}\mathcal{O}$ *such that* $\mathbf{W} =$ loc $\mathfrak{A}$, *but no ideal defining* $\mathbf{W}$ *can have fewer than* $n - m$ *generators.*

*Proof.* First suppose that $\mathbf{V}$ and $\mathbf{W}$ are germs of holomorphic subvarieties at the origin in $\mathbf{C}^k$ with $\dim \mathbf{V} = n$ but with $\mathbf{W}$ not necessarily a subvariety of $\mathbf{V}$ and with no restriction on the dimension of $\mathbf{W}$, and let $\mathbf{U} = \mathbf{V} \cup \mathbf{W}$. It will be demonstrated by induction on $n$ that there are $n$ germs $\mathbf{f}_j \in {}_k\mathcal{O}$ such that $\mathbf{W} = \{Z \in \mathbf{U} : \mathbf{f}_1(Z) = \cdots = \mathbf{f}_n(Z) = 0\}$. The first assertion of the corollary is an immediate consequence of this, since it just amounts to the special case in which $\mathbf{W} \subseteq \mathbf{V}$. For $n = 0$ the desired result is quite trivial, since $\mathbf{V} = \mathbf{0}$ so that $\mathbf{W} = \mathbf{U}$. For the inductive step consider a germ $\mathbf{U} = \mathbf{V} \cup \mathbf{W}$ with $\dim \mathbf{V} = n > 0$, and suppose that the desired result has been demonstrated for all lower dimensions. To each irreducible component $\mathbf{V}_i$ of the germ $\mathbf{V}$ associate an element $\mathbf{g}_i \in {}_k\mathcal{O}$ as follows. If $\mathbf{V}_i \subseteq \mathbf{W}$, simply take $\mathbf{g}_i = 0$. If $\mathbf{V}_i \nsubseteq \mathbf{W}$, so that $\mathbf{V}_i$ is necessarily an irreducible component of $\mathbf{U}$ as well, let $\mathbf{U}_i'$ be the union of all the other irreducible components of $\mathbf{U}$ and let $\mathbf{U}_i'' = \mathbf{U}_i' \cup (\mathbf{V}_i \cap \mathbf{W})$, noting that $\mathbf{U}_i''$ is a proper subvariety of $\mathbf{U}$. Then take for $\mathbf{g}_i$ any germ such that $\mathbf{g}_i | \mathbf{U} \neq 0$ and $\mathbf{g}_i \in$ id $\mathbf{U}_i''$. In both cases the germ $\mathbf{g}_i$ vanishes identically on $\mathbf{W}$ and on all the irreducible components of $\mathbf{V}$ other than $\mathbf{V}_i$, and if $\mathbf{V}_i \nsubseteq \mathbf{W}$, the zero locus of $\mathbf{g}_i$ on $\mathbf{V}_i$ is by Theorem 7 a subvariety of $\mathbf{V}_i$ of codimension one. The germ $\mathbf{f}_n = \sum_i \mathbf{g}_i \in {}_k\mathcal{O}$ then also vanishes on all of $\mathbf{W}$, so the subvariety $\mathbf{U}' = \{Z \in \mathbf{U} : \mathbf{f}_n(Z) = 0\}$ contains $\mathbf{W}$. But if $\mathbf{U}' = \mathbf{V}' \cup \mathbf{W}$, where $\mathbf{V}'$ is the union of those irreducible components of $\mathbf{U}'$ not contained in $\mathbf{W}$, it is clear that $\dim \mathbf{V}' = n - 1$. By the induction hypothesis there are $n - 1$ germs $\mathbf{f}_j \in {}_k\mathcal{O}$ such that $\mathbf{W} = \{Z \in \mathbf{U}' : \mathbf{f}_1(Z) = \cdots = \mathbf{f}_{n-1}(Z) = 0\}$; but then $\mathbf{W} = \{Z \in \mathbf{U} : \mathbf{f}_1(Z) = \cdots = \mathbf{f}_{n-1}(Z) = \mathbf{f}_n(Z) = 0\}$ as desired. The first assertion of the theorem has thereby been proved, and since the second assertion is an immediate consequence of Corollary 8, the proof is then concluded.

**Examples.** It is easy to see that the preceding theorem is the best possible result that can be expected to hold in general. For any integers $0 \leqq m \leqq n$ there are germs

**W** and **V** of holomorphic varieties of respective dimensions $m$ and $n$ such that $\mathbf{W} \subseteq \mathbf{V}$ and that it requires at least $n$ elements of ${}_{\mathbf{V}}\mathcal{O}$ to describe the subvariety **W**. Indeed the base point $\mathbf{0} \in \mathbf{V}$ requires at least $n$ elements of ${}_{\mathbf{V}}\mathcal{O}$ to describe as a consequence of Theorem 9, where $n = \dim \mathbf{V}$. If $\mathbf{V} = \mathbf{V}' \cup \mathbf{V}''$ is a reducible germ where both components $\mathbf{V}'$ and $\mathbf{V}''$ have dimension $n$ and $\mathbf{V}' \cap \mathbf{V}'' = \mathbf{0}$, and if $\mathbf{W} \subseteq \mathbf{V}'$ is an arbitrary subvariety of $\mathbf{V}'$, then it clearly requires $n$ germs $\mathbf{f}_1, \ldots, \mathbf{f}_n \in {}_{\mathbf{V}}\mathcal{O}$ to describe **W**, since the restrictions $\mathbf{f}_j|\mathbf{V}''$ describe the zero-dimensional subvariety $\mathbf{0} \in \mathbf{V}''$. That even more subtle phenomena are involved, that reducibility does not altogether explain what is really going on, is indicated by the following example, which has other interesting properties as well. Let $V$ be the holomorphic subvariety of $\mathbb{C}^{2n} = \mathbb{C}^n \times \mathbb{C}^n$ defined by

$$V = \{(Z, W) \in \mathbb{C}^n \times \mathbb{C}^n : w_i z_j = w_j z_i \text{ for } 1 \leqq i, j \leqq n\} \tag{1}$$

This subvariety can also be viewed as the set of all $2 \times n$ complex matrices of rank strictly less than two and is thus one of a class of algebraic varieties that have been fairly extensively investigated. Its relevance to the present considerations was pointed out by Grauert and Remmert. It is clear from (1) that

$$V \cap \{(Z, W) \in \mathbb{C}^n \times \mathbb{C}^n : z_1 \neq 0\}$$

$$= \left\{(Z, W) \in \mathbb{C}^n \times \mathbb{C}^n : z_1 \neq 0 \text{ and } w_j = \frac{w_1 z_j}{z_1} \text{ for } 2 \leqq j \leqq n\right\}$$

from which it follows readily that $V$ is a connected $(n + 1)$-dimensional complex manifold outside the subvariety $\{(Z, W) \in V: z_1 = 0\}$. A similar result evidently holds if $z_1$ is replaced by any other coordinate function $z_i$ or $w_i$, so that actually the origin is the singular point of $V$ and $V$ is an $(n + 1)$-dimensional irreducible holomorphic subvariety of $\mathbb{C}^{2n}$. Moreover the germ of this subvariety is irreducible at each of its points, although that is really only interesting at the singular point $O \in V$. Then for any complex parameter $t \in \mathbb{C}$, introduce the holomorphic subvariety $V(t)$ of $\mathbb{C}^{2n}$ defined by

$$V(t) = \{(Z, W) \in \mathbb{C}^n \times \mathbb{C}^n : w_j = tz_j \text{ for } 1 \leqq j \leqq n\} \tag{2}$$

noting that $V(t)$ is an $n$-dimensional complex manifold contained in $V$ and that $V(t_1) \cap V(t_2) = 0$ whenever $t_1 \neq t_2$. Now $V(t)$ is a subvariety of $V$ of codimension one, but it is easy to see that it cannot be defined in any neighborhood of the origin by fewer than $n$ functions. Indeed suppose that $f_1, \ldots, f_r$ are holomorphic functions in some open neighborhood $U$ of the origin in $V$ such that

$$V(t) \cap U = \{(Z, W) \in U : f_j(Z, W) = 0 \text{ for } 1 \leqq j \leqq r\}$$

It then follows that for any parameter $t_1 \neq t$,

$$0 = V(t) \cap V(t_1) \cap U$$

$$= \{(Z, W) \in V(t_1) \cap U : f_j(Z, W) = 0 \text{ for } 1 \leqq j \leqq r\}$$

and since $V(t_1)$ is an $n$-dimensional variety, it is a consequence of Theorem 9 that $r \geqq n$ as desired.

An extension of Theorem 5 in a rather different direction is the following useful result.

**10. THEOREM.** *If* $\mathbf{V}_1$ *and* $\mathbf{V}_2$ *are germs at the origin in* $\mathbb{C}^m$ *of pure dimensional holomorphic subvarieties and if* $\mathbf{W}$ *is any irreducible component of* $\mathbf{V}_1 \cap \mathbf{V}_2$, *then*

$$\operatorname{codim} \mathbf{W} \leqq \operatorname{codim} \mathbf{V}_1 + \operatorname{codim} \mathbf{V}_2$$

*or equivalently,*

$$\dim \mathbf{W} \geqq \dim \mathbf{V}_1 + \dim \mathbf{V}_2 - m$$

*Proof.* Represent $\mathbf{V}_j$ as the germ at the origin of a holomorphic subvariety $V_j$ of pure dimension $n_j$ in an open polydisc $\Delta(0; R) \subseteq \mathbb{C}^m$, where

$$V_j = \{Z \in \Delta(0; R) : f_1^j(Z) = \cdots = f_\nu^j(Z) = 0\}$$

for some holomorphic functions $f_k^j$ in $\Delta(0; R)$. To prove the desired result it is evidently only necessary to show that $\dim_A(V_1 \cap V_2) \geqq n_1 + n_2 - m$ at each point $A \in V_1 \cap V_2$, since that implies that $\dim W \geqq n_1 + n_2 - m$ if $W$ is any irreducible component of $V_1 \cap V_2$. For this purpose consider the translates of $V_j$ by vectors $-T$ where $T \in V_j \cap \Delta(0; R/2)$. These translates can be viewed as subvarieties

$$V_j(T) = \left\{Z \in \Delta\left(0; \frac{R}{2}\right) : f_1^j(Z + T) = \cdots = f_\nu^j(Z + T) = 0\right\}$$

and an open neighborhood of the origin on $V_j(T)$ is biholomorphic to an open neighborhood of the point $T \in V_j$. If $T \in \mathfrak{R}(V_2)$, then $V_2(T)$ is a complex submanifold of dimension $n_2$ in an open neighborhood of the origin and hence can be defined as the set of common zeros of $m - n_2$ coordinate functions near the origin in $\mathbb{C}^m$. It then follows from Corollary 8 that $\dim_0(V_1(A) \cap V_2(T)) \geqq n_1 + n_2 - m$. Now there are points $T \in \mathfrak{R}(V_2)$ arbitrarily near $A$, since $\mathfrak{R}(V_2)$ is dense in $V_2$, so since $\dim_0(V_1(A) \cap V_2(T))$ is an upper semicontinuous function of $T$ by Corollary 8, $\dim_A(V_1 \cap V_2) = \dim_0(V_1(A) \cap V_2(A)) \geqq n_1 + n_2 - m$ as desired, thereby concluding the proof.

**Example.** The preceding theorem of course holds whenever $\mathbf{V}_1$ and $\mathbf{V}_2$ are germs of holomorphic subvarieties of the germ of an $m$-dimensional regular holomorphic variety $\mathbf{V}$, and the corresponding global result is an immediate consequence; but it does not necessarily hold if the ambient variety $\mathbf{V}$ has singularities. For the $(n + 1)$-dimensional variety (1) considered earlier in this section, the subvarieties $V(t)$ defined by (2) are pure $n$-dimensional, but $\dim_0 V(t_1) \cap V(t_2) = 0$.

When considering germs of holomorphic varieties rather than of holomorphic subvarieties of $\mathbb{C}^m$ it is of course of some interest to have algebraic characterizations of dimension involving only the local ring of the variety, since by Corollary B17 a germ of a variety is entirely determined by the purely algebraic structure of its local ring. Two such characterizations will be discussed here; the first is the following almost immediate corollary of Theorem 9.

**11. THEOREM.** *If* $\mathbf{V}$ *is the germ of a holomorphic variety, then* $\dim \mathbf{V}$ *is the least integer n such that there are n germs* $\mathbf{f}_1, \ldots, \mathbf{f}_n$ *in* ${}_{\mathbf{V}}\mathcal{O}$ *generating an ideal* $\mathfrak{A} \subseteq {}_{\mathbf{V}}\mathcal{O}$ *for which* $\sqrt{\mathfrak{A}} = {}_{\mathbf{V}}\mathfrak{m}$.

*Proof.* It is an obvious consequence of Hilbert's zero-theorem on the variety $\mathbf{V}$, Corollary E4, that $\sqrt{\mathfrak{A}} = {}_{\mathbf{V}}\mathfrak{m}$ precisely where $\operatorname{loc} \mathfrak{A} = \mathbf{0}$, the base point of the germ $\mathbf{V}$. By applying Theorem 9 to the holomorphic subvariety $\mathbf{0} \subseteq \mathbf{V}$ of dimension zero, it follows that there are $n$ germs $\mathbf{f}_1, \ldots, \mathbf{f}_n$ in ${}_{\mathbf{V}}\mathcal{O}$ generating an ideal $\mathfrak{A}$ such that $\operatorname{loc} \mathfrak{A} = \mathbf{0}$, but no ideal $\mathfrak{A}$ defining $\mathbf{0}$ can have fewer than $n$ generators. That is just the desired result, thus concluding the proof.

**12. DEFINITION.** *A* **system of parameters** *for a germ* $\mathbf{V}$ *of a holomorphic variety is a set of n elements* $\mathbf{f}_1, \ldots, \mathbf{f}_n \in {}_{\mathbf{V}}\mathcal{O}$ *generating an ideal* $\mathfrak{A} \subseteq {}_{\mathbf{V}}\mathcal{O}$ *for which* $\sqrt{\mathfrak{A}} = {}_{\mathbf{V}}\mathfrak{m}$, *where* $n = \dim \mathbf{V}$. *Such a system of parameters is said to be* **regular** *if actually* $\mathfrak{A} = {}_{\mathbf{V}}\mathfrak{m}$.

It follows directly from Theorem 11 that any germ of a holomorphic variety does admit systems of parameters, but to determine which germs of varieties have regular systems of parameters requires some further analysis of the properties of systems of parameters. The elements of any system of parameters must of course belong to the maximal ideal ${}_{\mathbf{V}}\mathfrak{m}$. Upon recalling Definition E6, it is clear that *a system of parameters for* $\mathbf{V}$ *can be characterized more geometrically as n elements* $\mathbf{f}_1, \ldots, \mathbf{f}_n \in {}_{\mathbf{V}}\mathcal{O}$, *which are the coordinate functions of a germ of a finite holomorphic mapping* $\pi: \mathbf{V} \to \mathbf{C}^n$ *where* $n = \dim \mathbf{V}$; thus these $n$ elements $\mathbf{f}_1, \ldots, \mathbf{f}_n$ can be considered as playing the role of a local coordinate system or local parameter system on the germ $\mathbf{V}$, whence the terminology. Moreover, Theorem 11 can be restated as the assertion that $\dim \mathbf{V}$ *is the least integer n such that there exists a finite holomorphic mapping* $\pi: \mathbf{V} \to \mathbf{C}^n$. As observed in Theorem E11, the condition that $\pi: \mathbf{V} \to \mathbf{C}^n$ be the germ of a finite holomorphic mapping is equivalent to the condition that the induced algebra homomorphism $\pi^*: {}_n\mathcal{O} \to {}_{\mathbf{V}}\mathcal{O}$ be a finite homomorphism—that is, that the homomorphism $\pi^*: {}_n\mathcal{O} \to {}_{\mathbf{V}}\mathcal{O}$ exhibit ${}_{\mathbf{V}}\mathcal{O}$ as a finitely generated ${}_n\mathcal{O}$-module. The elements of a system of parameters are just the images $\mathbf{f}_j = \pi^*(\mathbf{z}_j)$ of the germs of the coordinate functions in $\mathbf{C}^n$. Since the germs $\mathbf{z}_j$ generate the maximal ideal ${}_n\mathfrak{m} \subseteq {}_n\mathcal{O}$, it is clear that the ideal $\mathfrak{A} \subseteq {}_{\mathbf{V}}\mathcal{O}$ generated by this system of parameters can be characterized as the ideal in ${}_{\mathbf{V}}\mathcal{O}$ generated by the image $\pi^*({}_n\mathfrak{m}) \subseteq {}_{\mathbf{V}}\mathcal{O}$—that is, as the ideal $\mathfrak{A} = {}_{\mathbf{V}}\mathcal{O} \cdot \pi^*({}_n\mathfrak{m})$. Alternatively by viewing ${}_{\mathbf{V}}\mathcal{O}$ as an ${}_n\mathcal{O}$-module and dropping any explicit reference to the homomorphism $\pi^*$, the subset $\mathfrak{A} \subseteq {}_{\mathbf{V}}\mathcal{O}$ can be characterized as the ${}_n\mathcal{O}$-submodule of ${}_{\mathbf{V}}\mathcal{O}$ spanned by ${}_n\mathfrak{m}$—that is, as the ${}_n\mathcal{O}$-submodule $\mathfrak{A} = {}_n\mathfrak{m} \cdot {}_{\mathbf{V}}\mathcal{O}$.

**13. LEMMA.** *For an ideal* $\mathfrak{A} \subseteq {}_{\mathbf{V}}\mathcal{O}$ *the following conditions are equivalent to one another:*

(i) $\sqrt{\mathfrak{A}} = {}_{\mathbf{V}}\mathfrak{m}$

(ii) ${}_{\mathbf{V}}\mathfrak{m}^{\nu} \subseteq \mathfrak{A} \subseteq {}_{\mathbf{V}}\mathfrak{m}$ *for some integer* $\nu > 0$.

(iii) ${}_{\mathbf{V}}\mathcal{O}/\mathfrak{A}$ *is a finite dimensional complex vector space.*

*Proof.* First suppose that the ideal $\mathfrak{A} \subseteq {}_V\mathcal{O}$ satisfies (i). If $\mathbf{f}_1, \ldots, \mathbf{f}_r \in {}_V\mathcal{O}$ is a set of generators for the maximal ideal ${}_V\mathfrak{m} \subseteq {}_V\mathcal{O}$, there are then integers $\nu_j > 0$ such that $\mathbf{f}_j^{\nu_j} \in \mathfrak{A}$. But any element $\mathbf{f} \in {}_v\mathfrak{m}$ can be written as a sum $\mathbf{f} = \sum_j \mathbf{x}_j \mathbf{f}_j$ for some $\mathbf{x}_j \in {}_V\mathcal{O}$, and if $\nu > \nu_1 + \cdots + \nu_r - r$, then any monomial in the multinomial expansion of $\mathbf{f}^\nu = (\sum_j \mathbf{x}_j \mathbf{f}_j)^\nu$ will involve a factor of $\mathbf{f}_j^{\nu_j}$ for some index $j$ and consequently $\mathbf{f}^\nu \in \mathfrak{A}$. Thus, ${}_V\mathfrak{m}^\nu \subseteq \mathfrak{A} \subseteq {}_V\mathfrak{m}$, so the ideal $\mathfrak{A}$ necessarily satisfies (ii). On the other hand, obviously any ideal $\mathfrak{A}$ satisfying (ii) also satisfies (i), so conditions (i) and (ii) are equivalent.

Next suppose that the ideal $\mathfrak{A} \subseteq {}_V\mathcal{O}$ satisfies (ii), and note that all ${}_V\mathcal{O}$-modules have natural structures as complex vector spaces, since $\mathbb{C} \subseteq {}_V\mathcal{O}$. For any integer $\mu > 0$ the finitely generated ${}_V\mathcal{O}$-module ${}_V\mathfrak{m}^\mu / {}_V\mathfrak{m}^{\mu+1}$ is annihilated by the ideal ${}_V\mathfrak{m}$ so is consequently a finitely generated ${}_V\mathcal{O}/{}_V\mathfrak{m}$-module. But ${}_V\mathcal{O}/{}_V\mathfrak{m} \cong \mathbb{C}$ and the structure of ${}_V\mathfrak{m}^\mu / {}_V\mathfrak{m}^{\mu+1}$ as an ${}_V\mathcal{O}/{}_V\mathfrak{m}$-module is the same as its structure as a complex vector space, so it is necessarily a finite-dimensional complex vector space. As complex vector spaces there is the obvious isomorphism

$$ {}_V\mathcal{O}/{}_V\mathfrak{m}^\nu \cong {}_V\mathcal{O}/{}_V\mathfrak{m} + {}_V\mathfrak{m}/{}_V\mathfrak{m}^2 + \cdots + {}_V\mathfrak{m}^{\nu-1}/{}_V\mathfrak{m}^\nu $$

and therefore ${}_V\mathcal{O}/{}_V\mathfrak{m}^\nu$ is also a finite-dimensional complex vector space. The natural homomorphism ${}_V\mathcal{O}/{}_V\mathfrak{m}^\nu \to {}_V\mathcal{O}/\mathfrak{A}$ of ${}_V\mathcal{O}$-modules or of complex vector spaces is of course surjective, and consequently ${}_V\mathcal{O}/\mathfrak{A}$ is a finite-dimensional complex vector space as well, showing that the ideal $\mathfrak{A}$ satisfies (iii). On the other hand, if the ideal $\mathfrak{A} \subseteq {}_V\mathcal{O}$ satisfies (iii), let $M_\nu$ be the image in ${}_V\mathcal{O}/\mathfrak{A}$ of the ideal ${}_V\mathfrak{m}^\nu \subseteq {}_V\mathcal{O}$ for any integer $\nu \geqq 1$. Here $M_\nu$ is a subspace of the finite-dimensional complex vector space ${}_V\mathcal{O}/\mathfrak{A}$, and since $M_\nu \supseteq M_{\nu+1}$, this sequence of subspaces is eventually stable in the sense that $M_{\nu_0} = M_{\nu_0+1}$ for some index $\nu_0$. However, since $M_{\nu_0+1} = {}_V\mathfrak{m} \cdot M_{\nu_0}$, it follows from Nakayama's lemma, Lemma A9, that $M_{\nu_0} = 0$. But that means that ${}_V\mathfrak{m}^{\nu_0} \subseteq \mathfrak{A} \subseteq {}_V\mathfrak{m}$ and hence that the ideal $\mathfrak{A}$ also satisfies (ii). Thus conditions (ii) and (iii) are equivalent, and that is enough to conclude the proof.

**14. THEOREM.** *If $\mathbf{f}_1, \ldots, \mathbf{f}_n$ is a system of parameters for a germ of a holomorphic variety $V$, defining a finite holomorphic mapping $\pi: V \to \mathbb{C}^n$ and generating an ideal $\mathfrak{A} \subseteq {}_V\mathcal{O}$, then $\dim_{\mathbb{C}} {}_V\mathcal{O}/\mathfrak{A}$ is the minimal number of generators of ${}_V\mathcal{O}$ as an ${}_n\mathcal{O}$-module with the module structure induced by $\pi^*$.*

*Proof.* Note that ${}_V\mathcal{O}/\mathfrak{A}$ is a finite-dimensional complex vector space as a consequence of the preceding lemma, since the condition that $\mathbf{f}_1, \ldots, \mathbf{f}_n$ be a system of parameters is that $\sqrt{\mathfrak{A}} = {}_V\mathfrak{m}$. Let $\mathbf{g}_1, \ldots, \mathbf{g}_\mu \in {}_V\mathcal{O}$ represent a basis for the finite dimensional vector space ${}_V\mathcal{O}/\mathfrak{A}$, where $\mu = \dim_{\mathbb{C}} {}_V\mathcal{O}/\mathfrak{A}$, and let $\mathbf{h}_1, \ldots, \mathbf{h}_\nu \in {}_V\mathcal{O}$ be generators of ${}_V\mathcal{O}$ as an ${}_n\mathcal{O}$-module, where $\nu$ is the minimal number of generators. Any element $\mathbf{f} \in {}_V\mathcal{O}$ can be written $\mathbf{f} = \sum_j \mathbf{x}_j \mathbf{h}_j$ for some germs $\mathbf{x}_j \in {}_n\mathcal{O}$, when ${}_V\mathcal{O}$ is viewed as an ${}_n\mathcal{O}$-module under the homomorphism $\pi^*: {}_n\mathcal{O} \to {}_V\mathcal{O}$ induced by $\pi: V \to \mathbb{C}^n$. Set $\mathbf{x}_j = c_j + \mathbf{m}_j$ where $c_j \in \mathbb{C}$ and $m_j \in {}_n\mathfrak{m}$, and note that $\mathbf{f} = \sum_j c_j \mathbf{h}_j + \mathbf{a}$ where $\mathbf{a} = \sum_j \mathbf{m}_j \mathbf{h}_j \in \mathfrak{A}$, since $\mathfrak{A} = {}_n\mathfrak{m} \cdot {}_v\mathcal{O}$. That shows that the elements $\mathbf{h}_j$ represent generators for the vector space ${}_V\mathcal{O}/\mathfrak{A}$ and consequently that $\nu \geqq \mu$. On the other hand, any element $\mathbf{f} \in {}_V\mathcal{O}$ can be written as $\mathbf{f} = \sum_j c_j \mathbf{g}_j + \mathbf{a}$ for some constants $c_j \in \mathbf{C}$ and some germ $\mathbf{a} \in \mathfrak{A}$. If $\mathfrak{B}$ is the submodule of the ${}_n\mathcal{O}$-module ${}_V\mathcal{O}$ spanned by the elements $\mathbf{g}_1, \ldots, \mathbf{g}_\mu$, it follows that ${}_V\mathcal{O} = \mathfrak{B} + \mathfrak{A} = \mathfrak{B} + {}_n\mathfrak{m} \cdot {}_V\mathcal{O}$, so by Nakayama's lemma, Lemma A9, necessarily $\mathfrak{B} = {}_V\mathcal{O}$. Consequently $\mu \geqq \nu$, and upon combining this with what has already been proved, it follows that $\mu = \nu$ and the proof is thereby concluded.

**15. THEOREM.** *A germ of a holomorphic variety admits a regular system of parameters if and only if it is the germ of a regular holomorphic variety.*

*Proof.* If $\mathbf{V}$ is the germ of a regular holomorphic variety of dimension $n$, then it can be viewed as the germ at the origin of the variety $\mathbf{C}^n$. The germs at the origin of the $n$ coordinate functions in $\mathbf{C}^n$ clearly generate the maximal ideal ${}_n\mathfrak{m}$, as is evident by considering the Taylor expansion of any element of ${}_n\mathfrak{m}$; hence, they form a regular system of parameters for $\mathbf{V}$. Conversely, suppose that $\mathbf{V}$ is the germ of an $n$-dimensional holomorphic variety admitting a regular system of parameters. This system of parameters will consist of the coordinate functions of a finite holomorphic mapping $\pi: \mathbf{V} \to \mathbf{C}^n$ and will generate the ideal $\mathfrak{A} = {}_V\mathfrak{m} \subseteq {}_V\mathcal{O}$. But then $\dim_{\mathbf{C}} {}_V\mathcal{O}/\mathfrak{A} = \dim_{\mathbf{C}} {}_V\mathcal{O}/{}_V\mathfrak{m} = 1$, and it follows from Theorem 14 that the induced homomorphism $\pi^*: {}_n\mathcal{O} \to {}_V\mathcal{O}$ exhibits ${}_V\mathcal{O}$ as an ${}_n\mathcal{O}$-module generated by a single element of ${}_V\mathcal{O}$. If $\mathbf{f} \in {}_V\mathcal{O}$ is such a generator, then the homomorphism $\phi: {}_n\mathcal{O} \to {}_V\mathcal{O}$ defined by setting $\phi(\mathbf{x}) = \mathbf{x}\mathbf{f}$ is surjective; hence ${}_V\mathcal{O} \cong {}_n\mathcal{O}/\mathfrak{B}$ for some ideal $\mathfrak{B} \subseteq {}_n\mathcal{O}$. The ideal $\mathfrak{B}$ must be equal to its own radical, since no power of a nonzero element of ${}_V\mathcal{O}$ can be zero, so that by Hilbert's zero-theorem $\mathfrak{B} = \text{id}\,\mathbf{W}$ for some germ $\mathbf{W}$ of a holomorphic subvariety of $\mathbf{C}^n$; and ${}_V\mathcal{O} \cong {}_W\mathcal{O}$ so that $\mathbf{V} = \mathbf{W}$ by Corollary B17. Since $\dim \mathbf{V} = n$, though, necessarily $\mathbf{V} = \mathbf{W} = \mathbf{C}^n$ by Theorem 3, and consequently $\mathbf{V}$ is the germ of a regular holomorphic variety. That concludes the proof.

While discussing finite holomorphic mappings in connection with systems of parameters, it is perhaps worth digressing briefly to examine some properties of dimension and finite holomorphic mappings. The first property is rather trivial, but nonetheless should be noted explicitly.

**16. THEOREM.** *If* $\pi: \mathbf{V} \to \mathbf{W}$ *is any germ of a finite holomorphic mapping, then* $\dim \mathbf{V} \leqq \dim \mathbf{W}$. *If* $\pi: \mathbf{V} \to \mathbf{W}$ *is the germ of a finite branched holomorphic covering, then* $\dim \mathbf{V} = \dim \mathbf{W}$.

*Proof.* If $\mathbf{f}_1, \ldots, \mathbf{f}_m$ is a system of parameters for $\mathbf{W}$ where $m = \dim \mathbf{W}$, and if $\pi: \mathbf{V} \to \mathbf{W}$ is a germ of a finite holomorphic mapping, then upon recalling Definition E6 it is clear that the $m$ germs $\pi^*(\mathbf{f}_1), \ldots, \pi^*(\mathbf{f}_m)$ in ${}_V\mathcal{O}$ generate an ideal $\mathfrak{A} \subseteq {}_V\mathcal{O}$ such that $\text{loc}\,\mathfrak{A} = 0$; but then it follows from Theorem 11 that $\dim \mathbf{V} \leqq m$ as desired. On the other hand, if $\pi: \mathbf{V} \to \mathbf{W}$ is the germ of a finite branched holomorphic covering, then $\dim \mathbf{V} \leqq \dim \mathbf{W}$ by the first part of the theorem, while since $\pi$ is a locally biholomorphic mapping over an open neighborhood of at least some point in the maximal dimensional component of $\mathfrak{R}(\mathbf{W})$, then clearly $\dim \mathbf{V} \geqq \dim \mathbf{W}$ as well. That suffices to conclude the proof.

The next result is not quite so obvious and so is perhaps more interesting. It is closely related to a result in algebraic geometry known as the theorem of the purity of the branch locus.

**17. THEOREM.** *If $\pi: V \to U$ is a finite branched holomorphic covering over an open subset $U \subseteq \mathbb{C}^n$ with branch locus $B_\pi \subset V$ and if $V$ is irreducible at a point $A \in B_\pi$, then $\dim_A B_\pi = \dim_{\pi(A)} \pi(B_\pi) = n - 1$.*

*Proof.* Recall from Theorem C13 that the branch locus $B_\pi$ is a holomorphic subvariety of $V$, and from Corollary C14 that its image $\pi(B_\pi)$ is a holomorphic subvariety of $U$. The mapping $\pi | B_\pi: B_\pi \to \pi(B_\pi)$ is of course a finite holomorphic mapping, so by Theorem E7 its restriction to any irreducible component of $B_\pi$ is a finite branched holomorphic covering. It is then clear from Theorem 16 that $\dim_A B_\pi \leqq \dim_{\pi(A)} \pi(B_\pi)$, the inequality arising from the possibility that $\dim_{A'} B_\pi > \dim_A B_\pi$ for some point $A' \in B_\pi$ with $\pi(A') = \pi(A)$. However, since $\pi(B_\pi)$ is a proper subvariety of $U$, necessarily $\dim_{\pi(A)}(\pi(B_\pi)) \leqq n - 1$, so in order to prove the desired result, it suffices just to show that $\dim_A B_\pi \geqq n - 1$. Suppose to the contrary that $\dim_A B_\pi < n - 1$. By restricting $V$ to a sufficiently small open neighborhood of $A$, it can be assumed that $A = \pi^{-1}(\pi(A))$ and hence that the order $o_\pi$ of the finite branched holomorphic covering $\pi$ is equal to the branching order $o_\pi(A)$, and $o_\pi = o_\pi(A) > 1$ since $A \in B_\pi$. Furthermore, $\dim_A B_\pi = \dim_{\pi(A)} \pi(B_\pi)$ in this case, and by taking for $U$ an appropriate sufficiently small neighborhood of $\pi(A)$ it can also be assumed that $\dim \pi(B_\pi) = \dim_{\pi(A)} \pi(B_\pi)$ and that $U$ is simply connected. Then the restriction $\pi: V - B_\pi \to U - \pi(B_\pi)$ is an unbranched covering space over the complement of a holomorphic subvariety $\pi(B_\pi)$ of a simply connected open subset $U \subseteq \mathbb{C}^n$, where $\dim \pi(B_\pi) < n - 1$. It is easy to see by induction on $\dim \pi(B_\pi)$ that each connected component of $V - B_\pi$ is mapped by $\pi$ homeomorphically onto $U - \pi(B_\pi)$. Indeed if $\dim \pi(B_\pi) = 0 < n - 1$, then $\pi(B_\pi)$ is a point in a simply connected open subset $U \subseteq \mathbb{C}^n$ with $n > 1$, and it is a rather standard result of topology that $U - \pi(B_\pi)$ is also simply connected, which implies the desired result. For the inductive step, whenever $\dim \pi(B_\pi) < n - 1$, then the regular part of the subvariety $\pi(B_\pi)$ is locally a topological submanifold of dimension $\leqq 2n - 3$ in $\mathbb{C}^n = \mathbb{R}^{2n}$, and it is again a rather standard result of topology that the complement of a submanifold of topological codimension $\geqq 3$ is locally simply connected. Therefore, $\pi: V - B_\pi \to U - \pi(B_\pi)$ can be extended to an unbranched covering over $U - \mathfrak{S}(\pi(B_\pi))$, and since $\mathfrak{S}(\pi(B_\pi))$ is a thin subset of $\pi(B_\pi)$ by Theorem E17 and that subvariety has even smaller dimension, the inductive step clearly follows. Now by Corollary E20 the closure of each of the $o_\pi$ connected components of $V - B_\pi$ is a holomorphic subvariety of $V$; but that contradicts the assumption that $V$ is irreducible at $A$, and with that contradiction the proof is concluded.

**18. COROLLARY.** *If $\pi: V \to U$ is a finite branched holomorphic covering over an open subset $U \subseteq \mathbb{C}^n$ with branch locus $B_\pi \subset V$ and if $V$ is irreducible at each of its points, then the subvarieties $B_\pi$ and $\pi(B_\pi)$ are of pure dimension $n - 1$.*

*Proof.* It follows from the preceding theorem that $\dim_A B_\pi = \dim_{\pi(A)} \pi(B_\pi) = n - 1$ at each point $A \in B_\pi$, thus immediately yielding the corollary.

**Example.** An irreducibility hypothesis is really necessary in Theorem 17. To see that, consider the holomorphic subvariety $V = V_1 \cup V_2 \subseteq \mathbf{C}^4$ defined by

$$V_1 = \{Z \in \mathbf{C}^4 : z_3 = z_1, z_4 = z_2\}$$

$$V_2 = \{Z \in \mathbf{C}^4 : z_3 = 2z_1, z_4 = 2z_2\}$$

The natural projection $\pi: \mathbf{C}^4 \to \mathbf{C}^2$ defined by $\pi(Z) = (z_1, z_2)$ for any point $Z \in \mathbf{C}^4$ clearly exhibits $V$ as a finite branched holomorphic covering of order two over $\mathbf{C}^2$, but the branch locus is $B_\pi = 0$ and hence has dimension zero rather than one.

The second purely algebraic characterization of the dimension of the germ of a holomorphic variety in terms of its local ring is expressed in terms of the prime ideals of the local ring, paralleling a corresponding characterization of dimension in algebraic geometry, and rests on the following notions.

**19. DEFINITION.** *The* **Krull dimension** *of a local ring $\mathcal{O}$ is the largest integer d for which there exist prime ideals $\mathfrak{p}_i \subset \mathcal{O}$ such that*

$$\mathfrak{p}_0 \subset \mathfrak{p}_1 \subset \cdots \subset \mathfrak{p}_d$$

*For any prime ideal $\mathfrak{p} \subset \mathcal{O}$ the* **depth** *of $\mathfrak{p}$ is the largest integer d for which there exist prime ideals $\mathfrak{p}_i \subset \mathcal{O}$ such that*

$$\mathfrak{p} \subset \mathfrak{p}_1 \subset \cdots \subset \mathfrak{p}_d$$

*and the* **height** *of $\mathfrak{p}$ is the largest integer h for which there exist prime ideals $\mathfrak{p}_i \subset \mathcal{O}$ such that*

$$\mathfrak{p} \supset \mathfrak{p}_1 \supset \cdots \supset \mathfrak{p}_h$$

It should be noted that all the inclusions in the preceding definitions are strict inclusions, and in particular that the ring $\mathcal{O}$ itself is not considered to be a prime ideal. A prime ideal in any local ring $\mathcal{O}$ always has a finite depth, since the ring $\mathcal{O}$ is Noetherian. As will shortly be seen, the local rings of germs of holomorphic varieties always have finite Krull dimensions, and any prime ideal in such a local ring has a finite height. It is clear that the maximal ideal $\mathfrak{m}$ in a local ring $\mathcal{O}$ is the only prime ideal of depth zero, and that the Krull dimension of $\mathcal{O}$ can also be described as the height of the maximal ideal $\mathfrak{m}$ of $\mathcal{O}$. The zero ideal in $\mathcal{O}$ is prime if and only if $\mathcal{O}$ is an integral domain. In that case the zero ideal is the only ideal of height zero, and the Krull dimension of $\mathcal{O}$ can be described as the depth of the zero ideal.

**20. THEOREM.** *If* **V** *is a germ of a holomorphic variety with local ring ${}_{\mathbf{V}}\mathcal{O}$, then the dimension of* **V** *is equal to the Krull dimension of ${}_{\mathbf{V}}\mathcal{O}$.*

*Proof.* As a preliminary observation note that whenever $\mathbf{W}' \subseteq \mathbf{W}''$ are irreducible germs of holomorphic subvarieties of **V** with $\dim \mathbf{W}' \leq \dim \mathbf{W}'' - 2$, then there is another irreducible germ $\mathbf{W}'''$ of a holomorphic subvariety of $\mathbf{V}$ such that $\mathbf{W}' \subset \mathbf{W}''' \subset \mathbf{W}''$. Indeed, since $\mathbf{W}' \subset \mathbf{W}''$, there must be a germ $\mathbf{f} \in {}_{\mathbf{V}}\mathcal{O}$ such that $\mathbf{f} \in \operatorname{id} \mathbf{W}'$ but $\mathbf{f} \notin \operatorname{id} \mathbf{W}''$. It follows from Theorem 7 that $\mathbf{W} = \operatorname{loc}\, {}_{\mathbf{W}''}\mathcal{O} \cdot (\mathbf{f}|\mathbf{W}'')$ is a holomorphic subvariety of $\mathbf{W}''$ of pure dimension $n''$, where $n'' = \dim \mathbf{W}''$, so any irreducible component $\mathbf{W}'''$ of $\mathbf{W}$ will evidently have the desired property.

Now choose an irreducible component $\mathbf{W}$ of $\mathbf{V}$ with $\dim \mathbf{W} = \dim \mathbf{V}$, and consider any chain of irreducible subvarieties $\mathbf{W}_i \subset \mathbf{W}$ such that

$$0 = \mathbf{W}_d \subset \mathbf{W}_{d-1} \subset \cdots \subset \mathbf{W}_1 \subset \mathbf{W}$$

It follows from Theorem 3 that any two consecutive subvarieties in this chain have dimensions that differ by at least one; hence, $d \leqq \dim \mathbf{W}$. On the other hand, if $d < \dim \mathbf{W}$, then at least one pair of consecutive subvarieties in this chain must have dimensions that differ by at least two, so in view of the preliminary observation above, this chain of subvarieties can be extended by inserting another subvariety appropriately. Thus $\dim \mathbf{W}$ is the length of a maximal chain of such subvarieties, and since $\mathfrak{p}_i = \operatorname{id} \mathbf{W}_i$ is prime precisely when $\mathbf{W}_i$ is irreducible, that serves to complete the proof.

**21. COROLLARY.** *If* $\mathbf{V}$ *is a germ of a holomorphic variety with local ring* ${}_{\mathbf{V}}\mathcal{O}$ *and if* $\mathfrak{p} \subset {}_{\mathbf{V}}\mathcal{O}$ *is a prime ideal with* $\mathbf{W} = \operatorname{loc} \mathfrak{p} \subset \mathbf{V}$, then

$$\dim \mathbf{W} = \operatorname{depth} \mathfrak{p} = \dim \mathbf{V} - \operatorname{height} \mathfrak{p}$$

*Proof.* The prime ideals $\bar{\mathfrak{p}}_i \subset {}_{\mathbf{W}}\mathcal{O}$ are precisely those ideals that can be expressed as $\bar{\mathfrak{p}}_i = \mathfrak{p}_i/\mathfrak{p}$ for prime ideals $\mathfrak{p}_i \subset {}_{\mathbf{V}}\mathcal{O}$ such that $\mathfrak{p}_i \supseteq \mathfrak{p}$. Thus chains of prime ideals $\bar{\mathfrak{p}}_i \subset {}_{\mathbf{W}}\mathcal{O}$ of the form $\bar{\mathfrak{p}}_0 \subset \bar{\mathfrak{p}}_1 \subset \cdots \subset \bar{\mathfrak{p}}_d$ correspond precisely to chains of prime ideals $\mathfrak{p}_i \subset {}_{\mathbf{V}}\mathcal{O}$ of the form $\mathfrak{p} \subseteq \mathfrak{p}_0 \subset \mathfrak{p}_1 \subset \cdots \subset \mathfrak{p}_d$. From Definition 19 and Theorem 20 it follows immediately that $\operatorname{depth} \mathfrak{p} = \dim \mathbf{W}$. On the other hand, for any chain of prime ideals $\mathfrak{p}_i \subset {}_{\mathbf{V}}\mathcal{O}$ of the form $\mathfrak{p} \supset \mathfrak{p}_1 \supset \cdots \supset \mathfrak{p}_h$, the germs $\mathbf{W}_i = \operatorname{loc} \mathfrak{p}_i$ are irreducible germs of holomorphic subvarieties of $\mathbf{V}$ of the form $\mathbf{W} \subset \mathbf{W}_1 \subset \cdots \subset \mathbf{W}_h$. Again any two consecutive subvarieties in this chain have dimensions differing by at least one, so that $\dim \mathbf{V} \geqq \dim \mathbf{W}_h \geqq \dim \mathbf{W} + h$. If $\dim \mathbf{V} > \dim \mathbf{W}_h$ or if two consecutive subvarieties in this chain have dimensions differing by at least two, then the chain can be extended by inserting another subvariety appropriately, so that the maximal length $h$ of such a chain is exactly $\dim \mathbf{V} - \dim \mathbf{W}$ and therefore that difference is equal to the height of $\mathfrak{p}$. That concludes the proof.

# H
# Holomorphic Functions on Varieties

Another category of semilocal results arises in the study of holomorphic functions themselves, rather than just germs of holomorphic functions, on varieties. Some of the elementary properties of holomorphic functions on open subsets of $\mathbb{C}^n$ hold in almost exactly the same form for functions on holomorphic varieties.

1. **THEOREM (identity theorem).** *If $f$ and $g$ are holomorphic functions on an irreducible holomorphic variety $V$ and if $f(Z) = g(Z)$ for all points $Z$ in a nonempty open subset of $V$, then $f(Z) = g(Z)$ for all points $Z$ in $V$.*

*Proof.* The regular locus $\mathfrak{R}(V)$ of the holomorphic variety $V$ is a dense open subset of $V$ and is itself a connected complex manifold as a consequence of Theorem E19. Since $f(Z) = g(Z)$ on a nonempty open subset of $\mathfrak{R}(V)$ as well, it follows immediately from the ordinary identity theorem, Theorem I-A3, that $f(Z) = g(Z)$ for all points $Z \in \mathfrak{R}(V)$. But since $f$ and $g$ are continuous on $V$ and $\mathfrak{R}(V)$ is dense in $V$, that implies that $f(Z) = g(Z)$ for all points $Z \in V$ and concludes the proof.

The irreducibility of $V$ was needed to show that the manifold $\mathfrak{R}(V)$ is connected. The irreducibility of $V$ is of course a necessary hypothesis, since two holomorphic functions can agree on an irreducible component of a reducible holomorphic variety $V$ without agreeing at all points of $V$.

2. **THEOREM (maximum modulus theorem).** *If $f$ is a holomorphic function on an irreducible holomorphic variety $V$ and if there is a point $A \in V$ such that $|f(Z)| \leqq |f(A)|$ for all points $Z$ in an open neighborhood of $A$, then $f(Z) = f(A)$ for all points $Z \in V$.*

*Proof.* Write the germ of the variety $V$ at the point $A$ as a union of irreducible germs of holomorphic varieties $\mathbf{V}_j$. Each germ $\mathbf{V}_j$ can be represented by a finite branched holomorphic covering of an open subset of some space $\mathbb{C}^{n_j}$. It then follows immediately from Theorem C10 that $f(Z) = f(A)$ for all points $Z \in V_j$ near $A$. But then $f(Z) = f(A)$ for all points $Z$ in an open neighborhood of $A$ in $V$, so by the identity theorem $f(Z) = f(A)$ for all points $Z \in V$. That concludes the proof.

It is clear that irreducibility is an essential hypothesis in the preceding theorem as well, but this hypothesis can be lessened in some useful direct applications of this theorem.

**3. COROLLARY.** *On a compact connected holomorphic variety all holomorphic functions are constant.*

*Proof.* Each irreducible component of a compact holomorphic variety is also compact, so any holomorphic function being continuous must attain its maximal value on each component. The maximum modulus theorem shows that the function must then be constant on each irreducible component and hence from the connectivity hypothesis must actually be constant throughout, as desired.

**4. COROLLARY.** *A compact holomorphic subvariety of* $\mathbb{C}^n$ *consists of only finitely many points.*

*Proof.* If $V$ is a compact holomorphic subvariety of $\mathbb{C}^n$, then each irreducible component $V_i$ of $V$ is also compact, so the coordinate functions in $\mathbb{C}^n$ must attain their maxima on $V_i$ at some points of $V_i$. But then by Theorem 2 these coordinate functions must be constant on $V_i$, so that $V_i$ consists of just a single point of $\mathbb{C}^n$. The germ of $V$ at each point is a finite union of irreducible germs of subvarieties by Theorem B7, so that the individual points comprising $V$ can have no limit points. Since $V$ is compact, it follows that $V$ consists of finitely many points of $\mathbb{C}^n$ as desired.

On the other hand, there are a number of the standard properties of holomorphic functions on open subsets of $\mathbb{C}^n$ that genuinely do not hold for holomorphic functions on arbitrary holomorphic varieties, one of the most basic being the extended form of Riemann's removable singularities theorem. This very important phenomenon will be discussed in some detail later, but it is perhaps worth pausing here to consider a simple example.

**Example.** Consider once again the holomorphic subvariety $V = \{(z_1, z_2) \in \mathbb{C}^2 : z_1^3 = z_2^2\}$ and the holomorphic mapping $\phi\colon \mathbb{C}^1 \to V$ defined by $\phi(t) = (t^2, t^3)$, recalling that $V$ is regular except for a singular point at the origin and that $\phi$ is a topological homeomorphism. The inverse of the mapping $\phi$ is a well-defined continuous function $f$ on $V$. It is holomorphic except at the origin, since $\phi$ is a locally biholomorphic map wherever $V$ is regular, but it cannot be holomorphic at the origin since otherwise $\phi$ would be biholomorphic and $V$ hence nonsingular. Thus this function $f$ is a surprisingly strong counterexample to Riemann's removable singularities theorem on $V$. It might also be noted that the function $f$ can be identified with the restriction to $V$ of the meromorphic function $z_2/z_1$, extended naturally at the origin on $V$. Thus $f$ is actually a continuous meromorphic function on $V$, but not holomorphic at the origin.

In discussing the ring ${}_V\mathcal{O}_U$ of holomorphic functions in an open subset $U$ of a holomorphic variety $V$, it is convenient to introduce as the **natural topology** the topology of uniform convergence on compact subsets of $U$, defined in the general case just as was done in the special case $V = \mathbb{C}^n$ in Definition I-F1. This topology can also be defined by the family of pseudonorms $\| \ \|_K$ associated to compact subsets $K$ of $U$ as in Definition I-F2. Many of the properties established for this topology in the special case $V = \mathbb{C}^n$ extend immediately to the general case; it is a metric topology, and ${}_V\mathcal{O}_U$ is a Hausdorff locally convex topological vector space satisfying the second axiom of countability. Completeness is less obvious in the general case, since the convergence of the restrictions to a subvariety $V \subseteq D$ of a sequence of holomorphic functions defined in an open set $D \subseteq \mathbb{C}^n$ says very little directly about any corresponding convergence of these functions in $D$, or even in any open neighborhood of $V$ in $D$. The study of the question of the completeness of these spaces ${}_V\mathcal{O}_U$ will be the main topic in the remainder of this section, although the auxiliary results established in the process have a considerable independent interest. To begin this study it is convenient to introduce the following extension of the notion of the germ of a holomorphic function at a point.

**5. DEFINITION.** *If $K$ is any closed subset of a holomorphic variety $V$, two functions $f_j \in {}_V\mathcal{O}_{U_j}$ holomorphic in open neighborhoods $U_j$ of $K$ are called* **equivalent at $K$** *if there is an open subset $W$ of $V$ such that $K \subseteq W \subseteq U_1 \cap U_2$ and $f_1|W = f_2|W$. This is an equivalence relation in the usual sense, an equivalence class is called a* **germ of a holomorphic function at K**, *and the set of all such germs is denoted by ${}_V\mathcal{O}_K$.*

If $K$ is a point of $V$, this reduces to the previously introduced notion of the germ of a holomorphic function at a point. As before, germs (equivalence classes) of holomorphic functions will be denoted by boldface letters; any function $f$ holomorphic in an open neighborhood of a closed subset $K$ of $V$ determines a germ $\mathbf{f}$ or $\mathbf{f}_K$ in ${}_V\mathcal{O}_K$. The set ${}_V\mathcal{O}_K$ has the natural structure of a ring and contains in a natural way the subring $\mathbb{C}$ of germs of constant functions; so it can be viewed as a complex algebra as well. Of course, if $K$ is too large, ${}_V\mathcal{O}_K$ may coincide with the subring $\mathbb{C}$ of the complex constants. Each germ $\mathbf{f} \in {}_V\mathcal{O}_K$ determines a continuous complex-valued function on the set $K$, for all representatives of $\mathbf{f}$ coincide on $K$, and any single representative is continuous in an open neighborhood of $K$. In some cases it is possible to identify this continuous function with the germ inducing it. In discussing this possibility it is convenient to say that a closed subset $K \subseteq V$ is **relatively irreducible** if it has arbitrarily small open neighborhoods that are themselves irreducible as holomorphic varieties. Note that when $V$ is a regular variety, a closed subset is relatively irreducible precisely when it is connected, while for a general variety $V$, the connectivity of a closed subset is a necessary but not sufficient condition for relative irreducibility. Note also that when $K$ is a subvariety of $V$, the relative irreducibility of $K$ and its irreducibility as a subvariety are quite independent notions. A reducible subvariety of a manifold may well be connected and hence relatively irreducible, while an irreducible subvariety $K \subseteq \mathbb{C}^n$ which is the intersection of two distinct subvarieties $V_1$ and $V_2$ of $\mathbb{C}^n$ is not relatively irreducible in $V_1 \cup V_2$. If $K$ is a relatively irreducible subset of a variety $V$ and $K$ has an interior point, then it follows immediately from the identity theorem that the continuous function on $K$ defined by a germ $\mathbf{f} \in {}_V\mathcal{O}_K$ uniquely determines that germ for any functions $f_j \in {}_V\mathcal{O}_{U_j}$ holomorphic in some open neighborhoods $U_j$ of $K$ and agreeing on $K$ must coincide in some connected open subneighborhood $U \subseteq U_1 \cap U_2$ of $K$. In this case a continuous function on $K$ corresponds to a germ in ${}_V\mathcal{O}_K$ precisely when it is the restriction to $K$ of some holomorphic function in an open neighborhood of $K$. The identification of elements of ${}_V\mathcal{O}_K$ with continuous functions on $K$ will be used quite freely and with no further comment in such a case, and it should not cause any confusion.

Note that if $K$ is any compact subset of $V$, it is possible to introduce a pseudonorm $\| \ \|_K$ on ${}_V\mathcal{O}_K$ by setting

$$\|\mathbf{f}\|_K = \sup_{Z \in K} |f(Z)| \tag{1}$$

for any element $\mathbf{f} \in {}_V\mathcal{O}_K$, where $f$ is the continuous function on $K$ determined by the germ $\mathbf{f}$. If $K$ is relatively irreducible in $V$ and $K$ has an interior point, this pseudonorm is clearly a norm.

**6. DEFINITION.** *If $K$ is a relatively irreducible compact subset of a holomorphic variety $V$ and $K$ has an interior point, then the* **natural topology** *on ${}_V\mathcal{O}_K$ is the metric topology defined by the norm* (1).

This topology extends naturally to a metric topology on any free module ${}_V\mathcal{O}_K^\nu$ by setting $\|\mathbf{F}\|_K = \max_i \|\mathbf{f}_i\|_K$ whenever $\mathbf{F} = (\mathbf{f}_1, \ldots, \mathbf{f}_\nu) \in {}_V\mathcal{O}_K^\nu$. It is evident that with this topology ${}_V\mathcal{O}_K$ is a Hausdorff locally convex topological vector space satisfying the first axiom of countability. Indeed ${}_V\mathcal{O}_K$ can be identified with a subalgebra of the uniform algebra $\mathscr{C}_K$ of all continuous complex-valued functions on the compact subset $K$ by associating to each germ in ${}_V\mathcal{O}_K$ the continuous function in $\mathscr{C}_K$ that it naturally determines. The algebra $\mathscr{C}_K$ with the uniform norm $\| \ \|_K$ defined as in (1) is a well-known Banach algebra. Even in the nicest of circumstances, though, ${}_V\mathcal{O}_K$ is not a closed subalgebra of $\mathscr{C}_K$.

**Example.** If $K = \bar{\Delta}(0; R) \subseteq \mathbb{C}^n$, then $\mathscr{C}_K$ is the Banach algebra of all complex-valued continuous functions on $K$ under the supremum norm defined as in (1), while ${}_n\mathcal{O}_K$ can be identified with the subset of $\mathscr{C}_K$ consisting of all those functions in $\mathscr{C}_K$ that are the restrictions to $K$ of functions holomorphic in an open neighborhood of $K$. If $\mathscr{A}_K \subseteq \mathscr{C}_K$ is the subset consisting of all those functions in $\mathscr{C}_K$ that are holomorphic in the interior $\Delta(0; R)$ of $K$, it is quite obvious that $\mathscr{A}_K$ is a closed subalgebra of $\mathscr{C}_K$ and contains ${}_n\mathcal{O}_K$. If $f \in \mathscr{A}_K$, then the function $f_\varepsilon$ defined by $f_\varepsilon(Z) = f((1-\varepsilon)Z)$ for $0 < \varepsilon < 1$ is clearly holomorphic in $\Delta(0; (1-\varepsilon)^{-1}R)$, so the restriction $f_\varepsilon | K$ belongs to the subset ${}_n\mathcal{O}_K \subseteq \mathscr{C}_K$. But obviously $f_\varepsilon | K$ converges to $f$ in the supremum norm (1) as $\varepsilon$ approaches zero, and therefore $f$ belongs to the closure of ${}_n\mathcal{O}_K$ in $\mathscr{C}_K$. Thus $\mathscr{A}_K$ is precisely the closure of ${}_n\mathcal{O}_K$ in $\mathscr{C}_K$; but clearly not all functions in $\mathscr{A}_K$ extend to holomorphic functions in a full open neighborhood of $K$, so that ${}_n\mathcal{O}_K$ is a proper subset of its closure $\mathscr{A}_K$.

Despite its obvious deficiencies, the natural topology on ${}_V\mathcal{O}_K$ is a very useful tool in a variety of circumstances. An important auxiliary result in the study of this topology is the following semilocal form of the Weierstrass division theorem.

**7. THEOREM.** *Let $\Delta(0;R) = \Delta' \times \Delta'' \subseteq \mathbb{C}^{n-1} \times \mathbb{C}^1$ be an open polydisc in $\mathbb{C}^n$ with closure $K = K' \times K''$, and let $\mathbf{h} \in {}_{n-1}\mathcal{O}_{K'}[z_n]$ be a monic polynomial of degree $\nu$ such that for any point $A' \in K'$ all $\nu$ roots of the polynomial $\mathbf{h}(A'; z_n)$ lie in $\Delta''$. Then there is a constant $M > 0$ such that any $\mathbf{f} \in {}_n\mathcal{O}_K$ can be written uniquely in the form $\mathbf{f} = \mathbf{g}\mathbf{h} + \mathbf{p}$ where $\mathbf{g} \in {}_n\mathcal{O}_K$, $\mathbf{p} = \sum_{j=0}^{\nu-1} \mathbf{p}_j z_n^j \in {}_{n-1}\mathcal{O}_{K'}[z_n]$, and*

$$\|\mathbf{g}\|_K \leqq M\|\mathbf{f}\|_K, \qquad \|\mathbf{p}_j\|_{K'} \leqq M\|\mathbf{f}\|_K \tag{2}$$

*Proof.* For any germ $\mathbf{f} \in {}_n\mathcal{O}_K$ choose a polydisc $\Delta(0; R+2\varepsilon)$ of polyradius $R+2\varepsilon = (r_1 + 2\varepsilon, \ldots, r_n + 2\varepsilon)$ with $\varepsilon > 0$ such that the germs $\mathbf{f}$ and $\mathbf{h}$ are represented by holomorphic functions $f$ and $h$ in this polydisc and that for any point $A' \in \Delta(0; R' + 2\varepsilon) \subseteq \mathbb{C}^{n-1}$ all $\nu$ roots of the polynomial $h(A', z_n)$ lie in the disc $\Delta'' = \Delta(0; r_n) \subseteq \mathbb{C}^1$. The integral formulas used in the proof of the Weierstrass division theorem, Theorem I-A5, when applied to the polydisc $\Delta(0; R + \varepsilon)$ yield holomorphic functions $g$ and $p$ in this polydisc such that $f = gh + p$ and that $p$ is a polynomial in $z_n$ of degree $\leqq \nu - 1$. It may be worth noting that the condition that $\mathbf{h}_0 \in {}_n\mathcal{O}_0$ be a Weierstrass polynomial in Theorem I-A5 is needed only to ensure that the zeros of that polynomial satisfy the conditions assumed here. To conclude the proof it is only necessary to verify the inequalities (2) for some constant $M$ not depending on $f$, since uniqueness follows readily just as in the proof of the Weierstrass division theorem. First the functions $p_j$ are given by integrals of the form

$$p_j(Z') = \frac{1}{2\pi i} \int_{|\zeta| = r_n + \varepsilon} \frac{h_j(Z', \zeta)}{h(Z', \zeta)} f(Z', \zeta)\, d\zeta$$

where $Z = (Z', z_n) \in \Delta(0; R' + \varepsilon) \times \Delta(0; r_n + \varepsilon)$ and $h_j(Z)$ are functions formed from the coefficients of this polynomial $h$ in a manner independent of the choice of the functions $f$. The denominator in this integral does not vanish for any point $(Z', \zeta)$ with $Z' \in \Delta(0; R' + \varepsilon)$ and $r_n \leqq |\zeta| \leqq r_n + \varepsilon$, so the path of integration can be moved to the circle $|\zeta| = r_n$ without changing the value of the integral. Then evidently $|p_j(Z')| \leqq M_j r_n \|f\|_K$ whenever $Z' \in K'$, where

$$M_j = \sup_{\substack{Z' \in K' \\ |\zeta| = r_n}} \frac{|h_j(Z', \zeta)|}{|h(Z', \zeta)|}$$

Choose a constant $m > 0$ such that $|h(Z)| \geqq m$ whenever $|z_k| = r_k$ for all $k$, as is possible since $h(Z) \neq 0$ on this compact subset. It then follows that

$$\begin{aligned}
|g(Z)| &= \left| \frac{f(Z) - p(Z)}{h(Z)} \right| \\
&\leqq \frac{1}{m} [|f(Z)| + \textstyle\sum |p_j(Z')| r_n^j] \\
&\leqq \frac{1}{m} \left[ 1 + \sum_{j=0}^{\nu-1} M_j r_n^{j+1} \right] \|f\|_K
\end{aligned}$$

whenever $|z_k| = r_k$ for all $k$. If $|z_1| < r_1$ but $|z_k| = r_k$ for all $k > 1$, then as a function of $z_1$ alone the above inequality holds on the boundary of the disc $|z_1| = r_1$; hence, by the maximum modulus theorem the same inequality holds for all $|z_1| < r_1$. Repeating the argument shows that this inequality holds on the entire boundary of the closed polydisc $K$ and hence on the polydisc $K$ itself. Therefore the desired inequalities follow, with $M = \max(M_0 r_n, \ldots, M_{\nu-1} r_n, (1 + \sum_j M_j r_n^{j+1})/m)$, and that suffices to conclude the proof.

Note that whenever $L \subseteq K$ are closed subsets of a holomorphic variety $V$, there is a natural restriction homomorphism $\rho_{LK}: {}_V\mathcal{O}_K \to {}_V\mathcal{O}_L$, and this is a continuous mapping in the natural topologies if $K$ and $L$ are compact and relatively irreducible and $L$ has an interior point. For any submodule $\mathcal{M}_K \subseteq {}_V\mathcal{O}_K^\nu$ the image $\rho_{LK}(\mathcal{M}_K) \subseteq {}_V\mathcal{O}_L^\nu$ is of course a submodule, and conversely for any submodule $\mathcal{M}_L \subseteq {}_V\mathcal{O}_L^\nu$ the inverse image $\rho_{LK}^{-1}(\mathcal{M}_L) = \{\mathbf{F} \in {}_V\mathcal{O}_K^\nu : \rho_{LK}(\mathbf{F}) \in \mathcal{M}_L\}$ is a submodule of ${}_V\mathcal{O}_K^\nu$. In special circumstances these submodules can be surprisingly closely related, as an application of the preceding theorem shows.

**8. THEOREM.** *Let $\mathcal{M}_0 \subseteq {}_n\mathcal{O}_0^\nu$ be any submodule at the origin in $\mathbb{C}^n$, and let $G_1, \ldots, G_\mu \in {}_n\mathcal{O}_U^\nu$ be any vectors of holomorphic functions in an open neighborhood $U$ of the origin such that the germs $\mathbf{G}_j \in {}_n\mathcal{O}_0^\nu$ generate $\mathcal{M}_0$. Then after a suitable nonsingular linear change of coordinates in $\mathbb{C}^n$, there exist an open polydisc $\Delta(0; R)$ with closure $K = \bar{\Delta}(0; R) \subseteq U$ and a constant $M > 0$ such that the germs $\mathbf{G}_j \in {}_n\mathcal{O}_K^\nu$ generate $\mathcal{M}_K = \rho_{OK}^{-1}(\mathcal{M}_O) \subseteq {}_n\mathcal{O}_K^\nu$ as an ${}_n\mathcal{O}_K$-module—indeed, such that any $\mathbf{F} \in \mathcal{M}_K$ can be written $\mathbf{F} = \sum_j \mathbf{h}_j \mathbf{G}_j$ for some germs $\mathbf{h}_j \in {}_n\mathcal{O}_K$ with $\|h_j\|_K \leqq M \|\mathbf{F}\|_K$. Moreover for any finite collection of submodules of free modules of various ranks there exists a single polydisc such that these results hold for all the modules simultaneously in that polydisc.*

*Proof.* The proof will be by induction on the dimension $n$. Since the theorem holds trivially in case $n = 0$, assume that it holds as stated for some dimension $n - 1 \geqq 0$. Consider first the special case $\nu = 1$ of the theorem in dimension $n$, and suppose given an ideal $\mathcal{M}_O \subseteq {}_n\mathcal{O}_O$ generated by the germs at the origin of some holomorphic functions $g_1, \ldots, g_\mu$ in an open neighborhood $U$ of the origin. After a suitable nonsingular linear change of coordinates in $\mathbb{C}^n$, it can be assumed that the germ $\mathbf{g}_1 \in {}_n\mathcal{O}_O$ is regular in the variable $z_n$; indeed, any finite set of such germs can simultaneously be made regular. Apply the Weierstrass preparation theorem to write $\mathbf{g}_1 = \mathbf{u}_1 \mathbf{p}_1$ where $\mathbf{u}_1$ is a unit in ${}_n\mathcal{O}_O$ and $\mathbf{p}_1$ is a Weierstrass polynomial of some degree $\delta$ in ${}_{n-1}\mathcal{O}_O[z_n]$. Then apply the Weierstrass division theorem to write $\mathbf{g}_j = \mathbf{u}_j \mathbf{p}_1 + \mathbf{p}_j$ for all $j > 1$ where $\mathbf{u}_j \in {}_n\mathcal{O}_O$ and $\mathbf{p}_j \in {}_{n-1}\mathcal{O}_O[z_n]$ are polynomials of degrees strictly less than $\delta$. Choose an open product neighborhood $V = V' \times \Delta'' \subseteq \mathbb{C}^{n-1} \times \mathbb{C}^1$ of the origin in $\mathbb{C}^n$, for some disc $\Delta'' = \Delta(0; r_n)$ of radius $r_n < 1$, with the properties that $\bar{V} \subseteq U$, that the germs $\mathbf{u}_j$, $\mathbf{p}_j$ can be represented by holomorphic functions $u_j$, $p_j$ in an open neighborhood of $\bar{V}$ with $g_1 = u_1 p_1$ and $g_j = u_j p_1 + p_j$ for $j > 1$, that $|u_1(Z)| \geqq m > 0$ for some constant $m$ and all points $Z \in \bar{V}$, and that for each point $A' \in \bar{V}'$ all $\delta$ roots of the polynomial $p_1(A', z_n)$ are contained in $\Delta''$. Here too it is clearly possible to choose a neighborhood $V$ having all these properties simultaneously for a finite number of sets of such functions $g_j$. Note that the polynomials $z_n^k p_j$ for $0 \leqq k \leqq \delta - 1$, $1 \leqq j \leqq \mu$, can be viewed as elements of the free ${}_{n-1}\mathcal{O}_{V'}$-

module consisting of all polynomials of degree strictly less than $2\delta$ in ${}_{n-1}\mathcal{O}_{V'}[z_n]$, and the inductive hypotheses can be applied to these elements. Therefore after a suitable nonsingular change of coordinates in $\mathbf{C}^{n-1}$ there exist an open polydisc $\Delta' \subseteq \mathbf{C}^{n-1}$ with closure $K' = \bar{\Delta}' \subseteq V'$ and a constant $M_1 > 0$ such that whenever $\mathbf{p}^* \in {}_{n-1}\mathcal{O}_{K'}[z_n]$ has its germ at the origin contained in the ${}_{n-1}\mathcal{O}_O$-module generated by $z_n^k p_j$, then $\mathbf{p}^* = \sum_{jk} \mathbf{h}_{jk}^* z_n^k \mathbf{p}_j$ for some germs $\mathbf{h}_{jk}^* \in {}_{n-1}\mathcal{O}_{K'}$ with $\|\mathbf{h}_{jk}^*\|_{K'} \leqq M_1 \|\mathbf{p}^*\|_{K'}$. Note further that the polydisc $\Delta = \Delta' \times \Delta''$ has closure $K = K' \times K'' \subseteq U$, and that Theorem 7 can be applied to the polynomial $\mathbf{p}_1$ in this set $K$ with an estimate (2) as in that theorem for some constant $M_2 > 0$. Now suppose that $\mathbf{f} \in {}_n\mathcal{O}_K$ is a germ for which $\mathbf{f} \in \mathcal{M}_O$—that is, for which $\mathbf{f} = \sum_j \mathbf{x}_j \mathbf{g}_j$ in ${}_n\mathcal{O}_O$ for some germs $\mathbf{x}_j \in {}_n\mathcal{O}_O$. By applying Theorem 7 write $\mathbf{f} = \mathbf{q}\mathbf{p}_1 + \mathbf{p}$ where $\mathbf{q} \in {}_n\mathcal{O}_K$ is a germ for which $\|\mathbf{q}\|_K \leqq M_2 \|f\|_K$ and $\mathbf{p} \in {}_{n-1}\mathcal{O}_{K'}[z_n]$ is a polynomial of degree strictly less than $\delta$ for which $\|\mathbf{p}\|_{K'} \leqq M_2 \|f\|_K$. Consider at first merely germs at the origin in $\mathbf{C}^n$, and apply the Weierstrass division theorem to write $\mathbf{x}_j = \mathbf{x}_j'\mathbf{p}_1 + \mathbf{y}_j$ for all $j > 1$, where $\mathbf{x}_j' \in {}_n\mathcal{O}_O$ and $\mathbf{y}_j \in {}_{n-1}\mathcal{O}_O[z_n]$ are polynomials of degrees strictly less than $\delta$. Then note that

$$\begin{aligned} \mathbf{p} = \mathbf{f} - \mathbf{q}\mathbf{p}_1 &= \sum_j \mathbf{x}_j \mathbf{g}_j - \mathbf{q}\mathbf{p}_1 \\ &= \mathbf{x}_1\mathbf{u}_1\mathbf{p}_1 + \sum_{j>1} (\mathbf{x}_j'\mathbf{p}_1 + \mathbf{y}_j)(\mathbf{u}_j\mathbf{p}_1 + \mathbf{p}_j) - \mathbf{q}\mathbf{p}_1 \end{aligned}$$

or equivalently

$$\mathbf{p} = \sum_j \mathbf{y}_j \mathbf{p}_j \tag{3}$$

where $\mathbf{y}_1 = \mathbf{x}_1\mathbf{u}_1 - \mathbf{q} + \sum_{j>1} [\mathbf{x}_j'\mathbf{u}_j\mathbf{p}_1 + \mathbf{x}_j'\mathbf{p}_j + \mathbf{y}_j\mathbf{u}_j]$. All entries in (3) with the possible exception of $\mathbf{y}_1$ are polynomials in ${}_{n-1}\mathcal{O}_O[z_n]$ of degree less than $\delta$; but then $\mathbf{y}_1\mathbf{p}_1 \in {}_{n-1}\mathcal{O}_O[z_n]$ so by the Weierstrass division theorem $\mathbf{y}_1 \in {}_{n-1}\mathcal{O}_O[z_n]$ and has degree strictly less than $\delta$. Then (3) asserts that the germ at the origin of the polynomial $\mathbf{p} \in {}_{n-1}\mathcal{O}_{K'}[z_n]$ is contained in the ${}_{n-1}\mathcal{O}_O$-module generated by $z_n^k \mathbf{p}_j$, so it follows from the inductive hypothesis as noted that $\mathbf{p} = \sum_j \mathbf{h}_j'\mathbf{p}_j$ for some polynomials $\mathbf{h}_j' \in {}_{n-1}\mathcal{O}_{K'}[z_n]$ of degree strictly less than $\delta$ with coefficients satisfying $\|\mathbf{h}_{jk}'\|_{K'} \leqq M_1 \|\mathbf{p}\|_{K'}$. Then

$$\begin{aligned} \mathbf{f} = \mathbf{q}\mathbf{p}_1 + \mathbf{p} &= \mathbf{q}\mathbf{p}_1 + \sum_j \mathbf{h}_j'\mathbf{p}_j \\ &= (\mathbf{q} + \mathbf{h}_1')\mathbf{p}_1 + \sum_{j>1} \mathbf{h}_j'(\mathbf{g}_j - \mathbf{u}_j\mathbf{p}_1) \\ &= \left(\mathbf{q} + \mathbf{h}_1' - \sum_{j>1} \mathbf{h}_j'\mathbf{u}_j\right)\mathbf{u}_1^{-1}\mathbf{g}_1 + \sum_{j>1} \mathbf{h}_j'\mathbf{g}_j \\ &= \sum_j \mathbf{h}_j\mathbf{g}_j \end{aligned}$$

Moreover for $j > 1$ note that $\|\mathbf{h}_j\|_K = \|\mathbf{h}_j'\|_K \leqq \delta M_1 M_2 \|f\|_K$, while

$$\|h_1\|_K \leq m^{-1}\left[M_2 + \delta M_1 M_2 + \sum_{j>1} \delta M_1 M_2 \|u_j\|_K\right] \|f\|_K$$

and that provides an estimate of the desired form.

The proof will be concluded by induction on the index $\nu$ for a fixed dimension $n$. The case $\nu = 1$ having just been handled, suppose that the theorem has been demonstrated for all indices strictly less than $\nu$ and consider a submodule $\mathcal{M}_O \subseteq {}_n\mathcal{O}_O^\nu$ generated by the germs at the origin of some holomorphic functions $G_1, \ldots, G_\mu \in {}_n\mathcal{O}_U^\nu$ in an open neighborhood $U$ of the origin. The first $\nu - 1$ entries in $G_j$ can be viewed as elements $G_j' \in {}_n\mathcal{O}_U^{\nu-1}$, and their germs at the origin generate a submodule $\mathcal{M}_O' \subseteq {}_n\mathcal{O}_O^{\nu-1}$. Furthermore, let $\mathcal{M}_O'' \subseteq {}_n\mathcal{O}_O^\nu$ be the submodule of $\mathcal{M}_O$ consisting of all elements in $\mathcal{M}_O$ having all but the last entry zero. Since ${}_n\mathcal{O}_O$ is Noetherian, the module $\mathcal{M}_O''$ will have finitely many generators $\mathbf{G}_1'', \ldots, \mathbf{G}_\lambda''$, each of which can be written $\mathbf{G}_k'' = \sum_j \mathbf{a}_{kj}\mathbf{G}_j$ for some germs $\mathbf{a}_{kj} \in {}_n\mathcal{O}_O$. After shrinking the neighborhood $U$ if necessary, it can be assumed that these germs can be represented by functions $G_k'' \in {}_n\mathcal{O}_U^\nu$ and $a_{kj} \in {}_n\mathcal{O}_U$, and that $G_k'' = \sum_j a_{kj} G_j$. The inductive hypotheses can be applied to the submodules $\mathcal{M}_O'$ and $\mathcal{M}_O''$, since the latter can be viewed as an ideal in ${}_n\mathcal{O}_O$. Therefore after a suitable nonsingular linear change of coordinates in $\mathbb{C}^n$, there will exist an open polydisc $\Delta(0; R) \subseteq \mathbb{C}^n$ with closure $K = \bar{\Delta}(0; R)$ such that the conclusions of the theorem hold in $K$ for these two sets of functions, with estimates given by constants $M'$ and $M''$. Now suppose that $\mathbf{F} \in {}_n\mathcal{O}_K^\nu$ is an element such that germ of $\mathbf{F}$ at the origin is contained in the module $\mathcal{M}_O$. Let $\mathbf{F}' \in {}_n\mathcal{O}_K^{\nu-1}$ consist of the first $\nu - 1$ entries of $\mathbf{F}$. It follows that the germ of $\mathbf{F}'$ at the origin is contained in the module $\mathcal{M}_O'$; therefore, $\mathbf{F}' = \sum_j \mathbf{h}_j' \mathbf{G}_j'$ for some germs $\mathbf{h}_j' \in {}_n\mathcal{O}_K$ with $\|h_j'\|_K \leq M' \|F'\|_K \leq M' \|F\|_K$. Then, however, $\mathbf{F} - \sum_j \mathbf{h}_j' \mathbf{G}_j \in \mathcal{M}_O''$, so that $\mathbf{F} - \sum_j \mathbf{h}_j' \mathbf{G}_j = \sum_k \mathbf{h}_k'' \mathbf{G}_k''$ for some germs $\mathbf{h}'' \in {}_n\mathcal{O}_K$ with $\|\mathbf{h}_k''\|_K \leq M'' \|\mathbf{F} - \sum_j \mathbf{h}_j' \mathbf{G}_j\|_K$. Altogether

$$\mathbf{F} = \sum_j \mathbf{h}_j' \mathbf{G}_j + \sum_{kj} \mathbf{h}_k'' \mathbf{a}_{kj} \mathbf{G}_j = \sum_j \mathbf{h}_j \mathbf{G}_j$$

where

$$\|\mathbf{h}_j\| \leq \left(M' + M'' \max_{k,j} \|a_{kj}\|_K\right) \|\mathbf{F}\|_K$$

The same result holds for any finite collection of such sets of $\nu$-tuples, and that suffices to conclude the proof.

As a brief digression from the natural course of the argument here, it is of interest to note one application of this theorem having the flavor of the discussion in section F. In that section several results were obtained showing that certain families of submodules of families of free modules are locally finitely generated. The families considered were all essentially constructed in terms of a given finite collection of holomorphic functions, but in some circumstances the finiteness is not really required.

**9. THEOREM.** *Let $\mathcal{G} = \{G\}$ be an arbitrary collection of holomorphic vectors of functions $G \in {}_n\mathcal{O}_U^\nu$ in a fixed open subset $U \subseteq \mathbb{C}^n$, and let $\mathcal{M}_A \subseteq {}_n\mathcal{O}_A^\nu$ be the submodule generated* *by the germs of these functions at any point $A \in U$. Then the family of submodules $\{\mathcal{M}_A\}$ of the family of free modules $\{{}_n\mathcal{O}_A^v\}$ is finitely generated in an open neighborhood of any point of $U$.*

*Proof.* For any given point $A \in U$ the submodule $\mathcal{M}_A \subseteq {}_n\mathcal{O}_A^v$ will be generated by the germs of a finite subset $G_{A1}, \ldots, G_{A\mu}$ of the functions in $\mathcal{G}$, since the ring ${}_n\mathcal{O}_A$ is Noetherian. It then follows immediately from Theorem 8 that there is an open neighborhood $U_A$ of the point $A$ such that the restriction of any $G \in \mathcal{G}$ can be written $G|U_A = \sum_{j=1}^{\mu} h_j(G_{Aj}|U_A)$ for some $h_j \in {}_n\mathcal{O}_{U_A}$. But that implies that the germs of $G_{A1}, \ldots, G_{A\mu}$ generate the module $\mathcal{M}_Z$ at any point $Z \in U_A$ and suffices to conclude the proof.

**10. COROLLARY.** *If $\mathcal{F} = \{f\}$ is an arbitrary collection of functions holomorphic in a fixed open subset $U \subseteq \mathbb{C}^n$, then the set $V = \{Z \in U : f(Z) = 0$ for all $f \in \mathcal{F}\}$ is a holomorphic subvariety of $U$.*

*Proof.* The function $f \in \mathcal{F}$ generate ideals in the local rings ${}_n\mathcal{O}_A$ at all points $A \in U$, and this family of ideals is finitely generated in an open neighborhood of each point of $U$ by Theorem 9. The desired result then follows immediately.

To return now to the discussion of the natural topology on the rings of germs of holomorphic functions, a useful application of Theorem 8 is the following.

**11. THEOREM (closure of modules theorem).** *For if any submodule $\mathcal{M}_O \subseteq {}_n\mathcal{O}_O^v$ at the origin in $\mathbb{C}^n$ and any open neighhorhood $U$ of the origin, after a suitable nonsingular linear change of coordinates in $\mathbb{C}^n$ there is an open polydisc $\Delta(0; R) \subseteq \mathbb{C}^n$ with closure $K \subseteq U$ such that the induced module $\mathcal{M}_K = \rho_{OK}^{-1}(\mathcal{M}_O) \subseteq {}_n\mathcal{O}_K^v$ is relatively closed in ${}_n\mathcal{O}_K^v$ in the natural topology.*

*Proof.* After shrinking $U$ if necessary, choose holomorphic functions $G_1, \ldots, G_\mu \in {}_n\mathcal{O}_U^v$ in the neighborhood $U$ such that the germs $\mathbf{G}_j \in {}_n\mathcal{O}_O^v$ generate the ${}_n\mathcal{O}_O$-module $\mathcal{M}_O$. After a suitable nonsingular linear change of coordinates in $\mathbb{C}^n$, there exist an open polydisc $\Delta = \Delta(0; R)$ with closure $K = \bar{\Delta}(0; R) \subseteq U$ and a constant $M > 0$ such that the conclusions of Theorem 8 hold in $K$ for the germs $\mathbf{G}_j \in {}_n\mathcal{O}_K^v$. Now given a sequence of germs $\mathbf{F}_i \in \mathcal{M}_K \subseteq {}_n\mathcal{O}_K^v$ converging in the natural topology of ${}_n\mathcal{O}_K^v$ to some germ $\mathbf{F} \in {}_n\mathcal{O}_K^v$, it is only necessary to show that $\mathbf{F} \in \mathcal{M}_K$. For this purpose note as a consequence of Theorem 8 that each germ $\mathbf{F}_i$ can be written $\mathbf{F}_i = \sum_{j=1}^{\mu} \mathbf{h}_{ij}\mathbf{G}_j$ for some germs $\mathbf{h}_{ij} \in {}_n\mathcal{O}_K$ with $\|\mathbf{h}_{ij}\|_K \leq M\|\mathbf{F}_i\|_K$. Since the germs $\mathbf{F}_i$ converge to $\mathbf{F}$ in norm, the constants $\|\mathbf{F}_i\|_K$ are bounded and hence all the constants $\|\mathbf{h}_{ij}\|_K$ are bounded. It then follows from Montel's theorem, Theorem I-F3, that after passing to a subsequence of $\{\mathbf{F}_i\}$ if necessary, the sequence of functions $h_{ij} \in {}_n\mathcal{O}_\Delta$ converges in the natural topology of ${}_n\mathcal{O}_\Delta$ to some function $h_j \in {}_n\mathcal{O}_\Delta$ for each fixed index $j$. In particular the sequence of functions $h_{ij}$ converge uniformly on $\Delta(0; R/2)$ to $h_j$. Thus whenever $Z \in \Delta(0; R/2)$, it follows that $F(Z) = \lim_i F_i(Z) = \lim_i \sum_j h_{ij}(Z)G_j(Z) = \sum_j h_j(Z)G_j(Z)$ and hence that the restriction $\rho_{OK}(\mathbf{F}) \in {}_n\mathcal{O}_O^v$ belongs to the submodule $\mathcal{M}_O$; but then $\mathbf{F} \in \mathcal{M}_K$ as desired and that concludes the proof.

It should be noted particularly that *the preceding theorem does not assert that the submodule $\mathcal{M}_K \subseteq {}_n\mathcal{O}_K^\vee$ is complete in the metric defining the topology, or closed in $\mathcal{C}_K^\vee$*, only that $\mathcal{M}_K$ is closed as a subset of the topological space ${}_n\mathcal{O}_K^\vee$. As already observed, ${}_n\mathcal{O}_K^\vee$ is not a closed subset of $\mathcal{C}_K^\vee$ and hence is not itself complete. Perhaps the most convenient form of the preceding result for the customary applications is the following.

**12. COROLLARY (closure of modules theorem).** *For any submodule $\mathcal{M}_O \subseteq {}_n\mathcal{O}_O^\vee$ at the origin in $\mathbb{C}^n$ and any open neighborhood $U$ of the origin, if $F_j \in {}_n\mathcal{O}_U^\vee$ are holomorphic functions such that the germs $\mathbf{F}_j \in {}_n\mathcal{O}_O^\vee$ lie in the submodule $\mathcal{M}_O$ and if the functions $\mathbf{F}_j$ converge in the natural topology of ${}_n\mathcal{O}_U^\vee$ to a function $F \in {}_n\mathcal{O}_U^\vee$, then the germ $\mathbf{F}_O \in {}_n\mathcal{O}_O^\vee$ also lies in the submodule $\mathcal{M}_O$.*

*Proof.* It follows from the preceding theorem that after a change of coordinates in $\mathbb{C}^n$ if necessary, there is an open polydisc $\Delta(0; R) \subset \mathbb{C}^n$ with closure $K \subset U$ such that $\mathcal{M}_K = \rho_{OK}^{-1}(\mathcal{M}_O) \subseteq {}_n\mathcal{O}_K^\vee$ is relatively closed in ${}_n\mathcal{O}_K^\vee$. Now since the germs $\mathbf{F}_j \in {}_n\mathcal{O}_O^\vee$ lie in the submodule $\mathcal{M}_O$, then the germs $\mathbf{F}_j \in {}_n\mathcal{O}_K^\vee$ lie in the submodule $\mathcal{M}_K$. Since the functions $F_j$ converge to $F$ in the topology of ${}_n\mathcal{O}_U^\vee$, the germs $\mathbf{F}_j \in {}_n\mathcal{O}_K^\vee$ necessarily converge to the germ $\mathbf{F} \in {}_n\mathcal{O}_K^\vee$, and consequently the latter germ must belong to $\mathcal{M}_K$. That means of course that the germ $\mathbf{F} \in {}_n\mathcal{O}_O^\vee$ must lie in $\mathcal{M}_O$ as desired, thereby concluding the proof of the corollary.

Some of the preceding considerations can now be used to show that the space of holomorphic functions on a holomorphic variety is indeed a Frechet space, thus accomplishing one of the goals set out for the present section. It is convenient first to establish the following preliminary result.

**13. LEMMA.** *Let $V$ and $W$ be holomorphic subvarieties of an open neighborhood $U$ of the origin in $\mathbb{C}^m$, $V$ being pure dimensional. Then after a nonsingular linear change of coordinates in $\mathbb{C}^m$ there exist an open polydisc $\Delta = \Delta(0; R)$ with closure $K = \bar{\Delta} \subseteq U$ and a constant $M > 0$ such that to each germ $\mathbf{f} \in {}_{V\cup W}\mathcal{O}_{K\cap(V\cup W)}$ for which $\mathbf{f}|K \cap W = 0$ there corresponds a germ $\mathbf{F} \in {}_m\mathcal{O}_K$ for which $\mathbf{F}|K \cap (V \cup W) = \mathbf{f}|K \cap (V \cup W)$ and $\|\mathbf{F}\|_K \leqq M \|\mathbf{f}\|_{K\cap(V\cup W)}$.*

*Proof.* If $V$ is of pure dimension $n$ and passes through the origin, then by Theorem G4 there is an $(m-n)$-dimensional linear subspace $L$ through the origin in $\mathbb{C}^m$ so that the germ at the origin of the intersection $L \cap V$ is the origin itself. It follows readily from Theorem E5 that after changing coordinates in $\mathbb{C}^m$ so that $L = \{Z \in \mathbb{C}^m : z_1 = \cdots = z_n = 0\}$ the natural projection $\mathbb{C}^m = \mathbb{C}^n \times \mathbb{C}^{m-n} \to \mathbb{C}^n$ exhibits the germ of $V$ at the origin as the germ of a finite branched holomorphic covering of some order $\nu$ over $\mathbb{C}^n$. After shrinking $U$ suitably, it can therefore be supposed that $U = U' \times U'' \subseteq \mathbb{C}^n \times \mathbb{C}^{m-n}$ and that the natural projection $\mathbb{C}^m \to \mathbb{C}^n$ exhibits $V$ as a finite branched holomorphic covering $\pi\colon V \to U'$ of pure order $\nu$. After shrinking $U$ even further if necessary, it can also be supposed that there is a holomorphic function $k \in {}_V\mathcal{O}_V$ that separates the sheets of this covering. From Theorem C6 applied to this function $k$, it follows that, for some fixed and nontrivial holomorphic function $d \in {}_n\mathcal{O}_{U'}$, to every holomorphic function $f \in {}_V\mathcal{O}_V$ there corresponds a unique polynomial $Q_f \in {}_n\mathcal{O}_{U'}[X]$ of degree at most $v - 1$ such that $d \cdot f = Q_f(k)$ in ${}_V\mathcal{O}_V$. The same construction can of course also be applied to every open neighborhood of the origin in $\mathbb{C}^n$ contained in $U'$, in the sense that every holomorphic function on $V \cap (\tilde{U}' \times U'')$ has a corresponding representation whenever $O \in \tilde{U}' \subseteq U'$. In particular, therefore, to every germ $\mathbf{f} \in {}_V\mathcal{O}_O$ there corresponds a unique polynomial $Q_\mathbf{f} \in {}_n\mathcal{O}_O[X]$ of degree at most $v - 1$ such that $\mathbf{d} \cdot \mathbf{f} = Q_\mathbf{f}(\mathbf{k})$ in ${}_V\mathcal{O}_O$. The mapping $\mathbf{f} \to Q_\mathbf{f}$ thus defined can be viewed as a homomorphism $\sigma\colon {}_V\mathcal{O}_O \to {}_n\mathcal{O}_O^v$ of ${}_n\mathcal{O}_O$-modules, where ${}_V\mathcal{O}_O$ has the natural structure of an ${}_n\mathcal{O}_O$-module induced by the holomorphic mapping $\pi\colon \mathbf{V} \to \mathbb{C}^n$ and the subspace of ${}_n\mathcal{O}_O[X]$ consisting of polynomials of order at most $v - 1$ is identified with ${}_n\mathcal{O}_O^v$.

Now let $\mathcal{N}_O \subseteq {}_V\mathcal{O}_O$ be the set of restrictions to $\mathbf{V}$ of those holomorphic functions on $\mathbf{V} \cup \mathbf{W}$ vanishing on $\mathbf{W}$, viewed as an ${}_n\mathcal{O}_O$-module, and consider the image submodule $\mathcal{M}_O = \sigma(\mathcal{N}_O) \in {}_n\mathcal{O}_O^v$. The latter submodule is of course finitely generated, so choose some germs $\mathbf{G}_j \in {}_m\mathcal{O}_O$ vanishing on $\mathbf{W}$ so that the images $Q_{\mathbf{g}_j} = \sigma(\mathbf{g}_j)$ of the restrictions $\mathbf{g}_j = \mathbf{G}_j|V$ generate the submodule $\mathcal{M}_O$. After shrinking the neighborhood $U$ even further if necessary, it can be supposed that $\mathbf{G}_j$ are the germs of some holomorphic functions $G_j \in {}_m\mathcal{O}_U$. Correspondingly the images $Q_{\mathbf{g}_j}$ will be the germs of the polynomials $Q_{g_j}$ associated to the functions $g_j = G_j|V$. Theorem 8 can then be applied to the ${}_n\mathcal{O}_O$-module $\mathcal{M}_O$ and the functions $Q_{g_j}$ representing generators of that module. After a nonsingular linear change of coordinates in $\mathbb{C}^n$ there exist an open polydisc $\Delta' = \Delta(0; R) \subseteq \mathbb{C}^n$ with closure $K' = \bar{\Delta}' \subseteq U'$ and a constant $M' > 0$ to which the conclusions of Theorem 8 apply. The polydisc $\Delta'$ can be shrunk further if necessary to ensure that there will exist a polydisc $\Delta'' = \Delta''(0; R'') \subseteq \mathbb{C}^{m-n}$ with closure $K'' \subseteq U''$ such that $V \cap (\Delta' \times U'') \subseteq \Delta' \times \Delta''$, and the polydisc $\Delta = \Delta' \times \Delta''$ will then have the desired properties. To verify that, consider any germ $\mathbf{f} \in {}_{V \cup W}\mathcal{O}_{K \cap (V \cup W)}$ for which $\mathbf{f}|K \cap W = 0$, and note that much as before there will be a polynomial $Q_\mathbf{f} \in {}_n\mathcal{O}_{K'}[X]$ of degree at most $v - 1$ such that $\mathbf{d} \cdot \mathbf{f} = Q_\mathbf{f}(\mathbf{k})$ in ${}_V\mathcal{O}_{K \cap V}$. The germ $Q_\mathbf{f} \in {}_n\mathcal{O}_{K'}^v$ is then an element of the induced module $\mathcal{M}_{K'}$, so by Theorem 8 there must exist germs $\mathbf{h}_j \in {}_n\mathcal{O}_{K'}$ such that $Q_\mathbf{f} = \sum \mathbf{h}_j Q_{\mathbf{g}_j}$ and $\|\mathbf{h}_j\|_{K'} \leq M' \|Q_\mathbf{f}\|_{K'}$. Then however as germs in ${}_V\mathcal{O}_K$ it is clear that $\mathbf{d} \cdot \mathbf{f} = Q_\mathbf{f}(\mathbf{k}) = \sum \mathbf{h}_j Q_{\mathbf{g}_j}(\mathbf{k}) = \mathbf{d} \cdot \sum \mathbf{h}_j \mathbf{g}_j$, so that $\mathbf{f} = \sum \mathbf{h}_j \mathbf{g}_j$. The germ $\mathbf{F} = \sum \mathbf{h}_j \mathbf{G}_j \in {}_m\mathcal{O}_K$ also has the properties that $\mathbf{F}|K \cap W = 0$ and $\mathbf{F}|K \cap V = \mathbf{f}|K \cap V$, so that $\mathbf{F}|K \cap (V \cup W) = \mathbf{f}$, and moreover $\|\mathbf{F}\|_K \leq \sum \|\mathbf{h}_j\|_{K'} \cdot \|\mathbf{G}_j\|_K \leq \sum M' \|Q_\mathbf{f}\|_{K'} \cdot \|\mathbf{G}_j\|_K$. Recall that the coefficients of the polynomials $Q_\mathbf{f}$ have values that are various symmetric functions of the values of the function $\mathbf{f}$ at various points of $V$. It is clear that $\|Q_f\|_{K'} \leq M'' \|\mathbf{f}\|_{K \cap (V \cup W)}$ for some constant $M'' > 0$ independent of f. Then set $M = M'M'' \sum \|\mathbf{G}_j\|_K$, and it follows that $\|\mathbf{F}\|_K \leq M \|\mathbf{f}\|_{K \cap (V \cup W)}$ as well, which suffices to conclude the proof.

**14. THEOREM.** *Let $V$ be a holomorphic subvariety of an open neighborhood $U$ of the origin in $\mathbb{C}^m$. Then there are open neighborhoods $U_1$, $U_2$ of the origin in $\mathbb{C}^m$, with closures $K_1$, $K_2$ for which $K_1 \subset K_2 \subset U$, and a constant $M > 0$ such that to each germ $\mathbf{f} \in {}_V\mathcal{O}_{K_2 \cap V}$ there corresponds a germ $\mathbf{F} \in {}_m\mathcal{O}_{K_1}$ for which $\mathbf{F}|K_1 \cap V = \mathbf{f}|K_1 \cap V$ and $\|\mathbf{F}\|_{K_1} \leq$*

$M\|\mathbf{f}\|_{K_1\cap V}$. (*The neighborhoods* $U_1$ and $U_2$ *can be taken as polydiscs in some linear coordinates in* $\mathbb{C}^m$, *though not necessarily in the same coordinates.*)

*Proof.* The proof will be by induction on the dimension of the subvariety $V$. The theorem holds quite trivially when $\dim V = 0$, so it is only necessary to demonstrate the inductive step. Assume therefore that the theorem holds for all subvarieties in $\mathbb{C}^m$ of dimensions strictly less than $n$, and consider a subvariety $V \subseteq U$ with $\dim V = n$. Write $V = V_0 \cup W$ where $V_0$ is of pure dimension $n$ and $\dim W < n$. By the inductive hypothesis on the subvariety $W$, there are open neighborhoods $U_1'$ and $U_2$ of the origin in $\mathbb{C}^m$ with closures $K_1' \subset K_2 \subset U$ and a constant $M' > 0$ such that for any germ $\mathbf{f} \in {}_V\mathcal{O}_{K_2\cap V}$ there corresponds to the restriction $\mathbf{f}|K_2 \cap W \in {}_W\mathcal{O}_{K_2\cap W}$ a germ $\mathbf{F}' \in {}_m\mathcal{O}_{K_1'}$ for which $F'|K_1' \cap W = f|K_1' \cap W$ and $\|\mathbf{F}'\|_{K_1} \leqq M'\|\mathbf{f}\|_{K_1\cap W}$. The germ $\mathbf{f}' = \mathbf{f}|K_1' \cap V - \mathbf{F}'|K_1' \cap V \in {}_V\mathcal{O}_{K_1'\cap V}$ vanishes on $K_1' \cap W$, so it follows from Lemma 13 that there exist an open neighborhood $U_1$ of the origin with closure $K_1 \subset U_1'$ and a constant $M'' > 0$ such that to every germ such as $\mathbf{f}'|K_1 \cap V \in {}_V\mathcal{O}_{K_1\cap V}$ there corresponds a germ $\mathbf{F}'' \in {}_m\mathcal{O}_{K_1}$ for which $\mathbf{F}''|K_1 \cap V = \mathbf{f}'|K_1 \cap V$ and $\|\mathbf{F}''\|_{K_1} \leqq M''\|\mathbf{f}'\|_{K_1\cap V}$. The germ $\mathbf{F} = \mathbf{F}'|K_1 + \mathbf{F}''$ then evidently has the properties that $\mathbf{F}|K_1 \cap V = \mathbf{f}|K_1 \cap V$ and $\|\mathbf{F}\|_{K_1} \leqq \|\mathbf{F}'\|_{K_1} + \|\mathbf{F}''\|_{K_1} \leqq M'\|\mathbf{f}\|_{K_1\cap W} + M''\|\mathbf{f}'\|_{K_1\cap V}$, while $\|f'\|_{K_1\cap V} \leqq \|\mathbf{f}\|_{K_1\cap V} + \|\mathbf{F}'|V\|_{K_1\cap V} \leqq \|\mathbf{f}\|_{K_1\cap V} + \|\mathbf{F}'\|_{K_1} \leqq (1 + M')\|\mathbf{f}\|_{K_1\cap V}$ so that altogether $\|\mathbf{F}\|_{K_1} \leqq M\|\mathbf{f}\|_{K_1\cap V}$ where $M = M' + M'' + M'M'' > 0$. In both the induction step and the application of Lemma 13 the new neighborhoods constructed can be taken to be open polydiscs in various linear coordinate systems in $\mathbb{C}^n$. That then suffices to complete the induction and thereby to conclude the proof of the theorem.

The preceding result is both of interest in itself and of considerable use in deriving other results; here it will be used just for the following basic property.

**15. THEOREM.** *The space of holomorphic functions on any holomorphic variety is a Frechet space in the natural topology.*

*Proof.* In view of the discussion earlier in this section of the natural topology on the space ${}_V\mathcal{O}_V$ of holomorphic functions on a holomorphic variety $V$, it is only necessary to demonstrate the completeness of ${}_V\mathcal{O}_V$; that follows quite easily from the preceding theorem. Indeed suppose that $\{f_n\}$ is a Cauchy sequence of holomorphic functions on $V$. That sequence converges uniformly on compact subsets of $V$, so the limit function is a well-defined continuous function $f$ on $V$. It is only necessary to show that $f$ is holomorphic on $V$. For that purpose consider an arbitrary point $O \in V$ and choose an open neighborhood $V_0$ of $O$ in $V$ so that $V_O$ can be represented by a holomorphic subvariety of an open neighborhood $U$ of $O$ in some space $\mathbb{C}^m$. Theorem 14 can be applied to this subvariety $V_O$; thus, there are open neighborhoods $U_1$ and $U_2$ of $O$ with closures $K_1 \subset K_2 \subset U$ and a constant $M > O$ so that any germ $\mathbf{g} \in {}_m\mathcal{O}_{K_2\cap V}$ extends to a germ $\mathbf{G} \in {}_m\mathcal{O}_{K_1}$ for which $\|\mathbf{G}\|_{K_1} \leqq M \cdot \|\mathbf{g}\|_{K_1\cap V}$. Choose a subsequence $\{f_{n_\nu}\}$ so that the germs $\mathbf{g}_\nu = (f_{n_\nu} - f_{n_{\nu-1}})|K_1 \cap V$ satisfy $\|\mathbf{g}_\nu\|_{K_1\cap V} < 2^{-\nu}$. The germ $\mathbf{f}_{n_1} = f_{n_1}|K_1 \cap V$ itself then extends to some germ $\mathbf{F}_{n_1} \in {}_m\mathcal{O}_{K_1}$ for which the size estimate will not be needed, but the germs $\mathbf{g}_\nu$ extend to germs $\mathbf{G}_\nu \in {}_m\mathcal{O}_{K_1}$ for which $\|G_\nu\|_{K_1} \leq M2^{-\nu}$. The germs

$$\mathbf{F}_{n_\nu} = \mathbf{F}_{n_1} + \mathbf{G}_2 + \cdots + \mathbf{G}_\nu \in {}_m\mathcal{O}_{K_1}$$

then converge uniformly on $U_1$ to a holomorphic function $F \in {}_m\mathcal{O}_{U_1}$, and since $\mathbf{F}_{n_\nu}|K_1 \cap V = \mathbf{f}_{n_\nu}$, it follows that $F|U_1 \cap V = f|U_1 \cap V$. Thus, $f$ is holomorphic on $V$ as desired, and the proof is thereby concluded.

# I

# Tangent Spaces

Any germ **V** of a holomorphic variety can be represented by a germ of a holomorphic subvariety at the origin in some space $\mathbb{C}^m$. From Theorem G3 it is clear that $m \geqq \dim \mathbf{V}$, and moreover that it is possible to take $m = \dim \mathbf{V}$ only when **V** is the germ of a regular variety. The minimal value of $m$ can be considered as another numerical invariant attached to the germ **V**, supplementing the basic invariant considered in the preceding section, the dimension of **V**, and measuring to some extent just how bad the singularity of **V** is. It is clearly a nontrivial invariant, at least to the extent of describing whether **V** is regular or singular, and indeed has a variety of uses. Just as in the case of dimension, it is of some interest to describe this new invariant directly in terms of the local ring ${}_{\mathbf{V}}\mathcal{O}$. That has the evident advantage of permitting an analysis of this invariant without requiring the examination of all possible ways of representing **V** as a holomorphic subvariety of $\mathbb{C}^m$. Moreover it leads to an interesting interpretation of the space $\mathbb{C}^m$ for this minimal value of $m$ as the analogue for general holomorphic varieties of the tangent space of a complex manifold as introduced in section I-C. With this as motivation, consider the following.

1. **DEFINITION.** *If* **V** *is the germ of a holomorphic variety with local ring* ${}_{\mathbf{V}}\mathcal{O}$, *a* **tangent vector** *to* **V** *is a linear mapping* $T: {}_{\mathbf{V}}\mathcal{O} \rightarrow \mathbb{C}$ *such that* $T(\mathbf{fg}) = \mathbf{f}(0)T(\mathbf{g}) + \mathbf{g}(0)T(\mathbf{f})$ *for any two germs* $\mathbf{f}, \mathbf{g} \in {}_{\mathbf{V}}\mathcal{O}$. *The set of all tangent vectors to* **V** *form a complex vector space called the* **tangent space** *to* **V** *and denoted by* $T(\mathbf{V})$. *The dimension of this vector space is called the* **tangential dimension** *of* **V** *and is denoted by* tdim **V**.

In this definition the complex numbers $\mathbf{f}(0)$, $\mathbf{g}(0)$ are the values of the germs **f**, **g** at the base point of the germ **V**, that point being conveniently denoted by $O$ unless otherwise specified. As a natural extension of this definition a **tangent vector** to a holomorphic variety $V$ at a point $A \in V$ is defined simply as a tangent vector to the germ $\mathbf{V}_A$ of the variety $V$ at the point $A$. The vector space consisting of all such tangent vectors is called the **tangent space** to the variety $V$ at the point $A$ and is denoted by either $T(\mathbf{V}_A)$ or $T_A(V)$. Correspondingly the dimension of this tangent space is called the **tangential dimension** of the variety $V$ at the point $A$ and is denoted by either tdim $\mathbf{V}_A$ or $\text{tdim}_A V$.

If $\mathbf{V}$ and $\mathbf{W}$ are two germs of holomorphic varieties and $\mathbf{F}: \mathbf{V} \to \mathbf{W}$ is a holomorphic mapping between them, then $\mathbf{F}$ induces a homomorphism of algebras $\mathbf{F}^*: {}_{\mathbf{W}}\mathcal{O} \to {}_{\mathbf{V}}\mathcal{O}$. For any tangent vector $T \in T(\mathbf{V})$, the composition $T \circ \mathbf{F}^*: {}_{\mathbf{W}}\mathcal{O} \to \mathbb{C}$ is clearly a linear mapping, and for any two germs $\mathbf{f}, \mathbf{g} \in {}_{\mathbf{W}}\mathcal{O}$,

$$(T \circ \mathbf{F}^*)(\mathbf{fg}) = T(\mathbf{F}^*(f) \cdot \mathbf{F}^*(g)) = \mathbf{f}(0) \cdot (T \circ \mathbf{F}^*(g)) + \mathbf{g}(0) \cdot (T \circ \mathbf{F}^*(f))$$

Thus $T \circ \mathbf{F}^*$ is a tangent vector to $\mathbf{W}$. The mapping that associates to any tangent vector $T \in T(\mathbf{V})$ the tangent vector $T \circ \mathbf{F}^* \in T(\mathbf{W})$ is called the **differential** of the holomorphic mapping $\mathbf{F}$ and is denoted by $d\mathbf{F}$. It is evident that $d\mathbf{F}: T(\mathbf{V}) \to T(\mathbf{W})$ is a linear mapping between these two tangent spaces. If $\mathbf{F}_1: \mathbf{V}_1 \to \mathbf{V}_2$ and $\mathbf{F}_2: \mathbf{V}_2 \to \mathbf{V}_3$ are any two holomorphic mappings, then for any tangent vector $T \in \mathrm{T}(\mathbf{V}_1)$ it follows that $d(\mathbf{F}_2 \circ \mathbf{F}_1)(T) = T \circ (\mathbf{F}_2 \circ \mathbf{F}_1)^* = T \circ \mathbf{F}_1^* \circ \mathbf{F}_2^* = d\mathbf{F}_2(d\mathbf{F}_1(T))$, so that $d(\mathbf{F}_2 \circ \mathbf{F}_1) = d\mathbf{F}_2 \circ d\mathbf{F}_1$. If $\mathbf{F}$ is the identity mapping, then $d\mathbf{F}$ is obviously also the identity mapping. It follows from this and the immediately preceding observation that $d\mathbf{F}$ is a linear isomorphism whenever $\mathbf{F}$ is a biholomorphic mapping.

It should be noted that for a regular holomorphic variety the definition of the tangent space introduced here coincides with the definition of the complex tangent space to a complex manifold as introduced in section I-C, a tangent vector at a point $A$ in a complex manifold evidently amounting to the same thing as a derivation at the point $A$. That leads to a useful explicit representation of the tangent space of a general holomorphic variety in terms of the following explicit representation of the complex tangent space to a complex manifold.

**2. THEOREM.** *For the complex manifold $\mathbb{C}^n$ the mapping*

$$T \in T_O(\mathbb{C}^n) \to \{T(\mathbf{z}_j) : 1 \leq j \leq n\} = \{T(\mathbf{z}_1), \ldots, T(\mathbf{z}_n)\} \in \mathbb{C}^n$$

*establishes a canonical isomorphism $T_O(\mathbb{C}^n) \simeq \mathbb{C}^n$, where $\mathbf{z}_j \in {}_n\mathcal{O}_O$ are the germs at the origin of the coordinate functions in $\mathbb{C}^n$.*

*Proof.* As in formula I-C(6) any tangent vector $T \in T_O(\mathbb{C}^n)$ is a linear mapping $T: {}_n\mathcal{O}_O \to \mathbb{C}$ of the form

$$T(\mathbf{f}) = \sum_{j=1}^{n} t_j \frac{\partial f}{\partial z_j}(0) \tag{1}$$

where $f$ is a holomorphic function in an open neighborhood of $O$ representing the germ $\mathbf{f} \in {}_n\mathcal{O}_O$ and $t_j = T(\mathbf{z}_j)$. Conversely for any complex constants $t_j$ the linear mapping $T: {}_n\mathcal{O}_O \to \mathbb{C}$ defined by (1) is a derivation for which $t_j = T(\mathbf{z}_j)$, and that suffices for the proof.

**3. THEOREM.** *If $\mathbf{V}$ is the germ of a holomorphic subvariety at the origin in $\mathbb{C}^m$ and $\mathbf{F}: \mathbf{V} \to \mathbb{C}^m$ is the inclusion mapping, then $d\mathbf{F}: T(\mathbf{V}) \to T_O(\mathbb{C}^m)$ is injective and its image is the linear subspace*

$$d\mathbf{F}(T(\mathbf{V})) = \{T \in T_O(\mathbb{C}^m) : T(\mathbf{g}) = 0 \text{ whenever } \mathbf{g} \in \operatorname{id} \mathbf{V} \subseteq {}_m\mathcal{O}_O\}$$

*Proof.* First note that whenever $T \in T(\mathbf{V})$ and $\mathbf{g} \in {}_m\mathcal{O}_O$, then $d\mathbf{F}(T)(\mathbf{g}) = T \circ \mathbf{F}^*(\mathbf{g}) = T(\mathbf{g} \circ \mathbf{F}) = T(\mathbf{g}|\mathbf{V})$, since the composition $\mathbf{g} \circ \mathbf{F}$ is really just the restriction of $\mathbf{g}$ to the subvariety $\mathbf{V}$. If $d\mathbf{F}(T) = 0$, then $T(\mathbf{g}|\mathbf{V}) = 0$ for all $\mathbf{g} \in {}_m\mathcal{O}_O$, and since ${}_{\mathbf{V}}\mathcal{O} = \{g|\mathbf{V} : \mathbf{g} \in {}_m\mathcal{O}_O\}$, it follows that $T = 0$; thus, the linear mapping $d\mathbf{F}$ is injective. Moreover, if $T \in T(\mathbf{V})$ and $\mathbf{g} \in \text{id}\,\mathbf{V} \subseteq {}_m\mathcal{O}_O$, then $d\mathbf{F}(T)(\mathbf{g}) = T(\mathbf{g}|\mathbf{V}) = 0$. On the other hand, whenever $T' \in T_0(\mathbf{C}^m)$ has the property that $T'(\mathbf{g}) = 0$ for all $\mathbf{g} \in \text{id}\,\mathbf{V} \subseteq {}_m\mathcal{O}_O$, then since ${}_{\mathbf{V}}\mathcal{O} \simeq {}_m\mathcal{O}_O|\text{id}\,\mathbf{V}$, it is clear that $T'$ induces a linear mapping $T\colon {}_{\mathbf{V}}\mathcal{O} \to \mathbf{C}$, and it is an utterly straightforward matter to verify that this induced mapping is actually a tangent vector $T \in T(\mathbf{V})$ for which $d\mathbf{F}(T) = T'$. That suffices to conclude the proof.

In some considerations it is convenient to identify the tangent space $T_0(\mathbf{C}^m)$ with the vector space $\mathbf{C}^m$ under the natural isomorphism provided by Theorem 2. In most considerations it is also convenient to identify the tangent space $T(\mathbf{V})$ of a germ of a holomorphic subvariety $\mathbf{V}$ at the origin in $\mathbf{C}^m$ with the linear subspace $d\mathbf{F}(T(\mathbf{V})) \subseteq T_0(\mathbf{C}^m)$, where $\mathbf{F}\colon \mathbf{V} \to \mathbf{C}^m$ is the inclusion mapping. That is justified by Theorem 3, and amounts merely to the standard practice of dropping explicit references to natural inclusion mappings. The combination of these two identifications provides the following useful explicit description of the tangent space to a germ of a holomorphic subvariety in $\mathbf{C}^m$.

**4. THEOREM.** *If* $\mathbf{V}$ *is the germ of a holomorphic subvariety at the origin in* $\mathbf{C}^m$, *then under the natural identifications*

$$T(\mathbf{V}) \simeq \left\{ T \in \mathbf{C}^m : \sum_{j=1}^{m} t_j \frac{\partial g}{\partial z_j}(0) = 0 \text{ for all } g \in \text{id}\,\mathbf{V} \subseteq {}_m\mathcal{O}_O \right\}$$

*Proof.* Under the identification described in Theorem 2, $T_0(\mathbf{C}^m) \simeq \mathbf{C}^m$ where a tangent vector $T \in T_0(\mathbf{C}^m)$ is identified with the vector $T = (t_1, \ldots, t_m) \in \mathbf{C}^m$ where $t_j = T(\mathbf{z}_j)$. As in (1) the tangent vector in $T_0(\mathbf{C}^m)$ corresponding to a vector $T = (t_1, \ldots, t_m) \in \mathbf{C}^m$ is the linear mapping $T\colon {}_m\mathcal{O}_O \to \mathbf{C}$ defined by $T(\mathbf{f}) = \sum_{j=1}^m t_j \partial f/\partial z_j(O)$. Under the further identification described in Theorem 3 the linear subspace $T(\mathbf{V}) \subseteq T_0(\mathbf{C}^m)$ corresponds to those tangent vectors $T \in T_0(\mathbf{C}^m)$ such that $T(\mathbf{g}) = 0$ for all $\mathbf{g} \in \text{id}\,\mathbf{V} \subseteq {}_m\mathcal{O}_O$, and the desired result follows immediately.

In the preceding theorem it is evidently not really necessary to consider all elements $\mathbf{g} \in \text{id}\,\mathbf{V}$, but only a basis for the ideal $\text{id}\,\mathbf{V} \subseteq {}_m\mathcal{O}_O$. That is to say, if $\mathbf{g}_1, \ldots, \mathbf{g}_r$ form a basis for the ideal $\text{id}\,\mathbf{V}$, then under the same identifications as before,

$$T(\mathbf{V}) \simeq \left\{ T \in \mathbf{C}^m : \sum_{j=1}^{m} t_j \frac{\partial g_k}{\partial z_j}(0) = 0 \text{ for } 1 \leq k \leq r \right\} \tag{2}$$

Thus in terms of the matrix

$$\frac{\partial(g_1, \ldots, g_r)}{\partial(z_1, \ldots, z_m)} = \left\{ \frac{\partial g_i}{\partial z_j} : 1 \leq i \leq r, i \leq j \leq m \right\}$$

an immediate consequence of the preceding theorem is the following observation.

**5. COROLLARY.** *If* $\mathbf{V}$ *is the germ of a holomorphic subvariety at the origin in* $\mathbb{C}^m$ *and* $\mathbf{g}_i \in {}_m\mathcal{O}_O$ *generate the ideal* $\mathrm{id}\,\mathbf{V} \subseteq {}_m\mathcal{O}_O$, *then*

$$\mathrm{tdim}\,\mathbf{V} = m - \mathrm{rank}\frac{\partial(g_1,\ldots,g_r)}{\partial(z_1,\ldots,z_m)}(0)$$

*Proof.* Theorem 4 can be restated as the assertion that

$$T(\mathbf{V}) \simeq \left\{T \in \mathbb{C}^m : \frac{\partial(g_1,\ldots,g_r)}{\partial(z_1,\ldots,z_m)}(0)\cdot T = 0\right\}$$

hence $\mathrm{tdim}\,\mathbf{V} = \dim T(\mathbf{V})$ obviously has the desired form.

**6. THEOREM.** *A germ* $\mathbf{V}$ *of a holomorphic variety can be represented by a germ of a holomorphic subvariety of* $\mathbb{C}^m$ *if and only if* $m \geq \mathrm{tdim}\,\mathbf{V}$.

*Proof.* If $\mathbf{V}$ can be represented by the germ of a holomorphic subvariety $V$ at the origin in $\mathbb{C}^m$, then it follows from Corollary 5 that $\mathrm{tdim}\,\mathbf{V} \leq m$. Moreover if $n = \mathrm{tdim}\,\mathbf{V} < m$, then it further follows from Corollary 5 that

$$\mathrm{rank}\frac{\partial(g_1,\ldots,g_r)}{\partial(z_1,\ldots,z_m)} = m - n$$

where the germs $\mathbf{g}_j \in {}_m\mathcal{O}_O$ generate $\mathrm{id}\,\mathbf{V}$. Some $m - n$ of the functions $g_j$ can thus be taken as part of a set of coordinate functions at the origin in $\mathbb{C}^m$, so after a biholomorphic change of coordinates in an open neighborhood of the origin in $\mathbb{C}^m$, it can be assumed that $\mathbf{z}_{n+1},\ldots,\mathbf{z}_m \in \mathrm{id}\,\mathbf{V} \subseteq {}_m\mathcal{O}_O$. But then $\mathbf{V} \subseteq \mathbb{C}^n$ where $\mathbb{C}^n \subseteq \mathbb{C}^m$ is the linear subspace spanned by the first $n$ coordinate axes, and that clearly suffices to conclude the proof.

The tangential dimension of a germ $\mathbf{V}$ of a holomorphic variety can thus be characterized as the least integer $m$ such that $\mathbf{V}$ can be represented by a germ of a holomorphic subvariety in $\mathbb{C}^m$. For this reason the tangential dimension of $\mathbf{V}$ is often called the **imbedding dimension** of $\mathbf{V}$. Imbeddings of a germ $\mathbf{V}$ in $\mathbb{C}^m$ with this smallest value of $m$ have sufficiently many special properties to warrant further discussion, for which it is convenient to introduce the following.

**7. DEFINITION.** *A germ* $\mathbf{V}$ *of a holomorphic subvariety at a point of* $\mathbb{C}^m$ *is said to be* **neatly imbedded** *if* $m = \mathrm{tdim}\,\mathbf{V}$.

Any germ of a holomorphic variety can of course always be represented by a neatly imbedded germ of a holomorphic subvariety at the origin in a space $\mathbb{C}^m$ of the appropriate dimension. When $\mathbf{V}$ is the germ of a holomorphic subvariety at the origin in $\mathbb{C}^m$, the tangent space $T(\mathbf{V})$ has been identified with a linear subspace $T(\mathbf{V}) \subseteq T_0(\mathbb{C}^m)$. The condition that $\mathbf{V}$ be neatly imbedded merely amounts to the condition that $T(\mathbf{V}) = T_0(\mathbb{C}^m)$. It is clear from Corollary 5 that *a germ of a holomor-*

*phic subvariety* **V** *at the origin in* $\mathbf{C}^m$ *is neatly imbedded if and only if* $\partial g_i/\partial z_j(0) = 0$, *where* $g_i$ *are any holomorphic functions in a neighborhood of the origin such that the germs* $\mathbf{g}_i \in {}_m\mathcal{O}_O$ *generate* id **V**. Equivalently, of course, *a germ of a holomorphic subvariety* **V** *at the origin in* $\mathbf{C}^m$ *is neatly imbedded if and only if* $\partial g/\partial z_j(0) = 0$ *for any representative* $g$ *of any germ* $\mathbf{g} \in$ id **V** *and any index* $j$. The significance of this notion will become more apparent after an analysis of the form that the differential of a holomorphic mapping takes under the various identifications that have been introduced here.

8. THEOREM. *If* $\mathbf{F}: \mathbf{C}^m \to \mathbf{C}^n$ *is the germ of a holomorphic mapping taking the origin to the origin, then under the natural identifications* $T_0(\mathbf{C}^m) \simeq \mathbf{C}^m$, $T_0(\mathbf{C}^n) \simeq \mathbf{C}^n$, *the differential* $d\mathbf{F}: \mathbf{C}^m \to \mathbf{C}^n$ *is the linear mapping described by the Jacobian matrix* $J_F(0)$.

*Proof.* Let $F$ be a holomorphic mapping from an open neighborhood of the origin in $\mathbf{C}^m$ into an open neighborhood of the origin in $\mathbf{C}^n$, representing the germ **F**, and write $F(Z) = (w_1(Z), \ldots, w_n(Z))$ in terms of its coordinate functions $w_j(Z)$. If $T = (t_1, \ldots, t_m) \in \mathbf{C}^m \simeq T_0(\mathbf{C}^m)$ and $S = (s_1, \ldots, s_n) = d\mathbf{F}(T) \in \mathbf{C}^n \simeq T_0(\mathbf{C}^n)$, then the tangent vector described by $T$ is represented by the linear differential operator $T(f) = \sum_j t_j \partial f/\partial z_j(0)$ and

$$s_i = d\mathbf{F}(T)(\mathbf{w}_i) = T \circ \mathbf{F}^*(\mathbf{w}_i) = T(\mathbf{w}_i(Z))$$

$$= \sum_j t_j \frac{\partial w_i}{\partial z_j}(0)$$

Thus $S = J_F(0) . T$ as desired, and the proof is thereby concluded.

Note that if **V** and **W** are germs of holomorphic subvarieties at the origins in the respective spaces $\mathbf{C}^m$ and $\mathbf{C}^n$, then any germ of a holomorphic mapping $\mathbf{G}: \mathbf{V} \to \mathbf{W}$ extends to a germ of a holomorphic mapping $\mathbf{F}: \mathbf{C}^m \to \mathbf{C}^n$. Indeed the coordinate functions $w_j$ in $\mathbf{C}^n$ restrict to holomorphic germs $\mathbf{w}_j | \mathbf{W} \in {}_{\mathbf{W}}\mathcal{O}$, and since $\mathbf{G}^*(\mathbf{w}_j|\mathbf{W}) \in {}_{\mathbf{V}}\mathcal{O}$, there will be some germs $\mathbf{f}_j \in {}_m\mathcal{O}_O$ such that $\mathbf{f}_j|\mathbf{V} = \mathbf{G}^*(\mathbf{w}_j|\mathbf{W})$. These germs $\mathbf{f}_j$ are then the coordinate functions of a germ of a holomorphic mapping $\mathbf{F}: \mathbf{C}^m \to \mathbf{C}^n$ such that $\mathbf{F}|\mathbf{V} = \mathbf{G}$ as desired. Of course, the mapping **F** is not uniquely determined, but nonetheless it does suffice to describe the differential $d\mathbf{G}$ as follows.

9. THEOREM. *If* **V** *and* **W** *are germs of holomorphic subvarieties at the origin in the respective spaces* $\mathbf{C}^m$ *and* $\mathbf{C}^n$, *if* $\mathbf{G}: \mathbf{V} \to \mathbf{W}$ *is any germ of a holomorphic mapping between them, and if* $\mathbf{F}: \mathbf{C}^m \to \mathbf{C}^n$ *is any extension of* **G**, *then under the natural identifications* $T(\mathbf{V}) \subseteq \mathbf{C}^m \simeq T_0(\mathbf{C}^m)$ *and* $T(\mathbf{W}) \subseteq \mathbf{C}^n \simeq T_0(\mathbf{C}^n)$, *the differential* $d\mathbf{G}: T(\mathbf{V}) \to T(\mathbf{W})$ *coincides with the restriction* $J_F(0)|T(\mathbf{V})$.

*Proof.* If $\mathbf{H_V}: \mathbf{V} \to \mathbf{C}^m$ and $\mathbf{H_W}: \mathbf{W} \to \mathbf{C}^n$ are the inclusion mappings, the condition that $\mathbf{F}: \mathbf{C}^m \to \mathbf{C}^n$ be an extension of $\mathbf{G}: \mathbf{V} \to \mathbf{W}$ merely means that $\mathbf{F} \circ \mathbf{H_V} = \mathbf{H_W} \circ \mathbf{G}$ and implies therefore that $d\mathbf{F} \circ d\mathbf{H_V} = d\mathbf{H_W} \circ d\mathbf{G}$. But the natural identifications considered here first involve identifying $T(\mathbf{V})$ with its image $d\mathbf{H_V}(T(\mathbf{V})) \subseteq T_0(\mathbf{C}^m)$ and correspondingly identifying $T(\mathbf{W})$ with its image $d\mathbf{H_W}(T(\mathbf{W})) \subseteq T_0(\mathbf{C}^n)$, and therefore with these identifications $d\mathbf{G}\colon T(\mathbf{V}) \to T(\mathbf{W})$ is just the restriction of the differential $d\mathbf{F}\colon T_0(\mathbb{C}^m) \to T_0(\mathbb{C}^n)$ to the subspace $T(\mathbf{V}) \simeq d\mathbf{H}_{\mathbf{V}}(T(\mathbf{V})) \subseteq T_0(\mathbb{C}^m)$. The second identifications $T_0(\mathbb{C}^m) \simeq \mathbb{C}^m$ and $T_0(\mathbb{C}^n) \simeq \mathbb{C}^n$ lead to the representation of $d\mathbf{F}$ as the linear mapping $J_{\mathbf{F}}(0)\colon \mathbb{C}^m \to \mathbb{C}^n$ as in Theorem 8, and the combination of these observations yields the desired result and thereby concludes the proof.

If in the preceding theorem the subvariety $\mathbf{V} \subseteq \mathbb{C}^m$ is neatly imbedded, then $T(\mathbf{V}) = T_0(\mathbb{C}^m)$ and consequently the differential $d\mathbf{G}\colon T(\mathbf{V}) \to T(\mathbf{W})$ is precisely the linear mapping described by the Jacobian $J_{\mathbf{F}}(0)$ of any extension $\mathbf{F}\colon \mathbb{C}^m \to \mathbb{C}^n$ of the mapping $\mathbf{G}$. In particular the Jacobian matrix $J_{\mathbf{F}}(0)$, or equivalently in view of Theorem 8 the differential $d\mathbf{F}$, is independent of the choice of extension $\mathbf{F}$ of the mapping $\mathbf{G}$. Therein lies the significance of neat imbeddings, as will be illustrated by the next observations.

**10. THEOREM.** *Two neatly imbedded germs of holomorphic subvarieties are equivalent as germs of holomorphic varieties if and only if they are equivalent as germs of holomorphic subvarieties.*

*Proof.* Suppose that $\mathbf{V}$ and $\mathbf{W}$ are germs of holomorphic subvarieties at the origin in the respective spaces $\mathbb{C}^m$ and $\mathbb{C}^n$. If they are equivalent as germs of holomorphic subvarieties, then they are obviously equivalent as germs of holomorphic varieties. On the other hand, if they are equivalent as germs of holomorphic varieties, then there is a biholomorphic mapping $\mathbf{G}\colon \mathbf{V} \to \mathbf{W}$. If $\mathbf{F}\colon \mathbb{C}^m \to \mathbb{C}^n$ is a holomorphic extension of $\mathbf{G}$ and $\mathbf{H}\colon \mathbb{C}^n \to \mathbb{C}^m$ is a holomorphic extension of $\mathbf{G}^{-1}$, the composition $\mathbf{H} \circ \mathbf{F}\colon \mathbb{C}^m \to \mathbb{C}^m$ restricts to the identity mapping $\mathbf{H} \circ \mathbf{F}|\mathbf{V}\colon \mathbf{V} \to \mathbf{V}$. The differential of this restriction is then nonsingular, and if $\mathbf{V}$ is neatly imbedded it follows immediately from Theorem 9 that $J_{\mathbf{H}\circ\mathbf{F}}(0) = J_{\mathbf{H}}(0)\cdot J_{\mathbf{F}}(0)$ is also a nonsingular mapping. Similarly if $\mathbf{W}$ is neatly imbedded, then $J_{\mathbf{F}\circ\mathbf{H}}(0) = J_{\mathbf{F}}(0)\cdot J_{\mathbf{H}}(0)$ is also nonsingular, so that $J_{\mathbf{F}}(0)$ itself is necessarily nonsingular. But that means that $\mathbf{F}\colon \mathbb{C}^m \to \mathbb{C}^n$ is the germ of a nonsingular holomorphic mapping, and since $\mathbf{F}(\mathbf{V}) = \mathbf{W}$, these germs are also equivalent as germs of holomorphic subvarieties. That suffices to conclude the proof.

This observation may help to clarify the relation between the notions of germs of holomorphic varieties and germs of holomorphic subvarieties. Any germ of a holomorphic variety is represented by a unique equivalence class of germs of neatly imbedded holomorphic subvarieties, but by a vast array of inequivalent germs of general holomorphic subvarieties. It is of course not reasonable to try to limit considerations solely to neatly imbedded germs of holomorphic subvarieties, since even though a holomorphic subvariety $V$ of an open subset $U \subseteq \mathbb{C}^m$ may represent a neatly imbedded subvariety at one point, it need not represent a neatly imbedded subvariety at any other point. An obvious example is that of a subvariety with a single singular point.

**11. THEOREM (inverse mapping theorem).** *If $\mathbf{G}\colon \mathbf{V} \to \mathbf{W}$ is a holomorphic mapping between germs of holomorphic varieties and if its differential $d\mathbf{G}\colon T(\mathbf{V}) \to T(\mathbf{W})$ is injective, then $\mathbf{G}$ is a biholomorphic mapping between $\mathbf{V}$ and its image $\mathbf{G}(\mathbf{V}) \subseteq \mathbf{W}$.*

*Proof.* It can be assumed that **V** and **W** are neatly imbedded germs of holomorphic subvarieties at the origins in the respective spaces $\mathbb{C}^m$ and $\mathbb{C}^n$, and that $\mathbf{G} = \mathbf{F}|\mathbf{V}$ where $\mathbf{F}: \mathbb{C}^m \to \mathbb{C}^n$ is the germ of a holomorphic mapping taking the origin to the origin. Since **V** and **W** are neatly imbedded, it follows from Theorem 9 that the differential $d\mathbf{G}: T(\mathbf{V}) \to T(\mathbf{W})$ is represented by the linear mapping $J_{\mathbf{F}}(0): \mathbb{C}^m \to \mathbb{C}^n$, so by the hypotheses of the present theorem the latter linear mapping is also injective; thus the Jacobian matrix $J_{\mathbf{F}}(0)$ is of rank $m$, so any mapping $F$ representing the germ **F** is nonsingular at the origin. It follows from Corollary C11 that the image under $F$ of an open neighborhood of the origin in $\mathbb{C}^m$ is a complex submanifold $M$ of an open neighborhood of the origin in $\mathbb{C}^n$, and an application of the ordinary inverse mapping theorem, Theorem C6, shows that $F: U \to M$ is a biholomorphic mapping if $U$ is sufficiently small. The restriction $\mathbf{G} = \mathbf{F}|\mathbf{V}$ is of course then a biholomorphic mapping between **V** and its image $\mathbf{G}(\mathbf{V}) \subseteq \mathbf{W}$, and the proof is thereby concluded.

While the preceding theorem shows that the *injectivity* of the differential $d\mathbf{G}: T(\mathbf{V}) \to T(\mathbf{W})$ has rather strong consequences, it should be noted that the *surjectivity* of the differential does not have any correspondingly strong consequences; in particular, it does not imply the surjectivity of the mapping $\mathbf{G}: \mathbf{V} \to \mathbf{W}$ itself. For example, whenever **V** is a neatly imbedded germ of a holomorphic subvariety at the origin in $\mathbb{C}^m$, the natural inclusion mapping $\mathbf{G}: \mathbf{V} \to \mathbb{C}^m$ has a surjective differential $d\mathbf{G}: T(\mathbf{V}) \to T_0(\mathbb{C}^m)$, but **G** is not surjective whenever **V** is a singular germ of a subvariety.

To describe a useful special case of the preceding theorem it is convenient to say that a holomorphic subvariety $V$ of a holomorphic variety $W$ is a **holomorphic retraction** of $W$ if there is a holomorphic mapping $R: W \to V$ such that $R|V: V \to V$ is the identity mapping; this is the holomorphic analogue of the familiar notion of a retraction between topological spaces. In case the ambient variety $W$ is regular, the condition that a subvariety $V \subseteq W$ be at least locally a holomorphic retraction of $W$ is easily described. On the one hand, it is clear that if $V$ is also regular, then it is always locally a holomorphic retraction of $W$; for an open neighborhood of any point $O \in V \subseteq W$ in $W$ can be represented by an open polydisc centered at $O$ in some space $\mathbb{C}^m$ in such a way that $V$ is the linear subspace $\mathbb{C}^n \subseteq \mathbb{C}^m$ in that neighborhood, and the natural projection mapping $\mathbb{C}^m \to \mathbb{C}^n$ provides the desired holomorphic retraction. On the other hand, this sufficient condition that $V$ be a holomorphic retraction of $W$ is necessary as well, as a consequence of the following result.

**12. COROLLARY.** *If a holomorphic subvariety $V$ of a complex manifold $W$ is a holomorphic retraction of $W$, then $V$ is necessarily a complex submanifold of $W$.*

*Proof.* Since the desired result is really purely local, it can be assumed that $W$ is an open neighborhood of the origin in some space $\mathbb{C}^m$, and by an appropriate choice of local coordinates near the origin in $\mathbb{C}^m$, it can further be assumed that the holomorphic subvariety $V \subseteq W$ is neatly imbedded in some complex submanifold $M = W \cap \mathbb{C}^n \subseteq W$ at the origin. If $F: V \to M$ is the inclusion mapping and $R: M \to V$

is the restriction to $M$ of a holomorphic retraction of $W$ to $V$, then the composition $R \circ F: V \to V$ is the identity mapping and hence its differential at the origin is nonsingular. However, $d(\mathbf{R} \circ \mathbf{F}) = d\mathbf{R} \circ d\mathbf{F}$ where $d\mathbf{F}: T_0(V) \to T_0(M)$ and $d\mathbf{R}: T_0(M) \to T_0(V)$, and since $\dim T_0(M) = \operatorname{tdim}_0(V) = \dim T_0(V)$, it follows that $d\mathbf{R}$ is also nonsingular. As a consequence of the preceding theorem the mapping $R$ must then be surjective, but for a retraction mapping that can be possible only when $M = V$ and $R$ is the identity mapping, thus showing that $V$ is regular at the origin and thereby concluding the proof.

It is apparent from the preceding discussion that the tangential dimension of a holomorphic variety may vary from point to point even when the variety itself is pure dimensional. The following simple observation about this variation is often quite useful.

**13. THEOREM.** *For any holomorphic variety $V$ and any integer $k$, the subset $\{Z \in V: \operatorname{tdim}_Z(V) > k\}$ is a holomorphic subvariety of $V$.*

*Proof.* It is of course enough just to show that the subset of interest here is a holomorphic subvariety of $V$ in an open neighborhood of any point $A \in V$. Some open neighborhood of $A$ in $V$ can be represented by a holomorphic subvariety of an open neighborhood $U$ of $O$ in a suitable complex space $\mathbb{C}^m$, and it follows from Cartan's theorem, Theorem F6, that if the neighborhood $U$ is sufficiently small, there will exist finitely many holomorphic functions $g_1, \ldots, g_r$ in $U$ such that the germs of these functions generate the ideal of the subvariety $V$ in the local ring ${}_m\mathcal{O}_Z$ at each point $Z \in U$. By Corollary 5,

$$\operatorname{tdim}_Z V = m - \operatorname{rank} \frac{\partial(g_1, \ldots, g_r)}{\partial(z_1, \ldots, z_m)}(Z)$$

at each point $Z \in V$. The subset $\{Z \in V \cap U : \operatorname{tdim}_Z(V) > k\}$ is then just the subset of $V \cap U$ consisting of all those points $Z \in V \cap U$ such that all $(m - k) \times (m - k)$ minors of the matrix $\partial(g_1, \ldots, g_r)/\partial(z_1, \ldots, z_m)$ vanish at $Z$ and hence is a holomorphic subvariety of $V \cap U$. That suffices to conclude the proof.

The preceding theorem can be used to complete in one important respect the discussion begun in section E of the singular locus $\mathfrak{S}(V)$ of a holomorphic variety $V$. For this purpose it is convenient first to note explicitly the following simple consequences of the preceding discussion of imbedding dimension.

**14. THEOREM.** *A germ* $\mathbf{V}$ *of a holomorphic variety is regular if and only if* $\operatorname{tdim} \mathbf{V} = \dim \mathbf{V}$.

*Proof.* As noted already it follows from Theorem G3 that the germ $\mathbf{V}$ is regular precisely when it can be represented by the germ of a holomorphic subvariety in $\mathbb{C}^n$ where $n = \dim \mathbf{V}$; but that is of course just the condition that $\operatorname{tdim} \mathbf{V} = \dim \mathbf{V}$, thus demonstrating the desired result.

**15. THEOREM.** *The singular locus of a holomorphic variety V is a holomorphic subvariety of V.*

*Proof.* Since the desired result is evidently a purely local one, it is only necessary to show that the singular locus $\mathfrak{S}(V)$ is a holomorphic subvariety in an open neighborhood $U$ of any point $A \in V$. If $U$ is sufficiently small, the intersection $U \cap V$ will have only finitely many irreducible components $V_1, \ldots, V_r$. Clearly

$$\mathfrak{S}(V) = \bigcup_{i \neq j} (V_i \cap V_j) \cup \bigcup_{i} \mathfrak{S}(V_i)$$

so it is actually enough to show that $\mathfrak{S}(V_i)$ is a holomorphic subvariety of $V_i$ for any irreducible component $V_i$ of $U \cap V$. The variety $V_i$ must be of pure dimension $n_i$ for some integer $n_i$, and it follows from Theorem 14 that $\mathfrak{S}(V_i) = \{Z \in V_i : \operatorname{tdim}_Z(V_i) > n_i\}$. But then Theorem 13 shows that $\mathfrak{S}(V_i)$ is a holomorphic subvariety of $V_i$, thus concluding the proof.

It is perhaps worth pointing out that as already demonstrated in Theorem E17, the singular locus $\mathfrak{S}(V)$ is a thin subset of $V$. Consequently $\mathfrak{S}(V)$ not only is a proper holomorphic subvariety of $V$ but moreover meets any irreducible component of $V$ in a proper holomorphic subvariety of that component. To turn next to a different aspect of the tangent space of a germ of a holomorphic variety, there is an alternative even more algebraic characterization of that tangent space. In considering this characterization it may be helpful to keep in mind the discussion of systems of parameters from section G. If ${}_{\mathbf{V}}\mathfrak{m}$ is the maximal ideal in the local ring ${}_{\mathbf{V}}\mathcal{O}$ of a germ $\mathbf{V}$ of a holomorphic variety, then as noted in Lemma G13 the quotient ${}_{\mathbf{V}}\mathfrak{m}/{}_{\mathbf{V}}\mathfrak{m}^2$ with its natural structure as a complex vector space is finite dimensional. The dual vector space is the space $\operatorname{Hom}_{\mathbb{C}}({}_{\mathbf{V}}\mathfrak{m}/{}_{\mathbf{V}}\mathfrak{m}^2, \mathbb{C})$ of all linear mappings $L: {}_{\mathbf{V}}\mathfrak{m}/{}_{\mathbf{V}}\mathfrak{m}^2 \to \mathbb{C}$.

**16. THEOREM.** *For any germ* $\mathbf{V}$ *of a holomorphic variety there is a canonical isomorphism between the tangent space* $T(\mathbf{V})$ *and the dual of the vector space* ${}_{\mathbf{V}}\mathfrak{m}/{}_{\mathbf{V}}\mathfrak{m}^2$.

*Proof.* In view of Definition 1 any tangent vector $T \in T(\mathbf{V})$ is a linear mapping $T: {}_{\mathbf{V}}\mathcal{O} \to \mathbb{C}$ and so by restriction determines a linear mapping $T|{}_{\mathbf{V}}\mathfrak{m}: {}_{\mathbf{V}}\mathfrak{m} \to \mathbb{C}$. Moreover $T(\mathbf{fg}) = f(0)T(\mathbf{g}) + g(0)T(\mathbf{f}) = 0$ whenever $\mathbf{f}, \mathbf{g} \in {}_{\mathbf{V}}\mathfrak{m}$, since the maximal ideal is characterized by the condition that $\mathbf{f}(0) = \mathbf{g}(0) = 0$, so the linear mapping $T$ annihilates the linear subspace ${}_{\mathbf{V}}\mathfrak{m}^2 \subset {}_{\mathbf{V}}\mathfrak{m}$ and hence can be viewed as a linear mapping $\phi(T): {}_{\mathbf{V}}\mathfrak{m}/{}_{\mathbf{V}}\mathfrak{m}^2 \to \mathbb{C}$. The mapping sending any tangent vector $T \in T(\mathbf{V})$ to the element $\phi(T) \in \operatorname{Hom}_{\mathbb{C}}({}_{\mathbf{V}}\mathfrak{m}/{}_{\mathbf{V}}\mathfrak{m}^2, \mathbb{C})$ is clearly a canonically defined linear mapping $\phi: T(\mathbf{V}) \to \operatorname{Hom}_{\mathbb{C}}({}_{\mathbf{V}}\mathfrak{m}/{}_{\mathbf{V}}\mathfrak{m}^2, \mathbb{C})$. If $\phi(T) = 0$, then the tangent vector $T: {}_{\mathbf{V}}\mathcal{O} \to \mathbb{C}$ vanishes on the subspace ${}_{\mathbf{V}}\mathfrak{m} \subseteq {}_{\mathbf{V}}\mathcal{O}$, and since $T(c) = 0$ automatically for any constant $c \in \mathbb{C} \subseteq {}_{\mathbf{V}}\mathcal{O}$ while any element $\mathbf{f} \in {}_{\mathbf{V}}\mathcal{O}$ can be written as the sum of the constant $\mathbf{f}(0)$ and an element of ${}_{\mathbf{V}}\mathfrak{m}$, it follows that $T(\mathbf{f}) = 0$ for any $\mathbf{f} \in {}_{\mathbf{V}}\mathcal{O}$. Thus $T = 0$ and $\phi$ is consequently an injective linear mapping. On the other hand, any element $L \in \operatorname{Hom}_{\mathbb{C}}({}_{\mathbf{V}}\mathfrak{m}/{}_{\mathbf{V}}\mathfrak{m}^2, \mathbb{C})$ can be viewed as a linear mapping $L: {}_{\mathbf{V}}\mathfrak{m} \to \mathbb{C}$ that vanishes on the subspace ${}_{\mathbf{V}}\mathfrak{m}^2 \subseteq {}_{\mathbf{V}}\mathfrak{m}$. Then for any $\mathbf{f} \in {}_{\mathbf{V}}\mathcal{O}$, set $T(\mathbf{f}) = L(\mathbf{f} - \mathbf{f}(0))$, and note that $\mathbf{f} - \mathbf{f}(0) \in {}_V\mathfrak{m}$ so that this is a well-defined mapping $T: {}_V\mathcal{O} \to \mathbb{C}$. It is obvious that the mapping $T$ is a linear mapping, and that $T|_V\mathfrak{m} = L$. Moreover whenever $\mathbf{f}, \mathbf{g} \in {}_V\mathcal{O}$, then

$$\begin{aligned} T(\mathbf{fg}) &= L[\mathbf{fg} - \mathbf{f}(0)\mathbf{g}(0)] \\ &= L[(\mathbf{f} - \mathbf{f}(0))(\mathbf{g} - \mathbf{g}(0)) + \mathbf{f}(0)(\mathbf{g} - \mathbf{g}(0)) + \mathbf{g}(0)(\mathbf{f} - \mathbf{f}(0))] \\ &= 0 + \mathbf{f}(0)L(\mathbf{g} - \mathbf{g}(0)) + \mathbf{g}(0)L(\mathbf{f} - \mathbf{f}(0)) \\ &= f(0)T(\mathbf{g}) + \mathbf{g}(0)T(\mathbf{f}) \end{aligned}$$

since $L$ is linear and vanishes on the subspace ${}_V\mathfrak{m}^2 \subseteq {}_V\mathfrak{m}$. Consequently $T \in T(\mathbf{V})$ and $L = \phi(T)$, so that $\phi$ is also a surjective linear mapping and the proof is thereby concluded.

**17. COROLLARY.** *For any germ* $\mathbf{V}$ *of a holomorphic variety*, tdim $\mathbf{V} = \dim_{\mathbb{C}} {}_V\mathfrak{m}/{}_V\mathfrak{m}^2$.

*Proof.* This is an immediate consequence of the preceding theorem, since $\dim_{\mathbb{C}} {}_V\mathfrak{m}/{}_V\mathfrak{m}^2 = \dim_{\mathbb{C}} \mathrm{Hom}_{\mathbb{C}}({}_V\mathfrak{m}/{}_V\mathfrak{m}^2, \mathbb{C})$.

**18. COROLLARY.** *For any germ* $\mathbf{V}$ *of a holomorphic variety*, tdim $\mathbf{V}$ *is equal to the minimal number of generators of the maximal ideal* ${}_V\mathfrak{m}$ *of the local ring* ${}_V\mathcal{O}$.

*Proof.* The germ $\mathbf{V}$ can be represented by the germ of a holomorphic subvariety at the origin in $\mathbb{C}^m$ where $m =$ tdim $\mathbf{V}$, and the $m$ coordinate functions in $\mathbb{C}^m$ when restricted to $\mathbf{V}$ clearly generate the maximal ideal ${}_V\mathfrak{m}$. On the other hand, if $\mathbf{f}_1, \ldots, \mathbf{f}_\nu$ generate the ideal ${}_V\mathfrak{m}$, then their residue classes span the complex vector space ${}_V\mathfrak{m}/{}_V\mathfrak{m}^2$; hence, $\nu \geq \dim_{\mathbb{C}} {}_V\mathfrak{m}/{}_V\mathfrak{m}^2 =$ tdim $\mathbf{V}$. That suffices to conclude the proof.

**19. COROLLARY.** *If* $\mathbf{F}: \mathbf{V} \to \mathbb{C}^n$ *is a holomorphic mapping from a germ of a holomorphic variety* $\mathbf{V}$ *to the germ of* $\mathbb{C}^n$ *at the origin, then* $\mathbf{F}$ *is a biholomorphic mapping between* $\mathbf{V}$ *and its image* $\mathbf{F}(\mathbf{V}) \subseteq \mathbb{C}^n$ *precisely when the coordinate functions of the mapping* $\mathbf{F}$ *are generators of the maximal ideal* ${}_V\mathfrak{m}$ *in the local ring of the germ* $\mathbf{V}$.

*Proof.* Let $\mathbf{f}_1, \ldots, \mathbf{f}_n$ be the coordinate functions of the mapping $\mathbf{F}$, and note that $\mathbf{f}_j \in {}_V\mathfrak{m} \subseteq {}_V\mathcal{O}$ since the mapping $\mathbf{F}$ by assumption takes $\mathbf{V}$ to the germ of $\mathbb{C}^n$ at the origin. First if the image $\mathbf{F}(\mathbf{V}) = \mathbf{W}$ is the germ of a holomorphic subvariety at the origin in $\mathbb{C}^n$ and $\mathbf{F}: \mathbf{V} \to \mathbf{W}$ is a biholomorphic mapping, then the induced homomorphism $\mathbf{F}^*: {}_W\mathcal{O} \to {}_V\mathcal{O}$ is an isomorphism between these local rings. The restrictions to $\mathbf{W}$ of the coordinate functions $z_1, \ldots, z_n$ in $\mathbb{C}^n$ evidently generate the maximal ideal ${}_W\mathfrak{m} \subset {}_W\mathcal{O}$, and consequently their images $\mathbf{F}^*(\mathbf{z}_j|\mathbf{W}) = \mathbf{f}_j$ generate the maximal ideal ${}_V\mathfrak{m} \subset {}_V\mathcal{O}$ as desired. Next suppose that the germs $\mathbf{f}_1, \ldots, \mathbf{f}_n$ generate the maximal ideal ${}_V\mathfrak{m} \subset {}_V\mathcal{O}$. In order to show that $\mathbf{F}$ is a biholomorphic mapping between $\mathbf{V}$ and its image $\mathbf{F}(\mathbf{V}) \subseteq \mathbb{C}^n$, it is sufficient by the inverse mapping theorem, Theorem 11, just to show that its differential $d\mathbf{F}: T(\mathbf{V}) \to T(\mathbb{C}^n)$ is injective. If $d\mathbf{F}(T) = 0$ for some tangent vector $T \in T(\mathbf{V})$, then $0 = d\mathbf{F}(T)(\mathbf{z}_j) = T(\mathbf{F}^*(\mathbf{z}_j)) = T(\mathbf{f}_j) = 0$ for $j = 1, \ldots, n$, and since $\mathbf{f}_1, \ldots, \mathbf{f}_n$ generate the maximal ideal ${}_V\mathfrak{m}$, the restriction $T|_{\mathbf{V}}\mathfrak{m} : {}_{\mathbf{V}}\mathfrak{m} \to \mathbb{C}$ must be the zero mapping. But it then follows from Theorem 16 that $T = 0$; hence, $dF$ is injective as desired and the proof is thereby concluded.

**Examples.** The preceding observations nicely lend themselves to the analysis of various singularities, the simplest example of which is the following. Consider the holomorphic mapping $F: \mathbb{C}^1 \to \mathbb{C}^n$ defined by

$$F(z) = (z^n, z^{n+1}, z^{n+2}, \ldots, z^{2n-1})$$

It is apparent that the germ of this mapping at the origin is finite and it then follows from Theorem E7 that its image is the germ $\mathbf{V}$ of a holomorphic subvariety at the origin in $\mathbb{C}^n$. Theorem G16 shows that $\dim \mathbf{V} = 1$, and since the singular locus is a proper holomorphic subvariety of $\mathbf{V}$, either $\mathbf{V}$ is regular or the origin is an isolated singular point of $\mathbf{V}$. The induced homomorphism $\mathbf{F}^*: {}_{\mathbf{V}}\mathcal{O} \to {}_1\mathcal{O}$ is injective by Corollary E9 and so can be used to identify the local ring ${}_{\mathbf{V}}\mathcal{O}$ of the germ $\mathbf{V}$ with a subring $\mathcal{R} \subseteq {}_1\mathcal{O}$. This subring $\mathcal{R}$ evidently consists of all germs in ${}_1\mathcal{O}$ of the form $\mathbf{f}(z^n, \ldots, z^{2n+1})$ for arbitrary elements $\mathbf{f} \in {}_n\mathcal{O}$, and its maximal ideal $\mathfrak{m} \subseteq \mathcal{R}$ corresponds to all such germs for arbitrary elements $\mathbf{f} \in {}_n\mathfrak{m} \subset {}_n\mathcal{O}$. For the present purposes it is enough to consider not the ring $\mathcal{R}$ itself but only its image $\tilde{\mathcal{R}}$ in the residue class ring ${}_1\mathcal{O}/{}_1\mathfrak{m}^{N+1}$ for some sufficiently large integer $N$. It is easily verified that $\tilde{\mathcal{R}}$ is the subring of ${}_1\mathcal{O}/{}_1\mathfrak{m}^{N+1}$ spanned by all polynomials in the arguments $z^n, \ldots, z^{2n+1}$, so that

$$\tilde{\mathcal{R}} = \{a_0 + a_n z^n + \cdots + a_N z^N\}/{}_1\mathfrak{m}^{N+1}$$

and correspondingly the image in $\tilde{\mathcal{R}}$ of the maximal ideal $\mathfrak{m} \subseteq \mathcal{R}$ is the ideal

$$\tilde{\mathfrak{m}} = \{a_n z^n + \cdots + a_N z^N\}/{}_1\mathfrak{m}^{N+1}$$

But then it is also easily verified that whenever $N \geqq 2n$,

$$\tilde{\mathfrak{m}}^2 = \{a_{2n} z^{2n} + \cdots + a_N z^N\}/{}_1\mathfrak{m}^{N+1}$$

and from this it follows immediately that

$$\operatorname{tdim} \mathbf{V} = \dim_{\mathbb{C}} {}_{\mathbf{V}}\mathfrak{m}/{}_{\mathbf{V}}\mathfrak{m}^2 \geqq \dim_{\mathbb{C}} \tilde{\mathfrak{m}}/\tilde{\mathfrak{m}}^2 = n$$

On the other hand, since $\mathbf{V} \subseteq \mathbb{C}^n$ necessarily $\operatorname{tdim} \mathbf{V} \leqq n$, and consequently $\operatorname{tdim} \mathbf{V} = n$. This therefore provides examples of one-dimensional germs of holomorphic varieties with any desired imbedding dimension. Furthermore beginning with the complex manifold $\mathbb{C}$, it is possible to construct a new one-dimensional holomorphic variety by replacing a neighborhood of the point $n \in \mathbb{C}$ by a neighborhood of the singular point with imbedding dimension $n$ just constructed and to do this for all integers $n = 1, 2, 3, \ldots$. The result is a one-dimensional holomorphic variety that cannot be represented as a holomorphic subvariety of any space $\mathbb{C}^n$, or indeed of any finite-dimensional complex manifold.

# J
# Holomorphic Vector Fields and Differential Equations

In the preceding section the discussion was focused on the tangent space to a holomorphic variety $V$ at a single point of $V$. But just as in the case of complex manifolds, it is often convenient to consider the union $T(V) = \bigcup_{Z \in V} T_Z(V)$ of the tangent spaces at all the points of $V$, the analogue of the tangent bundle of a complex manifold. Of course to be of other than purely formal interest this union $T(V)$ should also have a holomorphic structure, tying together the tangent spaces to $V$ at its various points in a manner reflecting the holomorphic structure of $V$. To describe such a structure, note that each point $A \in V$ has an open neighborhood $V_A$ in $V$ such that the variety $V_A$ can be represented by a holomorphic subvariety of an open subset $U_A$ in some space $\mathbb{C}^n$. Moreover if $V_A$ is sufficiently small, then by Cartan's theorem, Theorem F6, there are finitely many functions $g_j \in {}_n\mathcal{O}_{U_A}$ that generate the ideal of the subvariety $V_A$ at each point of $U_A$. The set

$$T(V_A) = \left\{(Z, T) \in U_A \times \mathbb{C}^n : g_j(Z) = \sum_k t_k \frac{\partial g_j}{\partial z_k}(Z) = 0 \text{ for all } j\right\} \tag{1}$$

is then a well-defined holomorphic subvariety of $U_A \times \mathbb{C}^n$. Moreover the natural projection $U_A \times \mathbb{C}^n \to U_A$ induces a holomorphic mapping $\pi$: $T(V_A) \to V_A$, since $V_A = \{Z \in U_A : g_j(Z) = 0 \text{ for all } j\}$, and it follows from Theorem I4, or more precisely from the version of that theorem expressed in equation I(2), that for any point $Z \in V_A$ the inverse image $\pi^{-1}(Z)$ can be identified with the tangent space $T_Z(V_A)$. There is thus an identification $T(V_A) \cong \bigcup_{Z \in V_A} T_Z(V_A)$ that can be used to impose the structure of a holomorphic variety on this portion of $T(V)$. This structure of course depends on the choice of a representation of $V_A$ as a holomorphic subvariety $V_A \subseteq U_A$. However note that if $V_A \subseteq U_A$ and $V'_A \subseteq U'_A$ are two choices of representatives of the same piece of a holomorphic variety as holomorphic subvarieties of open sets of $\mathbb{C}^n$ and $\mathbb{C}^{n'}$ respectively and if the neighborhoods $V_A$ and $V'_A$ are sufficiently small, then the natural biholomorphic mapping $G$: $V_A \to V'_A$ can be extended to a holomorphic mapping $F$: $U_A \to U'_A$, and similarly $G^{-1}$ can be extended to a holomorphic mapping $F'$: $U'_A \to U_A$. With the identifications made before, it follows from Theorem I9 that there is an induced holomorphic mapping $dF$: $T(V_A) \to T(V'_A)$ defined by

$$dF(Z, T) = (F(Z), J_F(Z)T) \quad \text{whenever} \quad (Z, T) \in T(V_A) \tag{2}$$

and similarly an induced holomorphic mapping $dF'$: $T(V_A') \to T(V_A)$, and that these mappings are inverses of each other. Consequently $T(V_A)$ and $T(V_A')$ are biholomorphic varieties, and it follows that the structure of a holomorphic variety introduced on $T(V)$ is really intrinsically defined, independently of the choices of representations of neighborhoods $V_A$ as holomorphic varieties.

1. DEFINITION. *The* **tangent variety** *of a holomorphic variety $V$ is the union $T(V) = \bigcup_{Z \in V} T_Z(V)$ with the structure of a holomorphic variety just described.*

It follows from the preceding description of the holomorphic structure on $T(V)$ that the natural projection $\pi$: $T(V) \to V$ defined by setting $\pi(T_Z) = Z$ is a holomorphic mapping. The tangent variety is thus like the tangent bundle of a complex manifold as discussed in section I-C in being a holomorphic family of vector spaces over $V$, though not in this case a regular holomorphic family since $T(V)$ need not be a regular variety. With the structure thus introduced on the tangent variety it is possible to introduce the following concept, a natural extension of the corresponding notion for complex manifolds.

2. DEFINITION. *A* **holomorphic vector field** *on a holomorphic variety $V$ is a holomorphic mapping $F$: $V \to T(V)$ such that $\pi(F(Z)) = Z$ for every point $Z \in V$, where $\pi$: $T(V) \to V$ is the natural projection from the tangent variety $T(V)$ to $V$. The set of all such holomorphic vector fields will be denoted by $\mathcal{O}_V(T(V))$.*

For any open subset $U \subseteq V$ the restriction to $U$ of a holomorphic vector field $F \in \mathcal{O}_V(T(V))$ is clearly a holomorphic vector field $F|U \in \mathcal{O}_U(T(U))$. The notion of the **germ of a holomorphic vector field** at a point $Z \in V$ is then the obvious one, paralleling Definition A1 of the germ of a holomorphic function. The germ at $Z$ of a holomorphic vector field $F$ will be denoted by $\mathbf{F}_Z$, and the set of all such germs will be denoted by $\mathcal{O}_Z(T(V))$. To examine the local properties of holomorphic vector fields, note that as before an open neighborhood $V_A$ of any point $A$ of the variety $V$ can be represented by a holomorphic subvariety of an open subset $U_A$ in some space $\mathbb{C}^n$, and if the neighborhood $V_A$ is sufficiently small, there will exist finitely many functions $g_j \in {}_n\mathcal{O}_{U_A}$ that generate the ideal of the subvariety $V_A$ at each point of $U_A$. The piece $T(V_A)$ of the tangent variety $T(V)$ can then be described as the holomorphic subvariety (1). In these terms a holomorphic vector field $F \in \mathcal{O}_{V_A}(T(V_A))$ can be viewed merely as a holomorphic mapping $F$: $V_A \to \mathbb{C}^n$ such that

$$\sum_k f_k(Z) \frac{\partial g_j}{\partial z_k}(Z) = 0 \qquad \text{for all } Z \in V_A \text{ and all } j \tag{3}$$

where $F(Z) = (f_1(Z), \ldots, f_n(Z))$. It is evident from this that $\mathcal{O}_{V_A}(T(V_A))$ has the natural structure of a module over the ring ${}_V\mathcal{O}_{V_A}$, since $fF \in \mathcal{O}_{V_A}(T(V_A))$ whenever $f \in {}_V\mathcal{O}_{V_A}$ and $F \in \mathcal{O}_{V_A}(T(V_A))$. Consequently $\{\mathcal{O}_Z(T(V))\}$ can be viewed as a family of modules over the family of rings ${}_V\mathcal{O}_Z$ as $Z$ varies over $V_A$. In the discussion of families of modules in section F only families of submodules of a family of free modules were actually considered. The family of modules $\{\mathcal{O}_Z(T(V))\}$ is of course really a family of submodules of the family of free modules of rank $n$ over the neighborhood $V_A$. But this representation depends on the particular representation of the neighborhood $V_A$ as a holomorphic subvariety of an open subset of $\mathbb{C}^n$, so in this case it is much more reasonable merely to view $\{\mathcal{O}_Z(T(V))\}$ as a family of abstract modules over the family of rings $\{{}_V\mathcal{O}_Z\}$. On the other hand, when viewed as a family of submodules of the family of free modules of rank $n$ over $V_A$, it follows from (3) that at each point $Z \in V_A$

$$\mathcal{O}_Z(T(V)) = \bigcap_j \mathscr{R}(\mathrm{g}_{j1}, \ldots, \mathrm{g}_{jn})$$

where $\mathrm{g}_{jk} \in {}_V\mathcal{O}_Z$ is the germ of the function $(\partial g_j/\partial z_k)|V$ at the point $Z$ and $\mathscr{R}(\mathrm{g}_{j1}, \ldots, \mathrm{g}_{jn})$ is the module of relations between these germs. Each of these families of modules of relations is finitely generated over an open neighborhood of any point of $V_A$ as a consequence of Oka's theorem on $V$, Corollary F9, and their intersection is then also finitely generated over an open neighborhood of any point of $V_A$ by Corollary F10. Thus *the family* $\{\mathcal{O}_Z(T(V))\}$ *of submodules of the family of free modules of rank n over* $V_A$ *is finitely generated over an open neighborhood of any point of* $V_A$.

As is no doubt quite familiar and evident from the preceding discussion, the family of modules $\{\mathcal{O}_Z(T(V))\}$ over a complex manifold $V$ is locally free. Thus there always exist an abundance of germs of holomorphic vector fields at any regular point of a holomorphic variety—indeed enough elements in $\mathcal{O}_Z(T(V))$ whenever $Z \in \mathscr{R}(V)$ so that their values at $Z$ span the full tangent space $T_Z(V)$. This last condition actually characterizes the regularity of the holomorphic variety $V$ at $Z$. At singular points the possible germs of holomorphic vector fields are very severely limited, as the discussion in the remainder of this section should make quite clear.

This discussion requires more knowledge of the local properties of holomorphic vector fields on complex manifolds, in particular of some local properties that amount to integrability conditions for holomorphic systems of ordinary differential equations. Although these conditions may be well known to some readers, they are an important application of the theory of holomorphic functions of several variables, albeit an application of only relatively elementary aspects of that theory, and so for the sake of completeness should be briefly treated here. Systems of differential equations arise when considering the problem of simplifying a germ of a holomorphic vector field in $\mathbb{C}^n$ by an appropriate local change of coordinates. A holomorphic vector field in an open neighborhood $U$ of the origin in $\mathbb{C}^n$ can be viewed just as a holomorphic mapping $E: U \to \mathbb{C}^n$. A biholomorphic mapping $H: U \to V$ taking $U$ to another open neighborhood of the origin and taking the origin to itself transforms the vector field represented by $E$ to the vector field represented by the mapping $F: V \to \mathbb{C}^n$ where

$$F(H(Z)) = J_H(Z)E(Z) \tag{4}$$

The simplest nonsingular vector field is that given by the mapping $E(Z) = E_1 = (1, 0, \ldots, 0)$ for all $Z \in U$. The question of which holomorphic germs of nonsingular vector fields can be transformed to this simplest of vector fields $E_1$ by a biholomorphic change of coordinates thus reduces to the question for which holomorphic germs **F** of nonsingular vector fields does there exist a germ **H** of a biholomorphic mapping in $\mathbb{C}^n$ satisfying the system of partial differential equations (4). For the vector field $E(Z) = E_1$, the system (4) has the explicit form

$$\frac{\partial h_j(Z)}{\partial z_1} = f_j(h_1(Z), \ldots, h_n(Z))$$

and so can be viewed as a system of ordinary differential equations involving holomorphic functions of the variable $z_1$ depending holomorphically on some auxiliary parameters $z_2, \ldots, z_n$. The basic existence theorem for such systems is the following.

3. **THEOREM.** *For any germs $\mathbf{f}_j \in {}_{1+n+r}\mathcal{O}$, $1 \leqq j \leqq n$, there exist unique germs $\mathbf{h}_j \in {}_{1+r}\mathcal{O}$ such that*

$$h_j(O, T) = O$$

$$\frac{\partial h_j}{\partial z}(z, T) = f_j(z, h_1(z, T), \ldots, h_n(z, T), T)$$

*in an open neighborhood of the origin in $\mathbb{C}^{1+r}$, where $f_j$* and *$h_j$ are any representatives of these germs.*

*Proof.* The germs $\mathbf{f}_j$ can be represented by power series

$$f_j(z, W, T) = \sum_{k,L} c^j_{kL}(T) z^k W^L$$

that converge in open neighborhoods of the closure $\bar{\Delta}(0; s)$ of some polydisc about the origin in $\mathbb{C}^{1+n+r}$, where $c^j_{kL}(T) = \sum_M c^j_{kLM} T^M$ is a power series in the variables $t_1, \ldots, t_r$, and it can be supposed that all these functions are uniformly bounded by some constant $C \geqq 0$ in the closed polydisc $\bar{\Delta}(0; s)$. The Cauchy inequalities of Theorem I-A5 show in this case that

$$|c^j_{kLM}| \leqq C s^{-k-|L|-|M|} \tag{5}$$

If there are any germs $\mathbf{h}_j$ satisfying the differential equations and initial conditions as in the statement of the theorem, they too can be represented by power series

$$h_j(z, T) = \sum_p x^j_p(T) z^p$$

that converge in some open neighborhood of the origin in $\mathbb{C}^{1+r}$, where $x^j_p(T) =$

$\sum_Q x^j_{pQ} T^Q$. The conclusions of the theorem are then that $x^j_0(T)$ vanishes identically —that is, that $x^j_{0Q} = 0$ for all multi-indices $Q$—and that

$$\sum_p (p+1)x^j_{p+1}(T)z^p = \sum_{k,L} c^j_{kL}(T)z^k\left(\sum_{p_1} x^j_{p_1}(T)z^{p_1}\right)^{l_1}\cdots\left(\sum_{p_n} x^j_{p_n}(T)z^{p_n}\right)^{l_n}$$

Comparing the terms independent of $z$ on the two sides of this identity shows that

$$x^j_1(T) = c^j_{0,0\ldots0}(T)$$

while comparing the coefficients of $z$ shows that

$$x^j_2(T) = c^j_{1,0\ldots0}(T) + \sum_\nu c^j_{0,0\ldots1\ldots0}(T)x^\nu_1(T)$$

where in $c^j_{0,0\ldots1\ldots0}$ the 1 occurs in the $\nu$th place, and so on. In general $x^j_p(T)$ appears equal to a polynomial function of the elements $c^k_q(T)$ for $q < p$ and of some corresponding elements $c^j_{lQ}(T)$, with all the coefficients of these polynomials being nonnegative integers. From using these formulas recursively it is evident that the elements $x^j_p(T)$ must be given explicitly by some formulas

$$x^j_p(T) = P^j_p(c^i_{kL}(T)) \tag{6}$$

where $P^j_p(T)$ are some polynomial functions with nonnegative rational coefficients. Thus, on the one hand, if the desired solutions $\mathbf{h}_j$ exist at all, they are clearly uniquely determined. On the other hand, in order to demonstrate the existence of these solutions, it is enough to show that the power series with coefficients determined by (6) converges in some open neighborhood of the origin. Now from (5) it follows that $|c^j_{kL}(T)| \leq |\tilde{c}^j_{kL}(T)|$ where

$$\tilde{c}^j_{kL}(T) = \sum_M Cs^{-k-|L|-|M|}|T|^M = Cs^{-k-|L|}\prod_{\nu=1}^{r}\left(1 - \frac{t_\nu}{s}\right)^{-1}$$

and since the coefficients of the polynomials $P^j_p(X)$ are all nonnegative, it then follows from (6) that $|x^j_p(T)| \leq \tilde{x}^j_p(T)$ where

$$\tilde{x}^j_p(T) = P^j_p(\tilde{c}^i_{kL}(T))$$

The functions $\tilde{x}^j_p(T)$ are the coefficients of the only possible solutions of the corresponding system of differential equations defined by the functions

$$\tilde{f}_j(z, W, T) = \sum_{k,L} \tilde{c}^j_{kL}(T)z^k W^L$$

$$= C\left(1 - \frac{z}{s}\right)^{-1}\prod_{\mu=1}^{n}\left(1 - \frac{w_\mu}{s}\right)^{-1}\prod_{\nu=1}^{r}\left(1 - \frac{t_\nu}{s}\right)^{-1}$$

For this simple a system of differential equations it is readily verified that there are solutions, satisfying the desired initial conditions, of the form

$$\tilde{h}_j(z, T) = s - s\left[1 - (n + 1)C \log\left(1 - \frac{z}{s}\right)\prod_\nu\left(1 - \frac{t_\nu}{s}\right)^{-1}\right]^{1/(n+1)}$$

They are holomorphic near the origin, so $\tilde{x}_p^j(T)$ are the coefficients of a convergent power series and hence so are $x_p^j(T)$. That suffices to conclude the proof.

This theorem shows that for any functions $f_j$ holomorphic near the origin in $\mathbb{C}^{1+n}$ there are uniquely determined holomorphic functions $h_j$ near the origin in $\mathbb{C}$ satisfying the system of ordinary differential equations $h_j'(z) = f_j(z, h_1(z), \ldots, h_n(z))$ and the initial conditions $h_j(0) = 0$. Moreover if the functions $f_j$ also depend holomorphically on some additional parameters in $\mathbb{C}^r$, then so do the solutions at least locally. With this result established, it is a very simple matter to extend it to allow more general initial conditions as follows.

4. **COROLLARY.**   *For any germs* $\mathbf{f}_j \in {}_{1+n+r}\mathcal{O}$, $1 \leq j \leq n$, *there are unique germs* $\mathbf{h}_j \in {}_{1+n+r}\mathcal{O}$ *such that*

$$h_j(0, W, T) = w_j$$

$$\frac{\partial h_j}{\partial z}(z, W, T) = f_j(z, h_1(z, W, T), \ldots, h_n(z, W, T), T)$$

*in an open neighborhood of the origin in* $\mathbb{C}^{1+n+r}$, *where* $f_j$ *and* $h_j$ *are any representatives of these germs.*

*Proof.*   It follows immediately from Theorem 3 that there are unique germs $\tilde{h}_j \in {}_{1+n+r}\mathcal{O}$ such that

$$\tilde{h}_j(0, W, T) = 0$$

$$\frac{\partial \tilde{h}_j}{\partial z}(z, W, T) = f_j(z, \tilde{h}_1(z, W, T) + w_1, \ldots, \tilde{h}_n(z, W, T) + w_n, T)$$

merely viewing both $W$ and $T$ as auxiliary parameters in that theorem. The functions $h_j(z, W, T) = \tilde{h}_j(z, W, T) + w_j$ then satisfy the conditions listed in the corollary, thus establishing the existence assertion. The argument can of course be reversed; any functions $h_j(z, W, T)$ satisfying the conditions listed lead to functions $\tilde{h}_j(z, W, T) = h_j(z, W, T) - w_j$ satisfying the appropriate conditions in Theorem 3, thus yielding the uniqueness assertion and concluding the proof.

For the immediate purpose here these observations will be used only to establish the following auxiliary result.

**5. COROLLARY.** *If $U$ is an open neighborhood of the origin in $\mathbb{C}^n$ and $F: U \to \mathbb{C}^n$ is a holomorphic mapping such that $F(0) \neq 0$, then after shrinking the neighborhood $U$ if necessary, there exists a biholomorphic mapping $G: U \to U'$ such that $J_G(Z)F(Z) = E_1$ for all points $Z \in U$, where $E_1 = (1, 0, \ldots, 0)$.*

*Proof.* By making a suitable nonsingular change of coordinates in the image space $\mathbb{C}^n$ of the mapping $F$, it can be assumed that $F(0) = E_1$, and then after shrinking $U$ if necessary, it can also be assumed that $f_1(Z) \neq 0$ for all $Z \in U$, where $F = (f_1, \ldots, f_n)$. The holomorphic mapping $G': U \to U'$ with coordinate functions

$$g_1'(Z) = \int_0^{z_1} f_1(t, z_2, \ldots, z_n)^{-1}\, dt, \qquad g_j'(z) = z_j \quad \text{for} \quad 2 \leqq j \leqq n$$

has Jacobian matrix

$$J_{G'}(Z) = \begin{pmatrix} f_1(Z)^{-1} & 0 & & 0 \\ 0 & 1 & \ddots & 0 \\ 0 & 0 & & 1 \end{pmatrix}$$

which is nonsingular at the origin, so after shrinking the neighborhood $U$ further, the mapping $G'$ will be biholomorphic. Moreover $F'(Z) = J_{G'}(Z)F(Z)$ has the property that $f_1'(Z) = 1$ for all points $Z \in U$. If $F'$ is viewed as a holomorphic mapping $F'': U' \to \mathbb{C}^n$ by setting $F''(Z') = F'(G^{-1}(Z'))$, and if the desired corollary is established for this mapping $F''$, then it also holds for the initial mapping $F$; for if, after shrinking $U$ further if necessary, there is a biholomorphic mapping $G''$: $U' \to U''$ such that $J_{G''}(Z')F''(Z') = E_1$ for all points $Z' \in U'$, then since the composite biholomorphic mapping $G = G'' \circ G' : U \to U''$ has Jacobian matrix $J_G(Z) = J_{G''}(G'(Z))J_{G'}(Z)$, it follows that $J_G(Z)F(Z) = J_{G''}(G'(Z))J_{G'}(Z)F(Z) = J_{G''}(G'(Z))F'(Z) = J_{G''}(Z')F''(Z') = E_1$ where $Z' = G'(Z)$.

It thus suffices to demonstrate the corollary just in the special case that $f_1(Z) = 1$ for all points $Z \in U$. Rather than constructing the mapping $G$, though, it is more convenient to construct what will be the inverse mapping $H = G^{-1}$, since that is a direct application of the preceding corollary. Write the coordinates in $\mathbb{C}^n$ as $W = (w_1, W')$ where $W' \in \mathbb{C}^{n-1}$, and it follows immediately from Corollary 4 that there are holomorphic functions $h_j$ in an open neighborhood $U'$ of the origin in $\mathbb{C}^n$ for $2 \leqq j \leqq n$ such that

$$h_j(0, W') = w_j \tag{7}$$

$$\frac{\partial h_j}{\partial w_1}(w_1, W') = f_j(w_1, h_2(w_1, W'), \ldots, h_n(w_1, W')) \tag{8}$$

in that neighborhood. Set $h_1(w_1, W') = w_1$. The resulting $n$ functions can be taken as the coordinate functions of a holomorphic mapping $H: U' \to \mathbb{C}^n$. It is evident from the definition of $h_1$ and the initial value conditions (7) that

$$J_H(0) = \begin{pmatrix} 1 & 0 & & 0 \\ * & 1 & \ddots & 0 \\ * & 0 & & 1 \end{pmatrix}$$

is nonsingular, so after shrinking the neighborhood $U'$ if necessary, $H$ will be a biholomorphic mapping $H: U' \to U$ taking the origin to the origin. As in the discussion preceding Theorem 3, it follows readily that $F(H(Z)) = J_H(Z)E_1$, that really being the content of formula (8) upon recalling that $w_1 = h_1(w_1, W')$. For the inverse biholomorphic mapping $G: U \to U'$, since $I = J_{GH}(Z) = J_G(H(Z))J_H(Z)$, then $J_G(H(Z))F(H(Z)) = J_G(H(Z))J_H(Z)E_1 = E_1$ as desired, thereby concluding the proof.

This corollary can now be used to complete the discussion at the beginning of this section dealing with holomorphic vector fields on holomorphic varieties. The basic result is the following.

**6. THEOREM (Rossi's theorem).** *If there exists a holomorphic vector field $F$ on a holomorphic variety $V$ such that $F(A) \neq 0$ at a point $A \in V$, then an open neighborhood of $A$ in $V$ is biholomorphic to a product $M \times W$ where $M$ is a regular one-dimensional variety.*

*Proof.* Suppose that $F$ is a holomorphic vector field on the variety $V$ and that $F(A) \neq 0$. An open neighborhood $V_A$ of $A$ in $V$ can be represented by a holomorphic subvariety of an open subset $U \subseteq \mathbb{C}^n$, and the piece $T(V_A)$ of the tangent variety $T(V)$ can correspondingly be represented as in (1). The restriction of the holomorphic vector field $F$ to the subset $V_A$ can then be viewed merely as a holomorphic mapping $F: V_A \to \mathbb{C}^n$ satisfying (3). The first step in the proof is to show that if the representative subvariety $V_A$ is suitably chosen, it can be supposed that $F(Z) = E_1$ for every point $Z \in V_A$. Indeed if the neighborhoods $V_A$ and $U$ are sufficiently small, there will exist a holomorphic mapping $\tilde{F}: U \to \mathbb{C}^n$ such that $\tilde{F}|V_A = F$, and after shrinking the neighborhood $U$ further if necessary, it follows from Corollary 5 that there is a biholomorphic mapping $H: U \to U'$ such that $J_H(Z)\tilde{F}(Z) = E_1$ for all points $Z \in U$. The image $V_A' = H(V_A)$ is then a holomorphic subvariety of $U'$ representing the same open neighborhood of $A$ in $V$ as $V_A$, and as in (2) the corresponding piece of the tangent variety of $V$ is $T(V_A') = dH(T(V_A))$. In this new representation of the tangent variety, the given holomorphic vector field is thus represented by the mapping $F': V_A' \to \mathbb{C}^n$ such that $(Z', F'(Z')) = dH(Z, F(Z)) = (H(Z), J_H(Z)F(Z))$ and hence such that $F'(Z') = E_1$ for all points $Z' \in V_A'$ as desired.

Now choose the represented subvariety $V_A \subseteq U$ so that $F(Z) = E_1$ throughout $V_A$, and let $g_j \in {}_n\mathcal{O}_U$ be holomorphic functions that generate the ideal of the subvariety $V_A$ at each point of $U$. It follows from (3) that $\partial g_j(Z)/\partial z_1 = 0$ for all $Z \in V_A$. The germ at the point $A$ of the function $\partial g_j/\partial z_1$ must thus lie in the ideal id $V_A \subseteq {}_n\mathcal{O}_A$, so that there are holomorphic functions $x_{jk}'$ in some polydisc $\Delta(A; R) \subseteq U$ such that

$$\frac{\partial g_j}{\partial z_1} = \sum_k x_{jk}' g_k \tag{9}$$

throughout $\Delta(A; R)$. Upon differentiating (9) note that

$$\frac{\partial^2 g_j}{\partial z_1^2} = \sum_k \left(\frac{\partial x'_{jk}}{\partial z_1}\cdot g_k + x'_{jk}\cdot\frac{\partial g_k}{\partial z_1}\right)$$
$$= \sum_k x''_{jk} g_k$$

where $x''_{jk} = \partial x'_{jk}/\partial z_1 + \sum_\ell x'_{j\ell} x'_{\ell k}$ are also holomorphic in $\Delta(A; R)$. The obvious induction argument then shows that for any $\nu \geq 0$,

$$\frac{\partial^\nu g_j}{\partial z_1^\nu} = \sum_k x^\nu_{jk} g_k$$

for some functions $x^\nu_{jk}$ that are holomorphic throughout $\Delta(A; R)$. At each point $B \in V_A \cap \Delta(A; R)$ it then follows that $\partial^\nu g_j/\partial z_1^\nu = \sum_k x^\nu_{jk}(B) g_k(B) = 0$ for all values $\nu$, so that the Taylor expansion of the function $g_j$ near $B$ must have the form

$$g_j(Z) = \sum_{\nu_1, \ldots, \nu_n} c^j_{\nu_1 \ldots \nu_n}(z_1 - b_1)^{\nu_1} \cdots (z_n - b_n)^{\nu_n}$$

where $c^j_{\nu_1 0 0} = 0$ for all values $\nu_1$. Consequently $g_j(z_1, b_2, \ldots, b_n) = g_j(B) = 0$ whenever $z_1 \in \Delta(b_1; s_1) \subseteq \Delta(a_1; r_1)$. That means of course that whenever $B \in V_A \cap \Delta(A; R)$, then $(z_1, b_2, \ldots, b_n) \in V_A$ for all such values of $z_1$, and it is clear from this observation that

$$V_A \cap \Delta(A; R) = \{Z \in \Delta(A; R) : g_j(a_1, z_2, \ldots, z_n) = 0 \text{ for all } j\}$$

That is just the desired result, and the proof is thereby concluded.

**7. COROLLARY.** *If there exist $m$ holomorphic vector fields $F_1, \ldots, F_m$ on a holomorphic variety $V$ such that the vectors $F_1(A), \ldots, F_m(A)$ are linearly independent at a point $A \in V$, then an open neighborhood of $A$ in $V$ is biholomorphic to a product $M \times W$ where $M$ is a regular $m$-dimensional variety.*

*Proof.* The proof will be by induction on $m$, with the case $m = 1$ being precisely the preceding theorem. For the induction step, if there are $m$ holomorphic vector fields $F_1, \ldots, F_m$ on $V$ such that $F_1(A), \ldots, F_m(A)$ are linearly independent, then since $F_1(A), \ldots, F_{m-1}(A)$ are also linearly independent, the induction hypothesis shows that an open neighborhood $V_A$ of $A$ in $V$ is biholomorphic to a product $M' \times W'$ where $M'$ is a regular $(m-1)$-dimensional variety. If $V_A$ is sufficiently small, it can thus be represented by a holomorphic subvariety of an open polydisc $\Delta(A; R) = \Delta(A'; R') \times \Delta(A''; R'') \subseteq \mathbb{C}^m \times \mathbb{C}^n$ in the form

$$V_A = \{Z = (Z', Z'') \in \Delta(A; R) : g_j(Z'') = 0\}$$

where $g_j$ are holomorphic functions in $\Delta(A''; R'')$ and generate the ideal of the subvariety $W'$ at each point of $\Delta(A''; R'')$. Then in view of (3) it follows readily that

$$T(V_A) = T(M') \times T(W')$$

The $m$ holomorphic vector fields $F_j$ can thus be viewed as pairs $F_j(Z) = (F_j'(Z'), F_j''(Z''))$ where $F_j': M' \to T(M') = \mathbb{C}^{m-1}$ and $F_j'': W' \to T(W')$ are holomorphic vector fields on $M'$ and $W'$, respectively. Since the values $F_1(A), \ldots, F_m(A)$ are linearly independent while $\dim T(M') = m - 1$, it follows that $F_j''(A'') \neq 0$ for some index $j$. But then an application of Theorem 6 shows that after shrinking the neighborhood $V_A$ further if necessary, $W' = M'' \times W$ where $M''$ is a regular one-dimensional variety. Therefore altogether $V_A \cong (M' \times M'') \times W$ where $M' \times M''$ is a regular $m$-dimensional variety, and that suffices to conclude the proof.

**8. COROLLARY.** *A holomorphic variety $V$ is regular at a point $A$ precisely when in some open neighborhood of $A$ on $V$ there are holomorphic vector fields $F_1, \ldots, F_n$ such that the values $F_1(A), \ldots, F_n(A)$ span $T_A(V)$.*

*Proof.* If the holomorphic variety $V$ is regular at $A$ and of dimension $m$ there, then $T_A(V)$ is $m$-dimensional and there clearly exist $m$ linearly independent holomorphic vector fields in some open neighborhood of $A$. On the other hand, if $\dim T_A(V) = m$ and if there exist holomorphic vector fields $F_1, \ldots, F_n$ near $A$ such that the values $F_1(A), \ldots, F_n(A)$ span $T_A(V)$, then it can be assumed that the values $F_1(A), \ldots, F_m(A)$ are linearly independent, and it follows from Corollary 7 that an open neighborhood $V_A$ of $A$ in $V$ has the form $V_A = M \times W$ where $M$ is a regular $m$-dimensional manifold. But since $\dim V_A \leq \dim T_A(V) = m$, necessarily $V_A = M$, and the proof is thereby concluded.

# K

# Holomorphic Extensions

The results obtained thus far in the analysis of holomorphic varieties can now be used to complete in some points the discussion of holomorphic extensions of holomorphic functions begun in section I-D. The considerations here will still be limited to results about holomorphic extensions of functions in subsets of $\mathbb{C}^n$, since a corresponding analysis of extension of holomorphic functions on arbitrary holomorphic varieties leads to rather different phenomena that will be treated later. Some corresponding extension properties of holomorphic subvarieties will also be discussed; in the case of subvarieties, though, there is really no difference between these extension properties for holomorphic subvarieties of subsets of $\mathbb{C}^n$ and for holomorphic subvarieties of arbitrary holomorphic varieties. The sort of extension theorem of interest here is that described in Theorem I-D4, which can be rephrased as the assertion that any function holomorphic in the complement of a holomorphic submanifold $V$ of codimension at least two in an open subset $D \subseteq \mathbb{C}^n$ can always be extended uniquely to a holomorphic function in all of $D$. As might now be expected, it is the codimension condition on $V$ rather than its regularity that is really the point. The proper general result is the following.

**1. THEOREM.** *If $V$ is a holomorphic subvariety of an open subset $D \subseteq \mathbb{C}^n$, if* $\dim V \leqq n-2$, *and if $f$ is a holomorphic function in $D - V$, then there is a unique holomorphic function $\tilde{f}$ on all of $D$ such that $\tilde{f}(Z) = f(Z)$ whenever $Z \in D - V$.*

*Proof.* The theorem will be proved by induction on $m = \dim V$. First when $m = 0$, then $V$ is actually a complex submanifold of any open subset of $\mathbb{C}^n$ containing it, so the desired result is just a special case of Theorem I-D4. Next for the inductive step suppose that the desired result has been established for all holomorphic subvarieties of dimension $m - 1$ or less, and consider a holomorphic subvariety $V$ of dimension $m$ in an open subset $D \subseteq \mathbb{C}^n$ where $n - 2 \geqq m$. The singular locus $\mathfrak{S}(V)$ is also a holomorphic subvariety of $D$ by Theorem I15—indeed, meets every irreducible component of $V$ in a proper subvariety of that component—so from Theorem G3 it follows that $\dim \mathfrak{S}(V) \leqq m - 1$. Now $V' = V \cap D'$ is a complex submanifold of the open subset $D' = D - \mathfrak{S}(V) \subseteq \mathbb{C}^n$ and $\dim V' = \dim V = m \leqq n - 2$, so it follows from Theorem I-D4 that every function $f$ holomorphic in

$D - V = D' - V'$ extends to a unique holomorphic function $\tilde{f}'$ in $D'$. But then, since $\dim \mathfrak{S}(V) \leqq m - 1$, it follows from the inductive hypothesis that $\tilde{f}'$ extends to a unique holomorphic function in all of $D$, and that suffices to conclude the proof.

If $V$ is a holomorphic subvariety of dimension $n - 1$ in an open subset $D \subseteq \mathbb{C}^n$, it is of course not the case that every function holomorphic in $D - V$ extends to a function holomorphic throughout $D$. Nonetheless there are various extension theorems applicable to this case as well, one of which is the following.

2. THEOREM. *If $V$ is an irreducible holomorphic subvariety of dimension $n - 1$ in an open subset $D \subseteq \mathbb{C}^n$, if $f$ is a holomorphic function in $D - V$, and if there is a point $A \in V$ such that $f$ extends to a holomorphic function in an open neighborhood of the point $A$, then $f$ extends to a holomorphic function in all of $D$.*

*Proof.* Note first of all that it is sufficient just to prove the theorem in the special case that $V$ is a regular holomorphic subvariety of $D$. Indeed if $V$ is an arbitrary irreducible subvariety of $D$ of dimension $n - 1$, then its singular locus $\mathfrak{S}(V)$ is a holomorphic subvariety of $D$ of dimension at most $n - 2$, while by Theorem E19 its regular locus $\mathfrak{R}(V)$ is connected; thus, $\mathfrak{R}(V)$ is an irreducible regular holomorphic subvariety of dimension $n - 1$ in the open subset $D' = D - \mathfrak{S}(V)$. If $f$ extends to a holomorphic function in an open neighborhood $E$ of the point $A$, then since $\mathfrak{R}(V)$ is dense in $V$, the set $E \cap \mathfrak{R}(V)$ is nonempty and hence $E$ can also be viewed as an open neighborhood of a point of $\mathfrak{R}(V)$. The special case of the theorem implies that $f$ extends to a holomorphic function in $D'$, and it then follows from Theorem 1 that there is a further extension to a holomorphic function in all of $D$ as desired.

Consider then the special case that $V$ is a regular holomorphic subvariety of $D$, and let $V_0$ be the subset of $V$ consisting of these points $A \in V$ such that $f$ extends to a holomorphic function in an open neighborhood of $A$. To complete the proof, it is clearly only necessary to show that $V_0 = V$. The complex manifold $V$ is connected by hypothesis, the subset $V_0$ is an open subset of $V$ by construction, and $V_0$ is nonempty also by hypothesis; thus, it is sufficient just to show that $V_0$ is closed in $V$. For that purpose consider an arbitrary point $A \in \bar{V}_0 \cap V$ and choose an open neighborhood $U$ of $A$ in $D$ such that $U$ is biholomorphic to an open polydisc $\Delta(0; R) \subseteq \mathbb{C}^n$ under a mapping taking $A$ to the origin and $V \cap U$ to the linear subset $L = \{Z \in \Delta(0; R) : z_n = 0\}$. In these terms $f$ is holomorphic in $\Delta(0; R) - L$ and extends to a holomorphic function in an open neighborhood of some point $B = (B', 0) \in L$. The function $\tilde{f}$ defined by

$$\tilde{f}(Z) = \frac{1}{2\pi i} \int_{|\zeta_n| = r_n/2} \frac{f(z_1, \ldots, z_{n-1}, \zeta_n)}{z_n - \zeta_n} \, d\zeta_n$$

is then holomorphic in $\Delta(0; R', r_n/2)$, where $R = (R', r_n)$. Moreover it coincides with $f$ in an open neighborhood of the polydisc $B' \times \Delta(0; r_n/2)$ since $f$ is holomorphic in such a neighborhood, and the identity theorem shows then that $\tilde{f}$ coincides with $f$ throughout $\Delta(0; R', r_n/2) - L$. Thus $\tilde{f}$ provides the desired extension of $f$ to 0, showing that $A \in V_0$ and hence that $V_0$ is closed. That suffices to conclude the proof.

Another extension theorem applicable for exceptional subvarieties of dimension $n-1$ in $\mathbb{C}^n$ is of course the extended form of Riemann's removable singularities theorem, Theorem I-D2. Indeed that and Theorem 1 are really the basic extension theorems for holomorphic functions, although there are various modifications of both results that are useful in some circumstances. Theorem 2 is just a modification of Theorem 1, and an illustrative example of a modification of Riemann's removable singularities theorem is the following.

3. THEOREM. *If $V$ is a holomorphic subvariety of an open subset $D \subseteq \mathbb{C}^n$, if $f$ is a holomorphic function in $D - V$, and if $f \in L^p(D)$ for some index $p$ in the range $2 \leqq p \leqq \infty$, then $f$ extends to a holomorphic function in all of $D$.*

*Proof.* Note that the holomorphic subvariety $V$, being a thin subset of $D$, is a set of Lebesgue measure zero in $D$ as a consequence of Corollary I-A9. The condition that $f$ be contained in $L^p(D)$ is thus meaningful even when $f$ is defined only in $D - V$. Note also that in the special case $p = \infty$ the present theorem reduces to the usual extended form of Riemann's removable singularities theorem, so this theorem can even be viewed as a generalization of the removable singularities theorem. It is clear from Theorems 1 and 2 that it is really sufficient just to prove the present theorem locally at a regular point of $V$ where that subvariety has dimension $n-1$; for then Theorem 2 shows that the function $f$ extends through any component of $V$ having dimension $n-1$, and the extension through all other components of $V$ follows immediately from Theorem 1. It can therefore be assumed that $D = \Delta(0; R)$ is an open polydisc in $\mathbb{C}^n$ and that $V = \{Z \in D : z_n = 0\}$. Moreover since $D$ has finite measure and consequently $L^p(D) \subseteq L^2(D)$ whenever $2 \leqq p \leqq \infty$, it can also be assumed merely that $f \in L^2(D)$. Now any function $f$ holomorphic in $D - V$ has for each fixed point $Z' \in \Delta(0; R') \subseteq \mathbb{C}^{n-1}$ a Laurent expansion in $z_n$ of the form $f(Z) = \sum_{\nu=-\infty}^{\infty} a_\nu(Z') z_n^\nu$ whenever $0 < |z_n| < r_n$, where $Z = (Z', z_n)$ and $R = (R', r_n)$ as usual, and the coefficients $a_\nu(Z')$ are evidently holomorphic functions in $\Delta(0; R')$. It follows from Fubini's theorem that when viewed as a function of $z_n$ alone, $f(Z) \in L^2(\Delta(0; r_n))$ for almost all points $Z' \in \Delta(0; R')$. If the desired result has already been demonstrated for the case $n = 1$, then for each $\nu < 0$ necessarily $a_\nu(Z') = 0$ almost everywhere in $\Delta(0; R')$, and continuity then implies that $a_\nu \equiv 0$ for $\nu < 0$ and consequently that $f$ is holomorphic in $\Delta(0; R)$ as desired. It is therefore finally enough just to prove this special case of the theorem when $n = 1$. If $f = \sum_{\nu=-\infty}^{\infty} a_\nu z^\nu \in L^2(\Delta(0; R))$, then of course $f \in L^2(A_\varepsilon)$ for the annulus $A_\varepsilon = \Delta(0; r) - \bar{\Delta}(0; \varepsilon)$ whenever $0 < \varepsilon < r$. The functions $z^\nu$ are readily seen to be orthogonal in $L^2(A_\varepsilon)$, so that

$$\infty > \|f\|_0^2 \geqq \|f\|_\varepsilon^2 = \sum_{\nu=-\infty}^{\infty} |a_\nu|^2 \|z^\nu\|_\varepsilon^2 \tag{1}$$

where $\|f\|_\varepsilon^2 = \int_{A_\varepsilon} |f(z)|^2 \, dx \, dy$. A simple calculation shows however that

$$\|z^\nu\|_\varepsilon^2 = \begin{cases} 2\pi(\log r - \log \varepsilon) & \text{if } \nu = -1 \\ \dfrac{\pi(r^{2\nu+2} - \varepsilon^{2\nu+2})}{\nu + 1} & \text{if } \nu \neq -1 \end{cases}$$

and therefore $\lim_{\varepsilon \to 0} \|z^{\nu}\|_{\varepsilon}^{2} = \infty$ whenever $\nu \leqq -1$. The inequality (1) can then persist as $\varepsilon$ tends to zero only if $a_{\nu} = 0$ whenever $\nu \leqq -1$, so the function $f$ must therefore be holomorphic in $\Delta(0; r)$ as desired. As already observed, that suffices to conclude the proof.

As noted at the beginning of this section, for holomorphic subvarieties there are extension theorems that are quite analogous to the preceding extension theorems for holomorphic functions. It is convenient to begin the discussion of these extension theorems by considering in the next two lemmas the following rather straightforward special case. Let

$$\Delta(0; R) = \Delta(0; R') \times \Delta(0; R'') \subseteq \mathbb{C}^{k} \times \mathbb{C}^{n-k} = \mathbb{C}^{n}$$

be an open polydisc decomposed nontrivially as a product as indicated, let $W$ be the $k$-dimensional regular holomorphic subvariety of $\Delta(0; R)$ defined by

$$W = \{Z = (Z', Z'') \in \Delta(0; R') \times \Delta(0; R''): Z'' = 0\}$$

and let $V$ be an irreducible $k$-dimensional holomorphic subvariety of $\Delta(0; R) - W$ such that

$$V \cap [\Delta(0; R') \times (\Delta(0; R'') - \bar{\Delta}(0; S''))] = \emptyset \tag{2}$$

for some polydisc $\Delta(0; S'') \subseteq \bar{\Delta}(0; S'') \subseteq \Delta(0; R'')$. The geometric situation is perhaps most easily kept in mind by referring to Figure 1; the assumption (2) is that the subvariety $V$ avoids the shaded region.

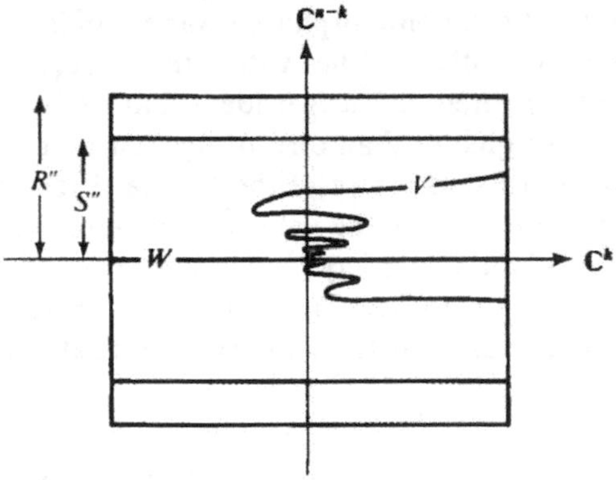


Figure 1

4. LEMMA. *In the situation described above, the subvariety $V$ meets every open neighborhood of the degenerate polydisc $A' \times \Delta(0; R'')$ for any point $A' \in \Delta(0; R')$.*

*Proof.* Intuitively the content of this lemma is the assertion that the subvariety $V$ must project essentially to the whole of the polydisc $\Delta(0; R')$ and so really cannot be quite as sketched in Figure 1. It is enough just to show that $V$ meets every open neighborhood of the degenerate polydisc $0 \times \Delta(0; R'')$, since there are biholomorphic mappings of the polydisc $\Delta(0; R')$ to itself taking any prescribed point $A' \in \Delta(0; R')$ to the origin. Consider first the special case $k = 1$, and suppose to the contrary of the desired result that there is a constant $\varepsilon > 0$ such that $|z_1| > \varepsilon$ whenever $Z = (z_1, z_2, \ldots, z_n) \in V$. For any index $2 \leqq j \leqq n$ either there is a point $B_j = (b_{j1}, \ldots, b_{jn}) \in V$ for which $b_{jj} \neq 0$ or $z_j$ is identically zero on $V$. If there is such a point $B_j$ consider for any integer $v_j > 0$ the restriction to $V$ of the function $(z_j/r_j)(z_1/r_1)^{-v_j}$, noting that this restriction is a holomorphic function $f_j$ on $V$ since $|z_1| > \varepsilon$ on $V$. By choosing $v_j$ sufficiently large, it can be supposed that $|f_j(B_j)| = |b_{jj}/r_j| \cdot |b_{j1}/r_1|^{-v_j} > 2$, since $b_{jj} \neq 0$ and $0 < |b_{j1}/r_1| < 1$. On the other hand, as a point $Z = (z_1, Z'') \in V$ approaches the boundary of $V$—that is to say, for points $Z$ outside an increasing sequence of compact subsets of $V$ exhausting $V$—either $|z_1| \to r_1$ or $\|Z''\| \to 0$. It is then evident from the form of the function $f_j$ that $|f_j(Z)| < 2$ whenever $Z$ is near enough the boundary of $V$. That is impossible by the maximum modulus theorem on $V$. There can thus be no such point $B_j$, so that $z_j$ is identically zero on $V$, and that is so for all indices $2 \leqq j \leqq n$, so that $V \subseteq \{Z \in \Delta(0; R) : z_2 = \cdots = z_n = 0\}$. Since both of these subvarieties are one-dimensional and the larger one is irreducible, this inclusion must actually be an identity. This subvariety does meet every open neighborhood of $0 \times \Delta(0; R'')$ though, contradicting the assumption and thereby proving the desired result in the special case $k = 1$.

The general case of the lemma then follows easily by induction on the dimension $k$, the initial case $k = 1$ having just been settled. Consider a $k$-dimensional subvariety $V$, choose any point $C = (c_1, \ldots, c_n) \in V$, and let $V_1 = V \cap \{z_1 = c_1, \ldots, z_{k-1} = c_{k-1}\}$. It is clear from Corollary G8 that each irreducible component of $V_1$ has dimension not less than one. On the other hand, $V_1 \cap \{z_k = c_k\} = V \cap \{z_1 = c_1, \ldots, z_k = c_k\}$ is a finite point set; for it is a holomorphic subvariety of $(c_1, \ldots, c_k) \times \Delta(0; R'')$ that is compact, since by hypothesis it lies within $(c_1, \ldots, c_k) \times \Delta(0; S'')$, hence is zero-dimensional. So by Theorem G7 it is evident that each irreducible component of $V_1$ has dimension not greater than one. Altogether then $V_1$ is pure one-dimensional, so by the one-dimensional case of the lemma there are points $Z \in V$ for which $z_k$ is arbitrarily close to zero; the value $c_k$ can be chosen as near zero as desired. Now by Theorem G7 the subvariety $V_{k-1} = V \cap \{z_k = c_k\}$ is of pure dimension $k - 1$, so from the case $k - 1$ of the lemma it follows that there are points $Z \in V_{k-1}$ for which $(z_1, \ldots, z_{k-1})$ is also arbitrarily near the origin. That then suffices to conclude the proof.

**5. LEMMA.** *In the situation described above, if the point set closure $\bar{V}$ of $V$ is a holomorphic subvariety in an open neighborhood of some point of $W$, then $\bar{V} \cap \Delta(0; R)$ is actually a holomorphic subvariety of the entire polydisc $\Delta(0; R)$.*

*Proof.* The restriction to $V$ of the natural projection $\mathbb{C}^n = \mathbb{C}^k \times \mathbb{C}^{n-k} \to \mathbb{C}^k$ is a holomorphic mapping $\pi: V \to \Delta(0; R')$. For any point $A' \in \Delta(0; R')$ the inverse image $\pi^{-1}(A') = V \cap (A' \times \Delta(0; R''))$ can be viewed as a holomorphic subvariety of the punctured polydisc $\Delta(0; R'') - 0$, and as a consequence of assumption (2) this subvariety must be contained within the smaller polydisc $\Delta(0; S'')$. The coordinate functions in $\mathbb{C}^{n-k}$ then evidently attain their maximum values on each irreducible component of $\pi^{-1}(A')$, so by the maximum modulus theorem for holomorphic varieties, Theorem H2, they must be constant on each component. Thus $\pi^{-1}(A')$ consists of a discrete set of points in $\Delta(0; R'') - 0$, and if there are infinitely many such points, their only limit point in $\bar{\Delta}(0; R'')$ is the origin 0. The germ $\pi_A$ of the mapping $\pi$ at any point $A \in V$ is thus necessarily finite, so as in Theorem E7 the restriction of $\pi_A$ to any irreducible component of the germ $\mathbf{V}_A$ is the germ of a finite branched holomorphic covering of its image. Actually since $\mathbf{V}_A$ is of pure dimension $k$, these images are necessarily germs of open subsets in $\mathbb{C}^k$, in view of Theorem G16, so that $\pi_A$ is itself the germ of a finite branched holomorphic covering of $\mathbb{C}^k$. Let $o_\pi(A)$ be the branching order of $\pi$ at $A$ as usual, and set $\nu(A') = \sum_{A \in \pi^{-1}(A')} o_\pi(A)$ for any point $A' \in \Delta(0; R')$. It is easy to see that if $\nu(A') \geqq \nu$ for some point $A' \in \Delta(0; R')$ and some integer $\nu$, then $\nu(Z') \geqq \nu$ for all points $Z'$ in an open neighborhood of $A'$ in $\Delta(0; R')$. Indeed if $A_1, \ldots, A_k \in \pi^{-1}(A')$ are any points for which $\sum_j o_\pi(A_j) = \mu \geqq \nu$ there is clearly an open polydisc $\Delta(A'; \varepsilon) \subseteq \Delta(0; R'')$ for which $\pi^{-1}(\Delta(A'; \varepsilon))$ contains $\mu$ disjoint finite branched holomorphic coverings of $\Delta(A'; \varepsilon)$, one passing through each point $A_j$. But then $\nu(Z') \geqq \mu \geqq \nu$ whenever $Z' \in \Delta(A'; \varepsilon)$.

With these preliminary observations and the accompanying notation, note that the hypothesis of the present lemma asserts the existence of an open polydisc $\Delta(B'; \delta) \subseteq \Delta(0; R')$ centered at some point $B' \in \Delta(0; R')$ such that $\bar{V}$ is a holomorphic subvariety in $\Delta(B'; \delta) \times \Delta(0; R'')$. Then evidently $\nu(A') < \infty$ for all points $A' \in \Delta(B'; \delta)$, since $\pi^{-1}(A')$ must be a discrete holomorphic subvariety of $\Delta(0; R'')$ rather than just of the punctured polydisc $\Delta(0; R'') - 0$. The projection $\pi(V)$ is dense in $\Delta(0; R')$ by Lemma 4, so it can be assumed that $B' \in \pi(V)$ and hence that $\nu(B') \geqq 1$. Furthermore by taking $\delta$ sufficiently small, it can also be assumed that $\nu(A') \geqq \nu(B')$ for all points $A' \in \Delta(B'; \delta)$. The subsets $\{A' \in \Delta(B'; \delta) : \nu(A') = \nu\}$ for the integers $\nu = 1, 2, 3, \ldots$ thus cover the polydisc $\Delta(B'; \delta)$, and it follows from the Baire category theorem that at least one of these subsets has a nonempty interior. There must consequently be an integer $\nu \geqq 1$ for which the set $U = \{A' \in \Delta(0; R') : \nu(Z') = \nu$ for all $Z'$ in an open neighborhood of $A'$ in $\Delta(0; R')\}$ is a nonempty open subset of the full polydisc $\Delta(0; R')$. It is clear that $\pi: \pi^{-1}(U) \to U$ is necessarily a finite branched holomorphic covering of $U$ of pure order $\nu$.

If, $U = \Delta(0; R')$ the desired result evidently holds, since then $V = \bar{V} \cap \Delta(0; R)$. Otherwise, to continue the argument, consider a point $A' \in \partial U \cap \Delta(0; R')$. If $\nu(A') > \nu$, then from the first part of the proof note that $\nu(Z') > \nu$ for all points $Z' \in \Delta(0; R')$ sufficiently near $A'$; but there are points $Z' \in U$ arbitrarily near $A'$, and for these points $\nu(Z') = \nu$, a contradiction. If $\nu(A') = \nu$, then the union of the germs of the subvariety $V$ at the points $\pi^{-1}(A')$ form a germ of a finite branched holomorphic covering of $\mathbb{C}^k$ of order $\nu$. Consequently for sufficiently small values $\varepsilon > 0$ the subset $\pi^{-1}(\Delta(A'; \varepsilon)) \subseteq V$ can be written as the disjoint union $\pi^{-1}(\Delta(A'; \varepsilon)) = V_1 \cup V_2$ of two holomorphic subvarieties of $\Delta(A'; \varepsilon) \times (\Delta(0; R'') - 0)$, and the restriction $\pi | V_1 : V_1 \to \Delta(A'; \varepsilon)$ is a finite branched holomorphic covering of order $\nu$. If $V_2 = \emptyset$, then clearly $\nu(Z') = \nu$ for all points $Z' \in \Delta(A'; \varepsilon)$ so that necessarily $A' \in U$, a contradiction. On the other hand, if $V_2 \neq \emptyset$, then applying Lemma 4 to this subvariety shows that $\pi(V_2)$ is dense in $\Delta(A'; \varepsilon)$ and hence that there are points $Z' \in \Delta(A'; \varepsilon) \cap U \cap \pi(V_2)$.

But for any such points clearly $v(Z') > v$, since $\pi^{-1}(Z')$ includes points in $V_2$ in addition to the $v$ points (counting multiplicity) in $V_1$, and that is a contradiction since $Z' \in U$. Thus it is also impossible that $v(A') = v$, so that necessarily $v(A') < v$. Again for sufficiently small values $\varepsilon > 0$ it is possible to write $\pi^{-1}(\Delta(A'; \varepsilon)) = V_1 \cup V_2$, where $V_1$ and $V_2$ are disjoint holomorphic subvarieties of $\Delta(A'; \varepsilon) \times (\Delta(0; R'') - 0)$ and the restriction $\pi | V_1 : V_1 \to \Delta(A'; \varepsilon)$ is a finite branched holomorphic covering of order $v(A')$. The set $V_2$ is nonempty—indeed, $\pi(V_2) \supseteq U \cap \Delta(A'; \varepsilon)$ since $v(Z') = v > v(A')$ whenever $Z' \in U \cap \Delta(A'; \varepsilon)$, but $V_2 \cap \pi^{-1}(A') = \emptyset$. Now for any given value $\eta > 0$ it is easy to see that by choosing $\varepsilon$ sufficiently small it can be supposed that $V_2 \subseteq \Delta(A'; \varepsilon) \times \Delta(0; \eta)$. Indeed if that were not the case there would exist a value $\eta > 0$ and a sequence of points $Z_\mu = (Z'_\mu, Z''_\mu) \in V_2$ such that $Z'_\mu \to A'$ while $\|Z''_\mu\| \geq \eta$, and by passing to a subsequence if necessary it can also be supposed that $Z''_\mu \to A''$ for some point $A'' \in \Delta(0; R'') - \Delta(0; \eta)$. The limit point $A = (A', A'')$ must be contained in $V_2$, since clearly $V_2$ is relatively closed in $\Delta(A'; \varepsilon) \times (\Delta(0; R'') - 0)$, but that is impossible since $V_2 \cap \pi^{-1}(A') = \emptyset$ as already noted. It then follows that for any sequence of points $Z'_\mu \in U$ such that $Z'_\mu \to A'$ there are points $Z_\mu = (Z'_\mu, Z''_\mu) \in V$ such that $Z'' \to 0$, by merely selecting any points $Z_\mu \in V_2$.

Consider then the finite branched holomorphic covering $\pi: \pi^{-1}(U) \to U$ of pure order $v$. To each coordinate function $z_j$ for $k + 1 \leq j \leq n$ there is associated as in Theorem C5 a monic polynomial $P_j(X) \in {}_k\mathcal{O}_U[X]$ of order $v$ such that $P_j(z_j)$ vanishes identically on $\pi^{-1}(U)$. The $v$ roots of the polynomial $P_j(X)$ over any point $Z' \in U$ are just the $j$th coordinates of the $v$ points $\pi^{-1}(Z') \in V$, counting multiplicities. Since $Z'' \neq 0$ whenever $Z = (Z', Z'') \in V$, it can be supposed by a suitable choice of coordinates in $\mathbb{C}^{n-k}$ that there is some point $Z' \in U$ such that $z_n \neq 0$ whenever $Z = (Z', z_{k+1}, \ldots, z_n) \in V$. The constant term $a_v \in {}_k\mathcal{O}_U$ of the polynomial $P_n(X)$ is then not identically zero in $U$. On the other hand, $a_v(Z'_\mu) \to 0$ for any sequence of points $Z'_\mu \in U$ converging to a point $A' \in \partial U \cap \Delta(0; R')$, since as demonstrated in the preceding paragraph there are points $Z_\mu = (Z'_\mu, Z''_\mu) \in V$ for which $Z''_\mu \to 0$ while $Z''_\mu \in \Delta(0; R'')$ and are hence uniformly bounded for any points $Z_\mu = (Z'_\mu, Z''_\mu) \in V$. Extending the function $a_v$ to the entire polydisc $\Delta(0; R')$ by setting $a_v(Z') = 0$ whenever $Z' \in \Delta(0; R') - U$ thus yields a continuous function on $\Delta(0; R')$, and one that is holomorphic wherever it is nonzero since it can be nonzero only at points of $U$. The extended function must be holomorphic in $\Delta(0; R')$ by Rado's theorem, Corollary I-Q6; since it is not identically zero, it follows that the complement of $U$ in $\Delta(0; R')$ is contained in the proper holomorphic subvariety $\{Z' \in \Delta(0; R'): a_v(Z') = 0\}$. All the coefficients of the polynomial $P_j(X)$ are holomorphic and uniformly bounded in $U$, since $Z'' \in \Delta(0; R'')$ whenever $Z = (Z', Z'') \in V$ for any point $Z' \in U$, so it follows from the extended form of Riemann's removable singularities theorem that these coefficients extend to holomorphic functions throughout $\Delta(0; R')$. Let $P_j(X) = P_j(Z'; X)$ also denote the polynomial with these extended functions as coefficients, and consider the holomorphic subvariety

$$\tilde{V} = \{Z = (Z', Z'') \in \Delta(0; R') \times \mathbb{C}^{n-k} : P_j(Z'; Z'') = 0 \text{ for } k + 1 \leq j \leq n\}$$

noting as observed in Theorem D1 that it too is a finite branched holomorphic covering of $\Delta(0; R')$ under the natural projection $\tilde{\pi}: \tilde{V} \to \Delta(0; R')$. It is clear that

$V \subseteq \tilde{V}$—indeed that $V - \pi^{-1}(D)$ is a connected component of $\tilde{V} - \tilde{\pi}^{-1}(\tilde{D})$ where $\tilde{D}$ is the union of the image of the branch locus of $\tilde{V}$ and of the subvariety $\{Z' \in \Delta(0; R') : a_v(Z') = 0\}$. It then follows directly from Theorem C15 that the closure of $V$ in $\Delta(0; R)$, which is the same as the closure of $V - \pi^{-1}(\tilde{D})$, is a holomorphic subvariety of $\Delta(0; R)$ as desired. That suffices to conclude the proof of the lemma.

It is worth noting here that in the preceding lemma the hypothesis that $\bar{V}$ is a holomorphic subvariety in an open neighborhood of some point of $W$ is really necessary in the proof. Indeed if $\{Z''_\mu\}$ is any sequence of points in $\Delta(0; S'') - 0$ such that $Z''_\mu \to 0$, then the subvariety $V = \bigcup_\mu \Delta(0; R') \times Z''_\mu$ satisfies all the other hypotheses of the lemma, but $V$ is clearly not a holomorphic subvariety in $\Delta(0; R)$. Now the special situation considered in the preceding lemmas and described in Figure 1 actually arises quite commonly, as the following result indicates.

**6. LEMMA.** *Let $D$ be an open subset in $\mathbb{C}^n$, $W$ be a pure-dimensional holomorphic subvariety of $D$, and $V$ be a pure-dimensional holomorphic subvariety of $D - W$ with* $\dim V = \dim W$. *Then to each point $A \in \mathfrak{R}(W) \cap \bar{V}$ there are an open neighborhood $U$ of $A$ in $D$ and local coordinates $z_1, \ldots, z_n$ in $U$ centered at $A$ such that in these coordinates the subvarieties $W \cap U$ and $V \cap U$ are in the special situation described in Figure* 1.

*Proof.* Choose an open neighborhood $U$ of $A$ in $D$ and a local coordinate system $z_1, \ldots, z_n$ in $U$ centered at $A$ such that $W \cap U$ is a $k$-dimensional linear subspace in these coordinates. It is fairly easy to see that after a nonsingular linear change of coordinates it can further be supposed that $W \cap L$ and $V \cap L$ are holomorphic subvarieties of dimension at most zero, where $L$ is the $(n - k)$-dimensional coordinate hyperplane $L = \{Z \in U : z_1 = \cdots = z_k = 0\}$. Actually since $W$ remains a subspace under such a change of coordinates, the intersection $W \cap L$ is just the origin in $L$, while since $V$ was only assumed to be a holomorphic subvariety outside $W$, the intersection $V \cap L$ is really only a holomorphic subvariety of the complement $L - O$ of the origin in $L$, in the new coordinate system. Indeed this assertion can readily be established by induction on the dimension $k$. The initial case $k = 0$ being quite trivial, it is only necessary to consider the inductive step. Note first that $W \cap U$ and $V \cap (U - W \cap U)$ have at most countably many irreducible components. Choose a point $A_v \neq A$ in each component, and let $l$ be a homogeneous linear function of the given local coordinates such that $l(A_v) \neq 0$ for all $A_v$. There certainly exists such a function, since for any individual $A_v$ the set of coefficients of those linear functions $l$ for which $l(A_v) \neq 0$ is a dense open subset of $\mathbb{C}^n$ and the intersection of countably many such subsets is nonempty by the Baire category theorem. After a nonsingular linear change of coordinates it can be supposed that $l(Z) = z_1$. The intersections of $W$ and $V$ with the coordinate hyperplane $z_1 = 0$ in $U$ are either empty or of pure dimension $k - 1$, by Theorem G7 and the observation that by construction $l$ does not vanish identically on any irreducible component of either subvariety. The inductive hypothesis can then be applied to these intersections; thus, after another nonsingular linear change of coordinates involving only the variables $z_2, \ldots, z_n$ it can be assumed that the set of points of $W$ and $V$ in the hyperplane $z_1 = 0$ at which $z_2 = \cdots = z_k = 0$ are holomorphic subvarieties of dimension at most zero, and that is just the desired result.

Now after yet another nonsingular linear change of coordinates involving only the variables $z_{k+1}, \ldots, z_n$, it can be supposed in addition to what has already been achieved that $W \cap U = \{Z \in U : z_{k+1} = \cdots = z_n = 0\}$. The intersection $V \cap L$ is at most zero-dimensional and so meets any compact subset of $L - O$ in finitely many points; thus, there are polydiscs $\bar{\Delta}(0; S'') \subseteq \Delta(0; R'') \subset L$ such that $V \cap [0 \times (\bar{\Delta}(0; R'') - \Delta(0; S''))] = \emptyset$. The compact subset $0 \times (\bar{\Delta}(0; R'') - \Delta(0; S'')) \subseteq L$ must indeed have an open neighborhood not meeting $V$, since $V$ is relatively closed in $U - W \cap U$, so there is a polydisc $\Delta(0; R') \subseteq \mathbb{C}^k$ sufficiently small that $V \cap [\Delta(0; R') \times (\Delta(0; R'') - \bar{\Delta}(0; S''))] = \emptyset$. That means that $W \cap U$ and $V \cap U$ are in the special form of interest, and thereby concludes the proof.

The basic extension theorem for holomorphic subvarieties now follows readily from the last two lemmas.

7. **THEOREM (theorem of Thullen, Remmert, and Stein).** *Let $D$ be an open subset of $\mathbb{C}^n$, $W$ be an irreducible holomorphic subvariety of $D$, and $V$ be a pure-dimensional holomorphic subvariety of $D - W$.*

(i) *If* $\dim V > \dim W$, *then* $\bar{V} \cap D$ *is a holomorphic subvariety of* $D$.
(ii) *If* $\dim V = \dim W$ *and* $\bar{V}$ *is a holomorphic subvariety in an open neighborhood of some point of* $W$, *then* $\bar{V} \cap D$ *is a holomorphic subvariety of* $D$.

*Proof.* (i) It is enough just to show that $\bar{V}$ is a holomorphic subvariety in an open neighborhood of any point $A \in \mathfrak{R}(W) \cap \bar{V}$; for then $\bar{V} \cap D$ is a holomorphic subvariety of $D - \mathfrak{S}(W)$, and since $\mathfrak{S}(W)$ is a holomorphic subvariety of $D$ with $\dim \mathfrak{S}(W) < \dim W$, the obvious induction yields the desired result. Consider then a point $A \in \mathfrak{R}(W) \cap \bar{V}$, choose an irreducible regular holomorphic subvariety $W'$ in an open neighborhood $U$ of $A$ in $D$ so that $W \cap U \subseteq W'$ and $\dim W' = \dim V$, and set $V' = V \cap U - V \cap W'$. It follows from Lemma 6 that after shrinking the neighborhood $U$ enough there will be a local coordinate system in $U$ in terms of which the subvarieties $W'$ and $V'$ are of the special form to which Lemma 5 applies. Since $\dim W < \dim W'$, there are points in $W'$ outside $W$, and near any such point $\bar{V}' = V'$ is a holomorphic subvariety; Lemma 5 then shows that $\bar{V}' \cap U$ is a holomorphic subvariety of $U$. By shrinking $U$ further if necessary, it can be supposed that $\bar{V}' \cap U$ has finitely many irreducible components $V_1, \ldots, V_r$; the complements $V_j' = V_j - V_j \cap W'$ remain irreducible, as is evident from Theorem E19 for instance, so that $V'$ has the irreducible components $V_1', \ldots, V_r'$. Now for the original subvariety $V$ let $V_0$ be any irreducible component of $V \cap U$. If $V_0 \subseteq W'$, then of course $V_0 = W' - W$ so that $\bar{V}_0 \cap U = W'$. If $V_0 \nsubseteq W'$, then $V_0 - V_0 \cap W'$ is an irreducible subvariety of $V'$ having the same dimension as $V'$, so that necessarily $V_0 - V_0 \cap W' = V_j'$ for some index $j$ and then $\bar{V}_0 \cap U = V_j$. It therefore follows that $\bar{V} \cap U$ is also a holomorphic subvariety of $U$ as desired—indeed, that either $\bar{V} \cap U = \bar{V}' \cap U$ or $\bar{V} \cap U = (\bar{V}' \cap U) \cup W'$.

(ii) It is enough to show that $\bar{V}$ is a holomorphic subvariety in an open neighborhood of any point $A \in \mathfrak{R}(W)$; for then $\bar{V} \cap D$ is a holomorphic subvariety of $D - \mathfrak{S}(W)$, and since $\dim \mathfrak{S}(W) < \dim W = \dim V$, an application of part (i) shows that $\bar{V} \cap D$ is a holomorphic subvariety of all of $D$. Let $X$ be the set of all points $A \in \mathfrak{R}(W)$ such that $\bar{V}$ is a holomorphic subvariety in some open neighborhood of the point $A$. The set $X$ is thus an open subset of $\mathfrak{R}(W)$, and it is an immediate consequence of the hypotheses in this case that $X$ is nonempty. On the other hand, for any point $A \in \bar{X} \cap \mathfrak{R}(W)$ it follows from Lemma 6 that there are an open neighborhood $U$ of $A$ in $D$ and a local coordinate system in $U$ in terms of which the subvarieties $W \cap U$ and $V \cap U$ are of the special form to which Lemma 5 applies. Since $X \cap U$ is nonempty, Lemma 4 shows that $\bar{V} \cap U$ is a holomorphic subvariety of $U$, from which it follows that $A \in X$ so that $X$ is also a closed subset of $\mathfrak{R}(W)$. The set $\mathfrak{R}(W)$ is connected since $W$ is an irreducible subvariety, by Corollary E20, and therefore $X = \mathfrak{R}(W)$ as desired. That concludes the proof of the theorem.

This theorem describes conditions under which a holomorphic subvariety $V$ defined only in the complement of another holomorphic subvariety $W$ extends through $W$ as a holomorphic subvariety, with the extension being given by the point set closure of $V$. The basic result is that $V$ automatically extends through any subvariety of strictly smaller dimension, and extends through any subvariety of the same dimension provided that it extends through at least one point on each irreducible component of the exceptional subvariety $W$; as noted earlier, these hypotheses are really necessary. The hypothesis that $W$ is irreducible is clearly just for convenience of statement, especially in connection with part (ii) of the theorem; the theorem can be applied to the components of $W$ successively if $W$ happens to be reducible. Similarly the hypothesis that $V$ is pure dimensional is also merely for convenience of statement; the theorem can be applied separately to the pure-dimensional components of $V$, provided of course that all are of the appropriate dimensions. It may be worth noting that it is not necessary to assume that $V$ has only finitely many irreducible components near any point of $W$; that follows automatically from the conclusion of the theorem though. Finally, it should be noted that the theorem extends trivially to the corresponding assertions for holomorphic subvarieties $W$ and $V$ of an arbitrary holomorphic variety $D$, for as was evident from the proof, the preceding theorem is really a local theorem, an arbitrary holomorphic variety $D$ can be represented locally as a holomorphic subvariety of an open subset $U$ in some space $\mathbb{C}^n$, and the theorem can be applied directly to the images of the subvarieties $W$ and $V$ in the subset $U \subseteq \mathbb{C}^n$.

# L

# Holomorphic Mappings

Thus far the only holomorphic mappings that have been examined in any detail are finite holomorphic mappings, but by this point enough machinery has been developed to permit the fairly easy derivation of a number of useful properties of holomorphic mappings in general. Again, though, attention will primarily be limited here to local properties of holomorphic mappings. The first basic notion to be used in this discussion is the following.

1. DEFINITION. *If $F: V \to W$ is a holomorphic mapping between two holomorphic varieties, the* **level set** *of $F$ at a point $A \in V$ is the subset*

$$L_A(F) = \{Z \in V : F(Z) = F(A)\} = F^{-1}(F(A))$$

It is obvious from the definition that the level set $L_A(F)$ is a holomorphic subvariety of $V$ passing through the point $A$. In an alternate terminology in common use the level set $L_A(F)$ is called the fibre of the mapping $F$ at the point $A$. The finite holomorphic mappings are precisely those holomorphic mappings for which the level sets are all finite point sets. More generally, the holomorphic mappings that are locally finite, in the sense that their germs at any points are the germs of finite holomorphic mappings, are precisely those holomorphic mappings for which the level sets are all zero-dimensional holomorphic subvarieties. That suggests that the dimensions of the level sets of holomorphic functions should be of some interest in the present case, as is indeed so. A useful preliminary observation about these dimensions is the following semicontinuity lemma.

2. LEMMA. *If $F: V \to W$ is a holomorphic mapping between two holomorphic varieties, then $\dim_Z L_Z(F)$ is an upper semicontinuous function of $Z \in V$.*

*Proof.* Since the function $\dim_Z L_Z(F)$ takes only integral values, upper semicontinuity just means that any point $A \in V$ has an open neighborhood $V_A$ such that $\dim_B L_B(F) \leq \dim_A L_A(F)$ whenever $B \in V_A$. Some open neighborhood $V_A$ of $A$ in $V$ can always be represented by a holomorphic subvariety

$$V_A = \{Z \in U : g_1(Z) = \cdots = g_r(Z) = 0\}$$

of an open neighborhood $U$ of the origin in $\mathbb{C}^m$, where $A$ corresponds to the origin and $g_j$ are holomorphic functions in $U$. If the neighborhood $U$ is sufficiently small, the mapping $F$ on $V_A$ can also be written as the restriction to $V_A$ of a holomorphic mapping $U \to \mathbb{C}^n$ described by coordinate functions $f_1, \ldots, f_n$ that are holomorphic in $U$. In these terms the level set of $F$ through a point $B \in V_A$ can be described locally as

$$L_B(F) \cap V_A = \{Z \in U : g_j(Z) = 0, f_k(Z) = f_k(B)\}$$

By translating this subvariety so that the point $B$ is taken to the origin, it is evident that the germ at the point $B$ of the subvariety $L_B(F)$ is biholomorphic to the germ at the origin of the subvariety

$$\{Z \text{ near } O: g_j(Z + B) = f_k(Z + B) - f_k(B) = 0\}$$

It follows immediately from the standard semicontinuity lemma for dimension, Lemma G6, that the dimension of the latter germ is an upper semicontinuous function of $B$, and that suffices to conclude the proof.

When investigating holomorphic mappings between complex manifolds, the rank of the mapping, the rank of the Jacobian matrix of the mapping when described in terms of any local coordinate systems, is a very convenient tool. It is possible to introduce a corresponding notion when considering holomorphic mappings between arbitrary holomorphic varieties, by utilizing the concept of the tangent space to an arbitrary holomorphic variety as treated in section I. That leads to a second basic notion to be used in the present discussion.

3. **DEFINITION.** *Let $F: V \to W$ be a holomorphic mapping between two holomorphic varieties and $dF_A: T_A(V) \to T_{F(A)}(W)$ be its differential at a point $A \in V$. The rank of the linear mapping $dF_A$ will also be referred to as the* **rank** *of the mapping $F$ at the point $A$ and will be denoted by* $\operatorname{rank}_A F$. *Correspondingly the* **corank** *of the mapping $F$ at the point $A$ will be defined as* $\operatorname{corank}_A F = \operatorname{tdim}_A V - \operatorname{rank}_A F$.

The rank of the mapping $F: V \to W$ at the point $A \in V$ is thus the dimension of the image of the linear mapping $dF_A: T_A(V) \to T_{F(A)}(W)$, while the corank is the dimension of the kernel of the linear mapping $dF_A$. Any two representatives of the germ of a holomorphic mapping have the same differential, so that the notions of the rank and corank of a germ of a holomorphic mapping between two germs of holomorphic varieties are obviously well defined. For mappings between complex manifolds the tangential dimension is constant on each connected component, being just the dimension of that component, so that introducing the corank as well as the rank is just a matter of notational convenience. But for mappings between general holomorphic varieties, where the tangential dimension can vary from point to point, these two notions are somewhat different, and as will become apparent, corank is actually the more useful and convenient concept.

To examine these notions more closely, consider a holomorphic mapping $F: V \to W$. As before some open neighborhood $V_A$ of a point $A$ in $V$ can be represented by a holomorphic subvariety

$$V_A = \{Z \in U : g_1(Z) = \cdots = g_r(Z) = 0\}$$

of an open neighborhood of the origin in $\mathbb{C}^m$, where $A$ corresponds to the origin and $g_j$ are holomorphic in $U$. Moreover by Cartan's theorem, Theorem F6, it can be supposed that the germs of the functions $g_j$ generate the ideal of the subvariety $V_A$ at each point of $U$. If the neighborhood $U$ is sufficiently small, the mapping $F$ on $V_A$ can also be written as the restriction to $V_A$ of a holomorphic mapping $\tilde{F}: U \to \mathbb{C}^n$ described by coordinate functions $f_1, \ldots, f_n$ that are holomorphic in $U$. Then as in Theorem I4 there is a canonical identification

$$T_Z(V) \cong \left\{t \in \mathbb{C}^m : \sum_j t_j \frac{\partial g_k}{\partial z_j}(Z) = 0,\ 1 \leqq k \leqq r\right\}$$

at each point $Z \in V_A$, and as in Theorem I9 the differential $dF_Z: T_Z(V) \to T_{F(Z)}(W)$ can be described in terms of this identification as the restriction to the subspace $T_Z(V) \subseteq \mathbb{C}^m$ of the linear mapping $J_{\tilde{F}}(Z): \mathbb{C}^m \to \mathbb{C}^n$ given by the Jacobian matrix of the mapping $\tilde{F}$ at the point $Z$. The kernel of the differential $dF_Z: T_Z(V) \to T_{F(Z)}(W)$ can therefore be identified with the linear subspace

$$\ker dF_Z \cong \left\{t \in \mathbb{C}^m : \sum_j t_j \frac{\partial g_k}{\partial z_j}(Z) = \sum_j t_j \frac{\partial f_l}{\partial z_j}(Z) = 0,\ 1 \leqq k \leqq r,\ 1 \leqq l \leqq n\right\} \tag{1}$$

That leads immediately to the following observation.

**4. THEOREM.** *If $F: V \to W$ is a holomorphic mapping between holomorphic varieties, then for any integer $\nu$*

$$\{Z \in V : \operatorname{corank}_Z F \geqq \nu\}$$

*is a holomorphic subvariety of $V$.*

*Proof.* Since the result is local, it suffices to prove it on an open neighborhood $V_A$ of any point $A \in V$. Since $\operatorname{corank}_Z F$ is just the dimension of the kernel of the linear mapping $dF_Z: T_Z(V) \to T_{F(Z)}(W)$ and that kernel is described by (1) when the neighborhood $V_A$ is as above, it follows that $\operatorname{corank}_Z F \geqq \nu$ precisely when the $(r + n) \times m$ matrix

$$\left\{\frac{\partial g_k}{\partial z_j}(Z), \frac{\partial f_l}{\partial z_j}(Z) : 1 \leqq k \leqq r,\ 1 \leqq l \leqq n;\ 1 \leqq j \leqq m\right\}$$

has rank $\leqq m - \nu$. That condition describes a holomorphic subvariety of $V_A$ as usual, thereby concluding the proof.

To continue the discussion with the same notation, the level set $L_A(F)$ through $A$ is just the holomorphic subvariety of $V_A$ consisting of those points $Z \in V_A$ for which $F(Z) = F(A)$ and hence can be described as

$$L_A(F) \cap V_A = \{Z \in U : f_l(Z) - f_l(A) = 0, g_k(Z) = 0, 1 \leqq k \leqq r, 1 \leqq l \leqq n\} \quad (2)$$

Although that serves to describe this subset geometrically, it is not good enough to describe the tangent space to this subvariety, a point of interest in the present discussion. The difficulty is that the functions in (2) may not serve to describe the ideal of the subvariety $L_A(F) \cap V_A$, but only an ideal having $\mathrm{id}_A(L_A(F) \cap V_A)$ as its radical. However it is always possible to add to the functions in (2) finitely many additional functions $h_1, \ldots, h_s$ holomorphic near $A$ so that all of these functions together generate $\mathrm{id}_A L_A(F) \cap V_A \subseteq {}_m\mathcal{O}_A$. Then under the canonical identifications as before

$$T_A(L_A(F)) \quad (3)$$

$$\cong \left\{ t \in \mathbb{C}^m : \sum_j t_j \frac{\partial h_i}{\partial z_j}(A) = \sum_j t_j \frac{\partial g_k}{\partial z_j}(A) = \sum_j t_j \frac{\partial f_l}{\partial z_j}(A) = 0, 1 \leqq i \leqq s, 1 \leqq k \leqq r, 1 \leqq l \leqq n \right\}$$

Comparing this with (1) leads directly to the following observation.

**5. LEMMA.** *If $F: V \to W$ is a holomorphic mapping between holomorphic varieties, then at any point $A \in V$*

$$\mathrm{tdim}_A L_A(F) \leqq \mathrm{corank}_A F$$

*Proof.* Represent an open neighborhood $V_A$ of the point $A$ as above. Then $\mathrm{corank}_A F$ is the dimension of the kernel of the linear mapping $dF_A: T_A(V) \to T_A(W)$ and that kernel can be described as in (1). On the other hand, $\mathrm{tdim}_A L_A(F) = \dim T_A(L_A(F))$ where $T_A(L_A(F))$ can be described as in (3); hence, $T_A(L_A(F)) \subseteq \mathrm{kernel}\, dF_A$ and the desired result follows immediately.

If it is really necessary to introduce some additional functions $h_j$ in order to secure the description (3), then it is quite conceivable that the inequality in the preceding lemma be a strict inequality; as will shortly be demonstrated, that may indeed be the case. Before that, though, it is convenient to carry the discussion a bit further, with one more useful if technical lemma before the first more major result.

**6. LEMMA.** *If $F: V \to W$ is a holomorphic mapping between holomorphic varieties, then*

$$\inf_{Z \in V} \dim_Z L_Z(F) = \inf_{Z \in V} \mathrm{corank}_Z F$$

*Proof.* It follows immediately from Lemma 5 that $\dim_Z L_Z(F) \leqq \operatorname{corank}_Z F$ at every point $Z \in V$, since of course $\dim_Z L_Z(F) \leqq \operatorname{tdim}_Z L_Z(F)$; hence,

$$\inf_{Z \in V} \dim_Z L_Z(F) \leqq \inf_Z \operatorname{corank}_Z F$$

To demonstrate the reversed inequality suppose to the contrary that

$$\nu = \inf_{Z \in V} \dim_Z L_Z(F) < \inf_Z \operatorname{corank}_Z F$$

and choose a point $A \in V$ at which $\dim_A L_A(F) = \nu$. The semicontinuity assertion of Lemma 2 then evidently implies that $\dim_Z L_Z(F) = \nu$ for all points $Z$ in an open neighborhood $V_A$ of $A$ in $V$. On the other hand, it follows readily from Theorem 4 that for some integer $\mu > \nu$, the subset $\{Z \in V_A : \operatorname{corank}_Z F \neq \mu\}$ is a proper holomorphic subvariety of $V_A$. Thus it is possible to find an open subset $U \subseteq \mathfrak{R}(V) \cap V_A$ such that $\dim_Z L_Z(F) = \nu$ and $\operatorname{corank}_Z F = \mu$ for all points $Z \in U$. Now the restriction of the mapping $F$ to the subset $U$ is then a holomorphic mapping of constant rank $n - \mu$ from the $n$-dimensional complex manifold $U$ into some space $\mathbb{C}^m$. But then as in Corollary I-C11 it follows that $\dim_Z L_Z(F) = \mu$ for all points $Z \in U$, which is a contradiction. That suffices to conclude the proof.

These observations can now be combined in the following very nice proof due to H. Holmann of a result first demonstrated by R. Remmert; this result extends and supersedes the semicontinuity assertion of Lemma 2.

**7. THEOREM.** *If $F: V \to W$ is a holomorphic mapping between holomorphic varieties, then for any integer $\nu$ the subset $\{Z \in V : \dim_Z L_Z(F) \geqq \nu\}$ is a holomorphic subvariety of $V$.*

*Proof.* Let $X = \{Z \in V : \dim_Z L_Z(F) \geqq \nu\}$ and introduce the descending sequence of sets $V = V_0 \supseteq V_1 \supseteq V_2 \supseteq \cdots$ defined inductively on $n$ by

$$V_n = \{Z \in V_{n-1} : \operatorname{corank}_Z(F|V_{n-1}) \geqq \nu\} \tag{4}$$

whenever $n \geqq 1$. From Theorem 4 it follows inductively on $n$ that each set $V_n$ is a holomorphic subvariety of $V_{n-1}$, so that the restriction $F|V_n$ is a well-defined holomorphic mapping of $V_n$ into $W$ for which the corank is well defined, and consequently this sequence is well defined and all the sets $V_n$ are holomorphic subvarieties of $V$. The first step in the proof is to demonstrate that

$$X = X \cap V_n = \{Z \in V_n : \dim_Z L_Z(F|V_n) \geqq \nu\} \tag{5}$$

for all $n \geqq 0$, and that too will be demonstrated by induction on $n$. When $n = 0$, condition (5) holds trivially since it just amounts to the definition of the set $X$. Assume that (5) holds for some index $n \geqq 0$ and view $F|V_n$ as a holomorphic mapping from the variety $V_n$ into $W$. Note that for any point $Z \in X$ necessarily

$$\nu \leqq \dim_Z L_Z(F|V_n) \leqq \operatorname{tdim}_Z L_Z(F|V_n) \leqq \operatorname{corank}_Z(F|V_n)$$

with the first inequality following from (5) and the last from Lemma 5. But then from (4) it follows that $Z \in V_{n+1}$, so that actually $X \subseteq V_{n+1}$. Since that is the case, it is evident that $\dim_Z L_Z(F|V_{n+1}) \geqq \nu$ whenever $Z \in X$, and therefore that (5) holds as well for the index $n + 1$ as desired. Now near any point $A \in V$ the descending sequence of holomorphic subvarieties $\{V_n\}$ will eventually be stable; that is to say, there must exist an open neighborhood $U$ of $A$ in $V$ and an index $n$ such that $U \cap V_n = U \cap V_{n+1}$. That means that $\operatorname{corank}_Z(F|V_n) \geqq \nu$ for all points $Z \in U \cap V_n$, and it then follows from Lemma 6 that $\dim_Z L_Z(F|V_n) \geqq \nu$ for all points $Z \in U \cap V_n$. But in view of (5) that in turn means that $U \cap V_n = U \cap X$, so that $X$ coincides with the holomorphic subvariety $V_n$ in the neighborhood $U$. That is the case for any point $A \in V$, so that $X$ is actually a holomorphic subvariety of $V$ and the proof is thereby concluded.

**Examples.** For any holomorphic mapping $F: V \to W$ between two holomorphic varieties and any integer $\nu$, it has been demonstrated in Theorems 4 and 7 that the sets $\{Z \in V : \operatorname{corank}_Z F \geqq \nu\}$ and $\{Z \in V : \dim_Z L_Z(F) \geqq \nu\}$ are holomorphic subvarieties of $V$. On the other hand, neither of the sets $\{Z \in V : \operatorname{rank}_Z F \geqq \nu\}$ and $\{Z \in V : \operatorname{tdim}_Z L_Z(F) \geqq \nu\}$ or of their complements are necessarily holomorphic subvarieties of $V$. To see that consider the holomorphic subvariety $V \subseteq \mathbb{C}^3$ defined by

$$V = \{Z \in \mathbb{C}^3 : z_1^2 - z_2^3 z_3 = 0\}$$

and the holomorphic mapping $F: V \to \mathbb{C}$ defined by $F(Z) = z_3$. From Theorems D1 and D2 it is readily verified that $V$ can be viewed as a two-sheeted finite branched holomorphic covering of the coordinate plane $\mathbb{C}^2$ of the variables $z_2$ and $z_3$—hence that $V$ has pure dimension two—and that the polynomial $z_1^2 - z_2^3 z_3$ generates the ideal of the subvariety $V$ at every point of $\mathbb{C}^3$. For any point $Z \in V$ the tangent space $T_Z(V)$ can thus be identified with the vector space

$$T_Z(V) = \{t \in \mathbb{C}^3 : 2z_1 t_1 - 3z_2^2 z_3 t_2 - z_2^3 t_3 = 0\} \tag{6}$$

Hence for any point $Z \in V$,

$$\operatorname{tdim}_Z V = \begin{cases} 3 & \text{if } z_1 = z_2 = 0 \\ 2 & \text{otherwise} \end{cases}$$

The singular locus $\mathfrak{S}(V)$ is consequently the subvariety $\mathfrak{S}(V) = \{Z \in V : z_1 = z_2 = 0\} = \{Z \in \mathbb{C}^3 : z_1 = z_2 = 0\}$, the coordinate axis of the variable $z_3$. The differential $dF_Z: T_Z(V) \to \mathbb{C}$ can correspondingly be identified with the linear mapping taking a vector $T = (t_1, t_2, t_3) \in T_Z(V)$ to $dF_Z(T) = t_3$. Now if $Z \in V$ is a point at which $z_1 = 0$, $z_2 \neq 0$, and hence $z_3 = 0$, it is clear from (6) that $t_3 = 0$ whenever $T \in T_Z(V)$, and therefore that rank $dF_Z = 0$. On the other hand, if $Z \in V$ is a point at which either $z_2 = 0$ or $z_1 z_2 \neq 0$, it is also clear from (6) that there exists a vector $T \in T_Z(V)$ for which $t_3 \neq 0$, and therefore that rank $dF_Z = 1$. These two observations show that $\{Z \in V : \operatorname{rank}_Z(F) \geqq 1\}$ consists of those points of $V$ lying outside the subvariety

$\{Z \in \mathbb{C}^3 : z_1 = z_3 = 0\} \subseteq V$ together with the origin, so that neither this set nor its complement is a holomorphic subvariety of $V$. Further for any point $A \in V$ the level set through that point is given by $L_A(F) = \{Z \in V : z_3 = a_3\} = \{Z \in V : z_1^2 = a_3 z_2^3, z_3 = a_3\}$ and is thus a one-dimensional holomorphic subvariety of $V$. If $a_3 = 0$, this subvariety is regular, while if $a_3 \neq 0$, it is easily verified that this subvariety has a singularity at $(0, 0, a_3)$ but is otherwise regular. Since $L_A(F)$ lies in the two-dimensional submanifold $\{Z \in \mathbb{C}^3 : z_3 = a_3\}$, its tangential dimension at this singular point must be two, so that for any point $A \in V$

$$\operatorname{tdim}_A L_A(F) = \begin{cases} 2 & \text{if } A = (0, 0, a_3) \text{ with } a_3 \neq 0 \\ 1 & \text{otherwise} \end{cases}$$

Thus $\{Z \in V : \operatorname{tdim}_Z L_Z(F) \geqq 2\}$ consists of those points in the singular locus $\mathfrak{S}(V) = \{Z \in \mathbb{C}^3 : z_1 = z_2 = 0\}$ except for the origin itself, so neither this set nor its complement is a holomorphic subvariety of $V$. Incidentally this example shows in addition that the inequality in Lemma 5 can indeed be a strict inequality at some points, since $\{Z \in V : \operatorname{corank}_Z L_Z(F) \geqq 2\}$ is a holomorphic subvariety of $V$ by Theorem 4.

Although the preceding results about the dimensions of the level sets of holomorphic mappings are possibly interesting enough in themselves, they take on added significance in the further analysis of properties of holomorphic mappings. The simplest class of holomorphic mappings, the finite holomorphic mappings, can be characterized as those holomorphic mappings for which the level sets are all of dimension zero. It might be expected that those holomorphic mappings for which the level sets have any constant dimension have some special properties; such mappings have indeed a quite special form, as follows.

**8. THEOREM.** *Let $V$ and $W$ be holomorphic varieties, with $V$ pure dimensional, and let $F: V \to W$ be a holomorphic mapping with the property that $\dim_Z L_Z(F) = \nu$ is a constant, independent of the point $Z \in V$. Then for any point $A \in V$ there are an open neighborhood $V_A$ of $A$ in $V$, a holomorphic subvariety $X_A$ of an open neighborhood of $F(A)$ in $W$, and an open subset $U \subseteq \mathbb{C}^\nu$, such that the restriction of $F$ to $V_A$ can be written as a composition $F|V_A = P \circ H$ where $H: V_A \to X_A \times U$ is a finite branched holomorphic covering and $P: X_A \times U \to X_A$ is the natural projection mapping.*

*Proof.* Since the level sets of $F$ are of dimension $\nu$ at any point, they must of course actually be of pure dimension $\nu$ at each point. The germ of the level set $L_A(F)$ at any point $A \in V$ can hence be represented as the germ of a finite branched holomorphic covering of $\mathbb{C}^\nu$, and this covering mapping can moreover be viewed as the restriction to the germ of the subvariety $L_A(F)$ of some germ $\mathbf{G}_A: \mathbf{V}_A \to \mathbb{C}^\nu$ of a holomorphic mapping taking the point $A \in V$ to the origin in $\mathbb{C}^\nu$. Let $\mathbf{H}_A: \mathbf{V}_A \to \mathbf{W}_{F(A)} \times \mathbb{C}^\nu$ be the germ of the holomorphic mapping defined by setting $H(Z) = (F(Z), G(Z))$ for all points $Z \in V$ near $A$, where $G$ represents the germ $\mathbf{G}_A$. Since $H^{-1}(F(A), 0)$ then consists of those points $Z \in L_A(F)$ near $A$ such that $G(Z) = 0$ while $G$ is a finite branched holomorphic covering, it is evident that $H^{-1}(F(A), 0)$ contains the point $A$ as an isolated point and hence that $\mathbf{H}_A$ is the germ of a finite holomorphic mapping. From Theorem E7 it follows that the restriction of $\mathbf{H}_A$ to each irreducible component of the germ $\mathbf{V}_A$ of the variety $V$ at $A$ is the germ of a finite branched holomorphic covering of its image, where that image is the germ of a holomorphic subvariety of $\mathbf{W}_{F(A)} \times \mathbf{C}^\nu$. Since $\mathbf{V}_A$ is pure dimensional, it is easy to see that $\mathbf{H}_A$ is itself the germ of a finite branched holomorphic covering of its image, which is the germ of a holomorphic subvariety of $\mathbf{W}_{F(A)} \times \mathbf{C}^\nu$. If $\mathbf{V}_A$ is not pure dimensional, it may happen that the image under $\mathbf{H}_A$ of one irreducible component of $\mathbf{V}_A$ is a proper holomorphic subvariety of the image of another component, and that cannot lead to a finite branched holomorphic covering since there is then no way that a dense open subset of any representative of $\mathbf{V}_A$ can be an ordinary covering mapping. That is a rather minor complication in many ways, and although the theorem as stated does require that $V$ be pure dimensional, the slightly weaker conclusion in which $H$ is a finite holomorphic mapping rather than a finite branched holomorphic covering does hold without that hypothesis. To continue the proof, upon choosing suitable representatives of all these germs, there thus exist connected open neighborhoods $V_A$ of the point $A$ in $V$, $W_{F(A)}$ of the point $F(A)$ in $W$, and $U$ of the origin in $\mathbf{C}^\nu$, together with a finite branched holomorphic covering $H: V_A \to H(V_A)$ where $H(Z) = (F(Z), G(Z))$ for any point $Z \in V_A$ and the image $H(V_A)$ is a holomorphic subvariety of $W_{F(A)} \times U$. The intersection $H(V_A) \cap (W_{F(A)} \times 0)$, which is of course the image under $H$ of the subvariety $G^{-1}(0) \subseteq V_A$, can be viewed as a holomorphic subvariety $X_A \subseteq W_{F(A)}$. For any point $B \in V_A$ the restriction of the covering $H$ to the subvariety $L_B(F) \cap V_A$ is necessarily itself a finite branched holomorphic covering

$$H|L_B(F) \cap V_A : L_B(F) \cap V_A \to H(V_A) \cap (F(B) \times U)$$

where its image $H(V_A) \cap (F(B) \times U)$ can be viewed as a holomorphic subvariety of $U$. Since the subvariety $L_B(F)$ is of dimension $\nu$ by hypothesis while $U$ is a connected open subset of $\mathbf{C}^\nu$, it follows that this image must indeed be all of $U$ and hence that $H(V_A) \cap (F(B) \times U) = F(B) \times U$. There must in particular exist some point $B_0 \in L_B(F) \cap V_A$ for which $H(B_0) = F(B) \times 0$, so that $F(B) = F(B_0)$ while $F(B_0) \in X_A$. Consequently it is actually the case that $H(V_A) = X_A \times U$. Finally since $H(Z) = (F(Z), G(Z))$, the restriction $F|V_A$ is merely the composition of the finite branched holomorphic covering $H: V_A \to X_A \times U$ and the natural projection $P: X_A \times U \to X_A$, thereby completing the proof of the theorem.

The preceding theorem actually describes the generic form of a holomorphic mapping between holomorphic varieties. To make this assertion precise it is convenient to introduce the notion of the **exceptional locus** $E(F)$ of a holomorphic mapping $F: V \to W$ between two holomorphic varieties. Write $V = \bigcup_i V_i$ where each $V_i$ is the union of those irreducible components $V_{ij}$ of $V$ for which $\dim V_{ij}$ and $\inf_{Z \in V_{ij}} \dim_Z L_Z(F|V_{ij})$ have some specified values. The subset

$$E(V_i) = \left\{ A \in V_i : \dim_A L_A(F) > \inf_{Z \in V_i} \dim_Z L_Z(F|V_i) \right\} \tag{7}$$

is by Theorem 7 a holomorphic subvariety of $V_i$—indeed, is evidently even a thin holomorphic subvariety of $V_i$. *The exceptional locus is then the subset*

$$E(F) = \left[\bigcup_i E(V_i)\right] \cup \left[\bigcup_{j \neq k} V_j \cap V_k\right] \tag{8}$$

*which is also a thin holomorphic subvariety of V.* What may at first glance seem unnecessary complications in this definition should at second glance appear to be a not unnatural attempt at a reasonable definition in which the exceptional locus is as small as possible. Each subvariety $V_i$ is pure dimensional, and outside the subvariety $E(V_i)$ the conclusion of Theorem 8 holds. Hence, the conclusion of Theorem 8 holds locally at each point of $V$ outside of $E(F)$, although the dimensions involved in that conclusion may differ on the various connected components of $V - E(F)$. Altogether then *at each point $A \in V - E(F)$ the mapping $F$ can be written locally as the composition $F = P \circ H$ of a finite branched holomorphic covering $H: V \to X \times \mathbb{C}^\nu$ and the natural projection $P: X \times \mathbb{C}^\nu \to X$,* where $X$ is a holomorphic subvariety of a neighborhood of $F(A)$ in $W$ and $\nu = \dim_A L_A(F)$. It is perhaps worth mentioning in passing that by Corollary I-C10 a holomorphic mapping $F: V \to W$ between complex manifolds having a Jacobian matrix of constant rank has an empty exceptional locus, but that a holomorphic mapping between complex manifolds with an empty exceptional locus need not have a Jacobian matrix of constant rank, even when it is just a finite holomorphic mapping.

If the variety $V$ in Theorem 8 is not pure dimensional, that theorem can at least be applied to each pure-dimensional piece. More generally the theorem can be applied to each irreducible component of $V$ or to any union of irreducible components of the same dimension, in view of the following simple auxiliary observation.

**9. LEMMA.** *If $V = V_1 \cup \cdots \cup V_r$ is a holomorphic variety where each $V_j$ is a union of irreducible components of $V$ and if $F: V \to W$ is a holomorphic mapping with the property that $\dim_Z L_Z(F) = \nu$ is a constant, independent of the point $Z \in V$, then the restrictions $F_j = F|V_j : V_j \to W$ also have the property that $\dim_Z L_Z(F_j) = \nu$ for all points $Z \in V_j$.*

*Proof.* For any point $Z \in V$ the holomorphic subvariety $L_Z(F) \subseteq V$ can be written as the union $L_Z(F) = L_Z(F_1) \cup \cdots \cup L_Z(F_r)$ where $L_Z(F_j) = L_Z(F) \cap V_j$ are holomorphic subvarieties of $L_Z(F)$; hence, $\dim_Z L_Z(F_j) \leq \dim_Z L_Z(F) = \nu$ for each index $j$. On the other hand, if $\dim_Z L_Z(F_j) = \mu < \nu$ for some index $j$, then it follows from the semicontinuity property of Lemma 2 that $\dim_T L_T(F_j) \leq \mu$ for all points $T \in V_j$ near $Z$. But there must exist some such point $T$ not contained in any other component $V_k$, and at such a point $\dim_T L_T(F_j) = \dim_T L_T(F) = \nu$, a contradiction. That thereby concludes the proof.

**10. THEOREM.** *If $F: V \to W$ is a holomorphic mapping between two holomorphic varieties and if $\dim_Z L_Z(F) = \nu$ is a constant independent of the point $Z \in V$, then each point $A \in V$ has arbitrarily small open neighborhoods $V_A$ in $V$ such that $F(V_A)$ is a holomorphic subvariety of an open neighborhood of $F(A)$ in $W$ and $\dim_{F(A)} F(V_A) = \dim_A V - \nu$.*

*Proof.* Choose any open neighborhood $V_A$ of $A$ in $V$ and write $V_A = \bigcup_i V_i$ where $V_i$ are the irreducible components of $V_A$. It is clear from Lemma 9 that the restrictions

$F_i = F|V_i: V_i \to W$ satisfy the hypotheses of Theorem 8, and it follows from that theorem that after shrinking the subvariety $V_i$ to a smaller neighborhood if necessary, the image $X_i = F_i(V_i)$ will be a holomorphic subvariety of an open neighborhood of $F(A)$ in $W$. Moreover the mapping $F_i$ is the composition $F_i = P_i \circ H_i$ of a finite branched holomorphic covering $H_i: V_i \to X_i \times U_i$ and the natural projection $P_i: X_i \times U_i \to X_i$, where $U_i$ is an open subset of $\mathbb{C}^\nu$. It follows from Theorem G16 that $\dim_A V_i = \dim_{H_i(A)}(X_i \times U_i)$, while it is evident upon considering just the regular points that $\dim_{H_i(A)}(X_i \times U_i) = \nu + \dim_{F_i(A)}(X_i)$; consequently, $\dim_{F_i(A)} X_i = \dim_A V_i - \nu$. Thus after shrinking the original neighborhood $V_A$ if necessary, the image $F(V_A) = \bigcup_i F(V_i) = \bigcup_i X_i$ is a finite union of holomorphic subvarieties of an open neighborhood of $F(A)$ in $W$ and hence is itself a holomorphic subvariety of that neighborhood, and $\dim_{F(A)} F(V_A) = \sup_i \dim_{F(A)} X_i = \sup_i(\dim_A V_i - \nu) = \dim_A V - \nu$, thereby concluding the proof.

What is in a sense the converse of the preceding theorem also holds, thus providing a characterization of those holomorphic mappings for which the images are holomorphic varieties locally. Before proving that though it is convenient to establish the following auxiliary result.

**11. LEMMA.** *If $F: V \to W$ is a holomorphic mapping taking some arbitrarily small open neighborhoods of a point $A$ in the variety $V$ onto open neighborhoods of the point $F(A)$ in the variety $W$ and if the variety $W$ is pure dimensional at the point $F(A)$, then*

$$\dim_A L_A(F) + \dim_{F(A)} W = \dim_A V$$

*Proof.* The proof will be by induction on the integer $n = \dim_{F(A)} W$. The initial case $n = 0$ is trivial, since then $V = L_A(F)$, so it is sufficient merely to demonstrate the inductive step. Suppose therefore that $W$ is of pure dimension $n$ at the point $F(A)$, and that the desired result holds for all varieties of dimension $< n$. Choose a nontrivial nonunit $\mathbf{f} \in {}_W\mathcal{O}_{F(A)}$ and a representative holomorphic function $f$ in an open neighborhood $W_{F(A)}$ of $F(A)$ in $W$, and let $g = f \circ F$ be the induced holomorphic function on $V_A = F^{-1}(W_{F(A)})$. The germ $\mathbf{g} \in {}_V\mathcal{O}_A$ is obviously a nonunit, since $g(A) = f(F(A)) = 0$. Moreover $\mathbf{g}$ is also nontrivial, since otherwise the function $f$ would vanish on the image $F(V_A)$, which by assumption is an open neighborhood of $F(A)$ in $W$, and hence the germ $\mathbf{f} \in {}_W\mathcal{O}_{F(A)}$ would necessarily be zero, contradicting the assumption that it is nonzero. The subvariety $W_1 = \{Z \in W_{F(A)} : f(Z) = 0\}$ is of pure dimension $n - 1$ at $F(A)$, by Theorem G7, while correspondingly for the subvariety $V_1 = \{Z \in V_A : g(Z) = 0\}$ necessarily $\dim_A V_1 = \dim_A V - 1$. Moreover since $V_1 = F^{-1}(W_1)$, then upon letting $F_1 = F|V_1 : V_1 \to W_1$ it is clear that $L_A(F_1) = L_A(F)$. It follows from the induction hypothesis that $\dim_A L_A(F_1) + \dim_{F(A)} W_1 = \dim_A V_1$, from which in view of the preceding observations the desired result follows immediately, thereby concluding the proof.

**12. THEOREM.** *Suppose that $F: V \to W$ is a holomorphic mapping between two holomorphic varieties $V$ and $W$, where $V$ is irreducible at a point $A \in V$. Then the images of some arbitrarily small open neighborhoods of $A$ in $V$ are holomorphic subvarieties of open*

***neighborhoods of $F(A)$ in $W$ if and only if $\dim_Z L_Z(F)$ is a constant, independent of the point $Z \in V$, in an open neighborhood of $A$ in $V$.***

*Proof.* If $\dim_Z L_Z(F)$ is constant in an open neighborhood of $A$, then it follows from Theorem 10 that the images of some arbitrarily small open neighborhoods of $A$ in $V$ are holomorphic subvarieties of open neighborhoods of $F(A)$ in $W$ as desired. Conversely suppose that the images of some arbitrarily small open neighborhoods of $A$ in $V$ are subvarieties of open neighborhoods of $F(A)$ in $W$. Upon restricting to a neighborhood of $A$ in $V$ and replacing $W$ by the image of this neighborhood, it can even be supposed that the images of some arbitrarily small open neighborhoods of $A$ in $V$ are actually open neighborhoods of $F(A)$ in $W$. If $V$ is irreducible at $A$ it is readily seen that $W$ must be irreducible at $F(A)$, so that $W$ is in particular pure dimensional at $F(A)$. It then follows from Lemma 11 that

$$\dim_A L_A(F) + \dim_{F(A)} W = \dim_A V \tag{9}$$

After setting $\nu = \dim_A L_A(F)$ and restricting to a still smaller open neighborhood of $A$ if necessary, it is possible by Theorem G11 to find a surjective finite holomorphic mapping $G: L_A(F) \to U$ where $U$ is an open subset of $\mathbb{C}^\nu$. Here $G$ can of course be taken as the restriction to $L_A(F)$ of a holomorphic mapping of some open neighborhood of $A$ in $V$ into the space $\mathbb{C}^\nu$. The mapping $H: V \to W \times U$ defined by $H(Z) = (F(Z), G(Z))$ then clearly has the property that $A$ is an isolated point of $H^{-1}(H(A))$, so that after restricting the neighborhood yet further if necessary, $H$ will be a finite holomorphic mapping. Since $V$ is irreducible at $A$, the mapping $H$ will be a finite branched holomorphic covering of the image, and since as a consequence of (9) necessarily $\dim_A V = \dim_{F(A)} W + \nu = \dim_{H(A)}(W \times U)$ that image must be all of $W \times U$, again after restricting the neighborhood as necessary. Thus it can be supposed that $H: V \to W \times U$ is a finite branched holomorphic covering. For any point $Z \in V$ the restriction of this covering to the subvariety $F(Z) \times U \subseteq W \times U$ is then a surjective finite holomorphic mapping $L_Z(F) \to F(Z) \times U$, and consequently by Theorem G16 it is clear that $\dim_Z L_Z(F) = \dim U = \nu$. Therefore $\dim_Z L_Z(F)$ is a constant near $A$ as desired, and the proof of the theorem is thereby concluded.

One implication in the preceding theorem, the implication also contained in Theorem 10, does not at all require the irreducibility hypothesis. The other implication really does require the irreducibility hypothesis though; if $V = V_1 \cup V_2$ is a reducible variety with components $V_1$ and $V_2$, and if $F: V \to W$ is a holomorphic mapping such that the restriction $F|V_1: V_1 \to W$ takes some arbitrarily small open neighborhoods of a point $A \in V_1 \cap V_2$ in $V_1$ onto open neighborhoods of $F(A)$ in $W$, then the level sets of the restriction $F|V_1$ must have constant dimension near $A$, but the restriction $F|V_2$ need not have that property. It is rather easy to see that this situation is essentially the only way in which that implication in the preceding theorem fails when $V$ is reducible at $A$. Before discussing this phenomenon further, though, it is convenient to establish two additional general properties of level sets, providing occasionally useful observations.

**13. LEMMA.** *If $F: V \to W$ is a holomorphic mapping between two holomorphic varieties, then at each point $A \in V$*

$$\dim_A L_A(F) + \dim_{F(A)} W \geqq \dim_A V$$

*If the inequality is actually an equality and $W$ is irreducible at $F(A)$, then $F$ maps some arbitrarily small open neighborhoods of $A$ in $V$ onto open neighborhoods of $F(A)$ in $W$.*

*Proof.* By restricting to a sufficiently small open neighborhood of the point $F(A)$ if necessary it can be assumed that $\dim W = \dim_{F(A)} W$. Let $V'$ be an irreducible component of $V$ such that $A \in V'$ and $\dim V' = \dim V$, and consider the restriction $F' = F|V' : V' \to W$. For any point $Z \in V' - E(F')$, where $E(F')$ is the exceptional locus of the mapping $F'$, the germ at $Z$ of the mapping $F'$ has as image the germ of a holomorphic subvariety of $W$ at $F(Z)$ and $\dim_Z L_Z(F') + \dim \mathbf{F}'(\mathbf{V}'_Z) = \dim_Z V'$. Since $\dim \mathbf{F}'(\mathbf{V}'_Z) \leqq \dim W$, it follows that

$$\dim_Z L_Z(F') + \dim W \geqq \dim_Z V' = \dim V \tag{10}$$

Now whenever $Z \in V' - E(F')$ does not lie in any irreducible component of $V$ other than $V'$, it is clear that $\dim_Z L_Z(F') = \dim_Z L_Z(F)$, and since there are such points $Z$ arbitrarily near $A$, it follows from the semicontinuity property of Lemma 2 that $\dim_A L_A(F) \geqq \dim_Z L_Z(F)$. Upon substituting this in (10) it further follows that

$$\dim_A L_A(F) + \dim_{F(A)} W \geqq \dim_Z L_Z(F') + \dim W \geqq \dim V \tag{11}$$

thereby demonstrating the first assertion of the lemma. Moreover if the extreme terms in (11) are actually equal, then clearly $\dim_A L_A(F) = \dim_Z L_Z(F')$. But then $\dim_A L_A(F') \leqq \dim_A L_A(F) = \dim_Z L_Z(F')$, from which upon recalling the definition of the exceptional locus and the semicontinuity property of Lemma 2 it is evident that $\dim_Z L_Z(F') = \dim_A L_A(F')$ for all points $Z \in V'$ near $A$. An application of Theorem 10 shows that there are arbitrarily small open neighborhoods $V'_A$ of $A$ in $V'$ such that the images $F(V'_A)$ are holomorphic subvarieties of open neighborhoods of $F(A)$ in $W$ and $\dim_{F(A)} F(V'_A) = \dim_A V'_A - \dim_A L_A(F') = \dim_{F(A)} W$. If $W$ is irreducible at $F(A)$, these images are actually open neighborhoods of $F(A)$ in $W$ and hence so are the larger sets $F(V_A)$ where $V_A = F^{-1}(F(V'_A)) \subseteq V$, therewith concluding the proof of the lemma.

The preceding lemma is occasionally useful in demonstrating that the image of a holomorphic variety is locally a holomorphic variety. That will be illustrated here by completing the discussion begun earlier of the role of the irreducibility hypothesis in Theorem 12. Suppose therefore that $F: V \to W$ is a holomorphic mapping with the property that the images of some arbitrarily small open neighborhoods of $A$ in $V$ are open neighborhoods of $F(A)$ in $W$. It will not be assumed that $V$ is irreducible at $A$, but by restricting to an irreducible component of $W$ and its inverse image, it can be supposed for this discussion that $W$ is irreducible at $F(A)$. If $V = V_1 \cup V_2 \cup \cdots$ where $V_j$ represent the irreducible components of $V$ at $A$, it follows from Lemma 13 that for each index $j$

$$\dim_A L_A(F|V_j) + \dim_{F(A)} W \geqq \dim_A V_j$$

while since $\dim_A L_A(F) = \sup_j \dim_A L_A(F|V_j)$ and $\dim_A V = \sup_j \dim_A V_j$, it follows from Lemma 11 that

$$\sup_j \dim_A L_A(F|V_j) + \dim_{F(A)} W = \sup_j \dim_A V_j$$

These two observations clearly imply that there exists some index $k$ for which $\dim_A L_A(F|V_k) = \sup_j \dim_A L_A(F|V_j)$ and $\dim_A V_k = \sup_j \dim_A V_j$, and for this index $k$,

$$\dim_A L_A(F|V_k) + \dim_{F(A)} W = \dim_A V_k$$

An application of the last part of Lemma 13 shows that $F|V_k$ takes some arbitrarily small open neighborhoods of $A$ in $V_k$ onto open neighborhoods of $F(A)$ in $W$, and it then follows from Theorem 12 that the level sets $L_Z(F|V_k)$ must have a constant dimension on $V_k$. Thus if $W$ is irreducible at $F(A)$, there must exist an irreducible component of $V$ of maximal dimension such that the images of this component are open neighborhoods of $F(A)$ in $W$ locally and the level sets of the restriction of $F$ to this component have constant dimensions; but of course nothing can be said about the other components of $V$.

One more point should be made to complete the preceding discussion. If $F: V \to W$ is a holomorphic mapping for which the level sets have a constant dimension, then each point $A \in V$ has arbitrarily small open neighborhoods $V_A$ such that $F(V_A)$ is a holomorphic subvariety of an open neighborhood $W_{F(A)}$ of $F(A)$ in $W$. Consequently for any such neighborhood $V_A$ the restriction $F|V_A: V_A \to F(V_A)$ is an open mapping *at the point* $A$, in the sense that the image of any open neighborhood of $A$ in $V_A$ contains an open neighborhood of $F(A)$ in $F(V_A)$. *It is not necessarily the case that some such restriction* $F|V_A : V_A \to F(V_A)$ *be an open mapping* though, even in the case that the level sets are of dimension zero. For instance, consider the subvariety $V = V_1 \cup V_2 \subseteq \mathbf{C}^4$ where $V_1 = \{Z \in \mathbf{C}^4 : z_1 = z_2 = 0\}$, $V_2 = \{Z \in \mathbf{C}^4 : z_3 = z_4 = 0\}$, and the subvariety $W = W_1 \cup W_2 \subseteq \mathbf{C}^3$ where $W_1 = \{W \in \mathbf{C}^3 : w_1 = 0\}$, $W_2 = \{W \in \mathbf{C}^3 : w_3 = 0\}$. The irreducible components $V_1, V_2, W_1, W_2$ are two-dimensional linear subspaces, with $V_1 \cap V_2$ consisting of just the origin and $W_1 \cap W_2$ being the one-dimensional linear subspace $w_1 = w_3 = 0$. The mapping $F: \mathbf{C}^4 \to \mathbf{C}^3$ defined by $F(z_1, z_2, z_3, z_4) = (z_1, z_2 + z_3, z_4)$ takes $V$ to $W$—indeed, restricts to a biholomorphic mapping taking $V_1$ to $W_1$ and another taking $V_2$ to $W_2$. It takes any open neighborhood of the origin in $V$ to an open neighborhood of the origin in $W$, but it does not take any sufficiently small open neighborhood of a point $A = (0, 0, a_3, 0)$ with $a_3 \neq 0$ to a set containing an open neighborhood of the image point $F(A) = (0, a_3, 0)$, since a small enough open neighborhood of $A$ will contain points of the component $V_1$ only, while any open neighborhood of $F(A)$ must contain points belonging to $W_2$ as well as points belonging to $W_1$. On the other hand, Lemma 13 shows that *a surjective holomorphic mapping* $F: V \to W$ *for which the level sets have a constant dimension is an open mapping if* $W$ *is irreducible at each point.*

# M
# Projective Spaces

Complex projective spaces are a familiar and important class of complex manifolds, arising naturally not just in algebraic geometry but also in complex analysis where they play a role in the investigation of local properties of holomorphic varieties and mappings to be pursued next. Although the general properties of projective spaces are no doubt quite well known, some special properties nicely illustrate the analytic results discussed in the preceding sections. So a discussion of projective spaces here may serve a purpose beyond just establishing notation and terminology.

One standard way to introduce complex projective $n$-space is to begin with the complement of the origin in $(n+1)$-space, the set $(\mathbb{C}^{n+1})^* = \{Z \in \mathbb{C}^{n+1} : Z \neq 0\}$, and to introduce an equivalence relation on this set by writing $Z \sim W$ whenever $Z$ and $W$ are points of $(\mathbb{C}^{n+1})^*$ such that $W = tZ$ for some nonzero complex number $t \in \mathbb{C}^*$. It is almost obvious that this is an equivalence relation in the usual sense, so it is possible to introduce the space $(\mathbb{C}^{n+1})^*/\sim$ of equivalence classes. To analyze this space more closely, introduce the open subsets $\tilde{U}_j = \{Z \in \mathbb{C}^{n+1} : z_j \neq 0\}$ for $0 \leq j \leq n$, it being customary in this context to write $Z = (z_0, z_1, \ldots, z_n)$, and note that $(\mathbb{C}^{n+1})^* = \bigcup_{j=0}^{n} \tilde{U}_j$ and that if a point $Z \in (\mathbb{C}^{n+1})^*$ lies in $\tilde{U}_j$ then so does the entire equivalence class $\{tz : t \in \mathbb{C}^*\}$ containing $Z$. It is a simple matter to verify that the holomorphic mapping $\tilde{H}_j \colon \tilde{U}_j \to \mathbb{C}^n$ defined by

$$\tilde{H}_j(Z) = \left(\frac{z_0}{z_j}, \frac{z_1}{z_j}, \ldots, \frac{z_{j-1}}{z_j}, \frac{z_{j+1}}{z_j}, \ldots, \frac{z_n}{z_j}\right) \tag{1}$$

takes $\tilde{U}_j$ onto $\mathbb{C}^n$, and that the inverse images of the points of $\mathbb{C}^n$ under this mapping are precisely the equivalence classes contained in $\tilde{U}_j$. Thus $\tilde{H}_j$ induces a one-to-one mapping $H_j \colon U_j \to \mathbb{C}^n$ between the space $U_j = \tilde{U}_j/\sim$ of equivalence classes in $\tilde{U}_j$ and $n$-space. The set $U_j$ can be viewed as a coordinate neighborhood on $(\mathbb{C}^{n+1})^*/\sim$ with local coordinates for the equivalence class containing a point $Z \in (\mathbb{C}^{n+1})^*$ given by $(t_{j1}, \ldots, t_{jn}) = \tilde{H}_j(Z)$. If $Z \in \tilde{U}_j \cap \tilde{U}_k$ for $j < k$, then it follows readily from (1) that $t_{jk} \neq 0$ and that the two sets of local coordinates representing the equivalence class containing $Z$ are related by

$$(t_{k1}, \ldots, t_{kn}) = t_{jk}^{-1}(t_{j1}, \ldots, t_{jj}, 1, t_{jj+1}, \ldots, t_{jk-1}, t_{jk+1}, \ldots, t_{jn})$$

Thus $H_k \circ H_j^{-1} : H_j(U_j \cap U_k) \to H_k(U_j \cap U_k)$ is a holomorphic mapping, and so similarly is $H_j \circ H_k^{-1}$, so that this coordinate covering $\{U_j, H_j\}$ describes a complex structure on the set $(\mathbb{C}^{n+1})^*/\sim$ as discussed in section I-C, and in the process of course describes a topological structure as well.

1. **DEFINITION.** *The n-dimensional* **complex projective space** *is the set $(\mathbb{C}^{n+1})^*/\sim$ with the structure of a complex manifold defined by the coordinate covering $\{U_j, H_j\}$. This manifold will be denoted by $\mathbb{P}_n(\mathbb{C})$, or just $\mathbb{P}_n$ for short.*

It is clear from the preceding discussion that the natural mapping $\tilde{H}$: $(\mathbb{C}^{n+1})^* \to \mathbb{P}^n$ that associates to each point $Z \in (\mathbb{C}^{n+1})^*$ its equivalence class $\tilde{H}(Z) \in (\mathbb{C}^{n+1})^*/\sim\, = \mathbb{P}^n$ is a holomorphic mapping. Note that any point $Z \in (\mathbb{C}^{n+1})^*$ is equivalent to the point $Z/\|Z\| \in \partial B(0; 1)$, so that the restriction of the mapping $\tilde{H}$ to the boundary of the unit ball in $\mathbb{C}^{n+1}$ is a surjective continuous mapping $\tilde{H}|\partial B(0; 1)$: $\partial B(0; 1) \to \mathbb{P}^n$. Since $\partial B(0; 1)$ is compact, it follows that *the projective space $\mathbb{P}^n$ is a compact complex manifold.* Although it will not be used here, it is perhaps worth noting that whenever $Z \in \partial B(0; 1)$, then $\tilde{H}^{-1}(\tilde{H}(Z)) = \{tZ : t \in \mathbb{C}, |t| = 1\}$, from which it can be demonstrated that the sphere $\partial B(0; 1)$ is a topological fibration over $\mathbb{P}^n$ with circles as fibres; that provides some information about the topological structure of $\mathbb{P}^n$. Of more interest here though, the compact complex manifold $\mathbb{P}^n$ contains as an open subset the coordinate neighborhood $U_0$ that is biholomorphic to $\mathbb{C}^n$, so that $\mathbb{P}^n$ can be viewed as a holomorphic compactification of $\mathbb{C}^n$. The complement $\mathbb{P}^n - \mathbb{C}^n$ consists precisely of those equivalence classes in $(\mathbb{C}^{n+1})^*$ lying in the linear subspace $L = \{Z \in \mathbb{C}^{n+1} : z_0 = 0\}$. The subset $L \cap (\mathbb{C}^{n+1})^* = \{(0, z_1, \ldots, z_n) \in (\mathbb{C}^{n+1})^*\}$ can of course be identified with $(\mathbb{C}^n)^*$, the restriction to $L$ of the equivalence relation in $(\mathbb{C}^{n+1})^*$ is the same equivalence relation in $(\mathbb{C}^n)^*$, and the restrictions to $L \cap U_j$ for $1 \leqq j \leqq n$ of the coordinate mappings are readily seen to amount to the standard coordinate mappings defining the complex structure of $\mathbb{P}^{n-1}$ on $(\mathbb{C}^n)^*/\sim$. Thus the complement $\mathbb{P}^n - \mathbb{C}^n$ is a complex submanifold of $\mathbb{P}^n$ biholomorphic to $\mathbb{P}^{n-1}$. In more picturesque if less precise terms, the compactification $\mathbb{P}^n$ of $\mathbb{C}^n$ arises by adjoining to $\mathbb{C}^n$ the projective space $\mathbb{P}^{n-1}$ at infinity. For the special case $n = 1$, this is the familiar way of viewing $\mathbb{P}^1$ as the compactification of $\mathbb{C}^1$ by adjoining the point $\mathbb{P}^0$ at infinity, thus identifying $\mathbb{P}^1$ with the Riemann sphere.

A customary terminology in this context is to call the coordinates $[z_0, \ldots, z_n]$ in $\mathbb{C}^{n+1}$ the *homogeneous coordinates* in $\mathbb{P}^n$, remembering that they are not really coordinates in the usual sense since all the $(n + 1)$-tuples $\{t[z_0, \ldots, z_n] : t \in \mathbb{C}^*\} = \{[tz_0, \ldots, tz_n] : t \in \mathbb{C}^*\}$ describe the same point in $\mathbb{P}^n$. The coordinates $(t_{01}, \ldots, t_{0n}) = H_0(Z)$ for the coordinate neighborhood $U_0$, which is normally merely identified with $\mathbb{C}^n$, are called the *inhomogeneous coordinates* in $\mathbb{P}^n$, remembering that they too are not really coordinates in $\mathbb{P}^n$ but just in the dense open subset $U_0 \cong \mathbb{C}^n$. A point $Z \in \mathbb{P}^n$ having homogeneous coordinates $[z_0, \ldots, z_n]$ thus has inhomogeneous coordinates $(z_1/z_0, \ldots, z_n/z_0)$, the latter being well defined only when $z_0 \neq 0$. A point having inhomogeneous coordinates $(z_1, \ldots, z_n)$ has homogeneous coordinates $[1, z_1, \ldots, z_n]$ or any scalar multiple thereof.

There are alternative ways of looking at complex projective spaces, two of which will be at least mentioned here. First note that the equivalence classes in $(\mathbb{C}^{n+1})^*$ representing the points of $\mathbb{P}^n$ are indeed precisely the one-dimensional linear subspaces through the origin in $\mathbb{C}^{n+1}$ with the common point at the origin removed; thus, *$\mathbb{P}^n$ can be viewed as the set of all complex lines through the origin in $\mathbb{C}^{n+1}$*. The generalization arising by considering instead of lines all the $k$-dimensional linear subspaces through the origin in $\mathbb{C}^{n+1}$ is the Grassmann manifold $G(k, n+1)$, which can also be given the structure of a compact manifold. The group $GL(n+1, \mathbb{C})$ of all $(n+1)\times(n+1)$ nonsingular complex matrices acting as a group of linear transformations on $\mathbb{C}^{n+1}$ permutes the set of all complex lines through the origin in $\mathbb{C}^{n+1}$ and consequently acts also as a group of one-to-one mappings on $\mathbb{P}^n$. A point $Z \in \mathbb{P}^n$ with homogeneous coordinates $[z_0, \ldots, z_n]$ is transformed by the matrix $A = \{a_{jk} : 0 \leqq j, k \leqq n\}$ to the point $W = AZ$ with homogeneous coordinates $[w_0, \ldots, w_n]$ given by $w_j = \Sigma_k a_{jk} z_k$. If $z_0 \cdot w_0 \neq 0$, the corresponding expression in terms of inhomogeneous coordinates $(z_1, \ldots, z_n)$ and $(w_1, \ldots, w_n)$ is

$$w_j = \frac{a_{j0} + \sum_{k=1}^n a_{jk} z_k}{a_{00} + \sum_{k=1}^n a_{0k} z_k}, \qquad 1 \leqq j \leqq n \tag{2}$$

showing that the resulting mapping is holomorphic. At points for which either $z_0$ or $w_0$ is zero, the mapping can be described only in terms of coördinates in other coordinate neighborhoods and is readily seen still to be holomorphic. Thus $\mathrm{GL}(n+1, \mathbb{C})$ *acts as a group of biholomorphic automorphisms of* $\mathbb{P}^n$, the action being described by (2) in inhomogeneous coordinates; these automorphisms are called **projective transformations.** It is clear from (2) that two matrices $A_1, A_2 \in \mathrm{GL}(n+1, \mathbb{C})$ describe the same projective transformation precisely when $A_2 = aA_1$ for some constant $a \in \mathbb{C}^*$, or equivalently precisely when $A_1$ and $A_2$ represent the same coset in the quotient group $\mathrm{GL}(n+1, \mathbb{C})/\mathbb{C}^*$ where $\mathbb{C}^*$ is identified with the subgroup $\{aI : a \in \mathbb{C}^*\} \subseteq \mathrm{GL}(n+1, \mathbb{C})$. Thus the projective transformations are really described by this quotient group, which will be called the **projective linear group** and will be denoted by $\mathrm{PL}(n, \mathbb{C})$. It is also clear that the group $\mathrm{GL}(n+1, \mathbb{C})$ or the group $\mathrm{PL}(n, \mathbb{C})$ acts transitively on $\mathbb{P}^n$. Thus if $K$ is the subgroup of $\mathrm{GL}(n+1, \mathbb{C})$ consisting of those linear transformations leaving fixed the line consisting of those points with homogeneous coordinates $[z_0, 0, \ldots, 0]$ for arbitrary $z_0 \in \mathbb{C}^*$, it follows that *$\mathbb{P}^n$ can be identified with the complex homogeneous space* $\mathrm{GL}(n+1, \mathbb{C})/K$, or equivalently with the complex homogeneous space $\mathrm{PL}(n, \mathbb{C})/[K/\mathbb{C}^*]$.

In studying projective spaces a special role is played by homogeneous polynomials in the homogeneous coordinates in $\mathbb{P}^n$—that is to say, by polynomials of the form

$$P(Z) = \sum_{|I|=d} a_I Z^I = \sum_{i_0+\cdots+i_n=d} a_{i_0\cdots i_n} z_0^{i_0}\cdots z_n^{i_n}$$

where all monomials have the same degree $d$. If $P$ is a homogeneous polynomial of degree $d$, then $P(tZ) = t^d P(Z)$ for any point $Z(\mathbb{C}^{n+1})^*$ and any constant $t \in \mathbb{C}^*$. Thus

$P(Z) = 0$ precisely when $P(tZ) = 0$ for all $t \in \mathbb{C}^*$, so the zeros of $P$ can be viewed as points in the projective space $\mathbb{P}^n$.

2. DEFINITION. *An* **algebraic subvariety** *of $\mathbb{P}^n$ is the set of common zeros of finitely many homogeneous polynomials.*

It is perhps worth noting that in this definition the polynomials are not required to be of the same degree. It is actually immaterial whether the polynomials involved are required to have the same degree or not. If $P$ is a homogeneous polynomial of degree $d$, then $z_0^\delta P(Z)$, $z_1^\delta P(Z)$, ..., $z_n^\delta P(Z)$ are homogeneous polynomials of degree $\delta + d$ and the set of zeros of $P(Z)$ is precisely the set of common zeros of these polynomials, so it is always possible to define an algebraic subvariety by a set of homogeneous polynomials all of which have the same degree. Rather more interesting though is the observation that the portion of the algebraic subvariety $V = \{Z \in \mathbb{P}^n : P_1(Z) = \cdots = P_r(Z) = 0\}$ contained in a coordinate neighborhood $U_j$ of $\mathbb{P}^n$ can be described in terms of the canonical coordinates $T = (t_1, \ldots, t_n) \in \mathbb{C}^n$ in that neighborhood as $V \cap U_j = \{T \in \mathbb{C}^n : P_i(t_1, \ldots, t_{j+1}, 1, t_{j+1}, \ldots, t_n) = 0, 1 \leq i \leq r\}$ and so is a holomorphic subvariety of $U_j$. Thus an algebraic subvariety of $\mathbb{P}^n$ is necessarily a holomorphic subvariety. Even more interesting since nontrivial is the converse assertion.

3. THEOREM (Chow's theorem). *The algebraic subvarieties of $\mathbb{P}^n$ are precisely the holomorphic subvarieties of the complex manifold $\mathbb{P}^n$.*

*Proof.* As just observed, it is obvious that an algebraic subvariety of $\mathbb{P}^n$ is a holomorphic subvariety. For the converse assertion consider a holomorphic subvariety $V \subseteq \mathbb{P}^n$, and let $V_0 = H^{-1}(V) \subseteq (\mathbb{C}^{n+1})^*$ be the inverse image of $V$ under the canonical holomorphic mapping $H: (\mathbb{C}^{n+1})^* \to \mathbb{P}^n$. The subset $V_0$ is thus a holomorphic subvariety of $(\mathbb{C}^{n+1})^*$—that is to say, is a holomorphic subvariety of $\mathbb{C}^{n+1}$ outside the origin—and must be invariant under the biholomorphic mappings $t: (\mathbb{C}^{n+1})^* \to (\mathbb{C}^{n+1})^*$ defined by multiplying the homogeneous coordinates by constants $t \in \mathbb{C}^*$. In particular the irreducible components of $V_0$ cannot consist of isolated points, since with any point $Z$ the subvariety $V_0$ must contain the entire line $\{tZ: t \in \mathbb{C}^*\}$ and so must all be of dimensions $> 0$. It then follows from the theorem of Thullen, Remmert, and Stein, Theorem K7, that the closure $\bar{V}_0$ of $V_0$ is a holomorphic subvariety at the origin in $\mathbb{C}^{n+1}$. Let $\mathbf{f}_1, \ldots, \mathbf{f}_m \in {}_{n+1}\mathcal{O}$ be a set of generators of the ideal of this subvariety at the origin, so that in some open polydisc $\Delta$ about the origin in $\mathbb{C}^{n+1}$ the subvariety $\bar{V}_0$ can be described as the set of common zeros of the holomorphic functions $f_j$ representing these germs. Write the power series expansion of $f_j$ at the origin in the form $f_j = \sum_{\nu=0}^{\infty} f_j^\nu$, where $f_j^\nu$ is a homogeneous polynomial of degree $\nu$, and note that whenever $Z \in \Delta \cap \bar{V}_0$, then $tZ \in \Delta \cap \bar{V}_0$ for all $t \in \mathbb{C}$ with $|t| \leq 1$; consequently, $0 = f_j(tZ) = \Sigma_\nu f_j^\nu(Z)t^\nu$. This latter power series in $t$ can only vanish identically in $\{t: |t| \leq 1\}\}$ when all of its coefficients are zero, so that actually $f_j^\nu(Z) = 0$ for all $\nu$ and $j$ whenever $Z \in \bar{V}_0$. The subvarieties $V_k = \{Z \in \Delta : f_j^\nu(Z) = 0$ for $1 \leq j \leq m, 0 \leq \nu \leq k\}$ are such that $V_0 \supseteq V_1 \supseteq \cdots$ and $\bigcap_k V_k = \Delta \cap \bar{V}_0$, so since ${}_{n+1}\mathcal{O}$ is Noetherian, $V_k = \Delta \cap \bar{V}_0$ for sufficiently large $k$.

But that means that $V$ can be described as the set of common zeros of the polynomials $f_j^\nu$ for $1 \leqq j \leqq m$, $0 \leqq \nu \leqq k$, so that $V$ is an algebraic subvariety of $\mathbb{P}^n$ as desired, and the proof is thereby concluded.

The only nontrivial part of the proof of the preceding theorem is really the application of the theorem of Thullen, Remmert, and Stein to the holomorphic subvariety $V_0 = H^{-1}(V) \subseteq (\mathbb{C}^{n+1})^*$, showing that it extends to a holomorphic subvariety $\bar{V}_0 \subseteq \mathbb{C}^{n+1}$. The latter subvariety is a **holomorphic** cone in $\mathbb{C}^{n+1}$, a holomorphic subvariety of $\mathbb{C}^{n+1}$ with the property that with any point $Z \in \mathbb{C}^{n+1}$ it also contains all the points $tZ$ for all $t \in \mathbb{C}$. Any subset of $\mathbb{C}^{n+1}$ that is the set of common zeros of a finite number of homogeneous polynomials is of course a holomorphic cone in $\mathbb{C}^{n+1}$, and the last part of the proof of the preceding theorem consists in showing that conversely any holomorphic cone in $\mathbb{C}^{n+1}$ is the set of common zeros of a finite number of homogeneous polynomials. Chow's theorem can thus be rephrased as the assertion that *the holomorphic subvarieties of* $\mathbb{P}^n$ *are precisely the images under the natural holomorphic mapping* $H: (\mathbb{C}^{n+1})^* \to \mathbb{P}^n$ *of holomorphic cones in* $\mathbb{C}^{n+1}$, *and the latter can be characterized as the sets of common zeros of finitely many homogeneous polynomials.* It is should also be noted, as a simple application of Corollary H10 and the above observations, that the set of common zeros of any collection of homogeneous polynomials, not just of finite collections, is a holomorphic cone. That assertion can also be proved without reference to Corollary H10, by using the observation that a holomorphic cone is determined uniquely by its germ at the origin.

# N
# Proper Holomorphic Mappings and Modifications

Even in the local study of holomorphic varieties there naturally arise a number of results that are basically of a global nature. One of the most useful and important of such results is the following.

1. **THEOREM (Remmert's proper mapping theorem).** *If $F: V \to W$ is a proper holomorphic mapping between holomorphic varieties, then its image $F(V)$ is a holomorphic subvariety of $W$ and* $\dim F(V) = \sup_i \sup_{Z \in V_i}[\dim V_i - \dim_Z L_Z(F|V_i)]$ *where $V_i$ are the irreducible components of $V$.*

*Proof.* The proof will be by induction on $n = \dim V$. The case $n = 0$ holds quite trivially, so it is only necessary to demonstrate the inductive step. Assume therefore that the theorem holds as stated whenever $\dim V < n$, and consider a proper holomorphic mapping $F: V \to W$ where $\dim V = n$. The restriction $F|V_i: V_i \to W$ to an irreducible component $V_i$ of $V$ is also a proper holomorphic mapping, and if the theorem holds for each of these restrictions, then it clearly holds for the given mapping $F$ since the assumption that $F$ is proper implies that only finitely many of the image varieties $F(V_i)$ meet any open subset of $W$ having compact closure. It can thus also be assumed that the $n$-dimensional variety $V$ is irreducible. Let $\nu = \inf_{Z \in V} \dim_Z L_Z(F)$ and introduce the exceptional locus $E(F) = \{Z \in V: \dim_Z L_Z(F) > \nu\}$, recalling from the discussion in section $L$ that $E(F)$ is a proper holomorphic subvariety of $V$. Thus for each irreducible component $E_i$ of $E(F)$ necessarily $\dim E_i < n$, while of course $\dim_Z L_Z(F|E_i) \geqq \nu$ at each point $Z \in E_i$. Since $\dim E(F) < n$ and the restriction $F|E(F)$ is also a proper holomorphic mapping, it follows from the inductive hypothesis that $Y = F(E(F))$ is a holomorphic subvariety of $W$ with $\dim Y = \sup_i \sup_{Z \in E_i}[\dim E_i - \dim_Z L_Z(F|E_i)] < n - \nu$. The inverse image $X = F^{-1}(Y)$ is a holomorphic subvariety of $V$, and the restriction $F|V - X$: $V - X \to W - Y$ is also a proper holomorphic mapping. Now for every point $Z \in V - X \subseteq V - E(F)$ it is necessarily the case that $\dim_Z L_Z(F) = \nu$, so that by Theorem L10 each point $Z \in V - X$ has an open neighborhood $V_Z$ for which $F(V_Z)$ is a holomorphic subvariety of an open neighborhood of $F(Z)$ with $\dim F(V_Z) = n - \nu$. It follows immediately from this and the fact that $F|V - X$ is proper that the image $F(V - X)$ is a holomorphic subvariety of pure dimension $n - \nu$ in $W - Y$.

Since dim $Y <$ dim $F(V - X)$, the theorem of Thullen, Remmert, and Stein, Theorem K7, shows that the closure of $F(V - X)$ in $W$ is a holomorphic subvariety $W_1$ of $W$. Since $F$ is proper, the image $F(V)$ is necessarily a closed subset of $W$, and consequently $W_1 \subseteq F(V)$. Since $V$ is irreducible and $X$ must be a proper subvariety of $V$ by the preceding observations about dimensions, the subset $V - X$ is dense in $V$, and consequently also $F(V) \subseteq W_1$. Therefore the image $F(V) = W_1$ is a holomorphic subvariety of $W$ of dimension $n - \nu$ as desired, and the proof is thereby concluded.

One common application of Remmert's proper mapping theorem is the observation that *the image of a compact holomorphic variety $V$ under any holomorphic mapping $F: V \to W$ is a holomorphic subvariety of $W$*, since any continuous mapping from a compact Hausdorff space to another topological space is necessarily proper. For some purposes an extension of the proper mapping theorem, due to N. Kuhlmann and H. Whitney, is quite useful. For this extension a mapping $F: V \to W$ between two topological spaces will be called **semiproper** if for any point $B \in W$ there are an open neighborhood $W_B$ of $B$ in $W$ and a compact subset $K \subseteq V$ such that $F(V) \cap W_B = F(K) \cap W_B$. If $W$ is locally compact, it is clear that any proper mapping $F: V \to W$ is semiproper; but there are a great many semiproper mappings that are not proper, for instance the natural projection mapping $\mathbb{C}^2 \to \mathbb{C}^1$. The following sometimes useful observation indicates other ways in which semiproper mappings may arise.

**2. THEOREM.** *If $V$ and $W$ are separable locally compact Hausdorff spaces, then any closed mapping $F: V \to W$ is semiproper.*

*Proof.* Suppose that $F: V \to W$ is a closed, continuous, but not semiproper mapping. That $F$ is closed means that the image under $F$ of any closed subset of $V$ is a closed subset of $W$. That $F$ is not semiproper means that for some point $B \in W$ there is a basis $\{W_B^\nu\}$ for the open neighborhoods of $B$ in $W$ such that for each index $\nu$ there is no compact subset $K$ of $V$ for which $F(K) \cap W_B^\nu = F(V) \cap W_B^\nu$. Choose a sequence of compact subsets $K_\nu \subset V$ such that $K_{\nu+1}$ contains an open neighborhood of $K_\nu$ and $\bigcup_\nu K_\nu = V$. It then follows that for each index $\nu$ there must be some point $B_\nu \in F(V) \cap W_B^\nu$ such that $B_\nu \notin F(K_\nu)$. Now $B_\nu = F(A_\nu)$ for some point $A_\nu \in V$, and $A_\nu \notin K_\nu$ so the sequence $\{A_\nu\}$ is a closed subset of $V$. Since $F$ is a closed mapping, the image sequence $\{B_\nu\}$ is also a closed subset of $W$. However, since $B$ is a limit point of this sequence, it must be the case that $B = B_{\nu_0}$ for some index $\nu_0$; but the same argument can be applied to the sequence from which the term $B_{\nu_0}$ is omitted, so that actually $B_\nu = B$ for infinitely many distinct indices. That is clearly impossible though, since if $B = B_{\nu_1} = f(A_{\nu_1})$, then $A_{\nu_1} \in K_\nu$ for all sufficiently large indices $\nu$ and in particular for some large index $\nu_2$ for which $B = B_{\nu_2}$, showing that $B_{\nu_2} \in f(K_{\nu_2})$ in contradiction to the construction of this sequence. That suffices to conclude the proof.

It may be worth noting in passing that not every semiproper mapping is necessarily closed. For example, if $W$ is a locally compact space, $G: V_1 \to W$ is any mapping that is not closed, $V$ is the disjoint union of $V_1$ and $W$, and $F: V \to W$ is the mapping defined by setting $F(Z) = G(Z)$ whenever $Z \in V_1$ and $F(Z) = Z$ whenever $Z \in W \subseteq V$, then $F$ is clearly semiproper but not closed. The proper mapping theorem extends to hold for semiproper mappings. The extension of course includes the original theorem as an immediate corollary, so that to some extent providing a separate proof of Theorem 1 is wasted effort; but that result is so important, and the proof in that special case is sufficiently more direct and transparent, that it seemed worth including.

**3. THEOREM (theorem of Kuhlmann and Whitney).** *If $F: V \to W$ is a semiproper holomorphic mapping between holomorphic varieties, then its image $F(V)$ is a holomorphic subvariety of $W$ and* $\dim F(V) = \sup_i \sup_{Z \in V_i}[\dim V_i - \dim_Z L_Z(F|V_i)]$ *where $V_i$ are the irreducible components of $V$.*

*Proof.* Perhaps the principal obstacle to a direct extension of the proof of Theorem 1 to this more general situation is that the restriction of a semiproper mapping $F: V \to W$ to a closed subset of $V$ is not necessarily itself semiproper. The semiproper hypothesis must be used from the beginning to restrict consideration to a compact subset of $V$ for the local analysis of the image $F(V)$ near a point of $W$, and in these considerations it is convenient to use a rather more complicated induction parameter. If $F: V \to W$ is any semiproper holomorphic mapping between holomorphic varieties and $V_i$ are the irreducible components of $V$, then following Whitney set $\lambda_i(F) = \inf_{Z \in V_i}[\dim V_i - \dim_Z L_Z(F|V_i)]$ and $\lambda(F) = \inf_i \lambda_i(F)$; the proof will be by induction on the integer $\mu(F) = \dim V - \lambda(F)$. It should be noted that in view of the hypothesis that $F$ is semiproper, the analysis of the image $F(V)$ near a point of $W$ can be reduced to a consideration of the finitely many components $V_i$ of $V$ meeting a suitably chosen compact subset of $V$, so that it can be supposed without loss of generality that $V$ is itself finite-dimensional.

To begin the proof, consider at first a semiproper holomorphic mapping $F: V \to W$ for which $\mu(F) = 0$. In this case $\dim V = \lambda(F) = \inf_i \inf_{Z \in V_i}[\dim V_i - \dim_Z L_Z(F|V_i)] \leqq \inf_i \dim V_i \leqq \sup_i \dim V_i = \dim V$, so that all these inequalities must actually be equalities; that means that $V$ is pure dimensional and $\dim_Z L_Z(F|V_i) = 0$ for every point $Z$ in every component $V_i$. The mapping $F$ is therefore a finite holomorphic mapping at each point $Z \in V$, so by Theorem E7 the image under $F$ of some open neighborhood $V_Z$ of $Z$ in $V$ is a holomorphic subvariety $F(V_Z)$ of an open neighborhood of $F(Z)$ in $W$. Since $F$ is semiproper, for any point $B \in W$ there are an open neighborhood $W_B$ of $B$ in $W$ and a compact subset $K \subseteq V$ such that $F(V) \cap W_B = F(K) \cap W_B$. For each point $Z \in X = F^{-1}(B) \cap K$ select some open neighborhood $V_Z$ for which $F(V_Z)$ is a holomorphic subvariety of an open neighborhood of $B = F(Z)$ in $W$. The closed subset $X \subseteq K$ is also compact and so will be covered by the neighborhoods $V_{Z_i}$ for but finitely many points $Z_i \in X$. It is then possible to choose a subneighborhood $W'_B \subseteq W_B$ so small that $F(V_{Z_i}) \cap W'_B$ is a holomorphic subvariety of $W'_B$ for each of these points $Z_i$ and $F[K - \bigcup_i (K \cap V_{Z_i})] \cap W'_B = \emptyset$, since $K - \bigcup_i (K \cap V_{Z_i})$ is compact. So $F(V) \cap W'_B = \bigcup_i [F(V_{Z_i}) \cap W'_B]$

is a holomorphic subvariety of $W_B'$ as desired. By recalling that $V$ is pure dimensional and that $\dim_Z L_Z(F|V_i) = 0$ in this case, and noting that $\dim F(V_{Z_i}) = \dim V$ as in Theorem G16, the theorem is therefore established in this case.

For the inductive step in the proof assume that the theorem holds as stated for all semiproper holomorphic mappings $F: V \to W$ with $\mu(F) < n$, and consider a semiproper holomorphic mapping $F: V \to W$ for which $\mu(F) = n$. For any point $B \in W$ choose an open neighborhood $W_B$ of $B$ in $W$ and a compact subset $K \subseteq V$ for which $F(V) \cap W_B = F(K) \cap W_B$, and choose an open neighborhood $U$ of $K$ for which $\bar{U}$ is also compact. For each component $V_i$ of $V$ meeting $U$ set $V_i' = \{Z \in U \cap V_i : \dim_Z L_Z(F|V_i) \geqq n + \dim V_i - \dim V\}$, where it follows from an application of Theorem L7 that $V_i'$ is a holomorphic subvariety of $U \cap V_i$, and put $V' = \bigcup_i V_i'$. At each point $Z \in V_i$ it is the case that $\dim V_i - \dim_Z L_Z(F|V_i) \geqq \lambda_i(F) \geqq \lambda(F) = \dim V - \mu(F) = \dim V - n$ and hence that $\dim_Z L_Z(F|V_i) \leqq n + \dim V_i - \dim V$. Thus actually $\dim_Z L_Z(F|V_i) = n + \dim V_i - \dim V$ at each point $Z \in V_i'$, so that $V_i'$ consists precisely of the level sets of this dimension and consequently $\dim_Z L_Z(F|V_i') = n + \dim V_i - \dim V$ at each point $Z \in V_i'$. It then follows from Theorem L10 that each point $Z \in V_i'$ has an open neighborhood in $V_i'$ that is mapped by $F$ to a holomorphic subvariety of an open neighborhood of $F(Z)$ in $W$, and that the dimension of this image is $\dim V_i' - n - \dim V_i + \dim V \leqq \dim V - n$. By arguing then as in the preceding paragraph of the proof, upon choosing an open subneighborhood $U'$ of $K$ with $\bar{U}' \subseteq U$ there will exist an open subneighborhood $W_B'$ of $B$ in $W$ for which the image $Y_i' = F(V_i' \cap U') \cap W_B'$ is either the empty set or a holomorphic subvariety of $W_B'$ with $\dim Y_i' \leqq \dim V - n$. Since only finitely many components $V_i$ meet $U$, after shrinking $W_B'$ further if necessary, the entire image $Y' = \bigcup_i Y_i' = F(V' \cap U') \cap W_B'$ will be either the empty set or a holomorphic subvariety of $W_B'$ with $\dim Y' \leqq \dim V - n$. Introduce the open subset $U'' = U' \cap F^{-1}(W_B')$ of $V$ and the holomorphic subvariety $X'' = U'' \cap F^{-1}(Y')$ of $U''$; it is clear that the restriction $F|U'' - X'' : U'' - X'' \to W_B' - Y'$ is itself a semiproper holomorphic mapping. Furthermore each point $Z \in V_i \cap (U'' - X'')$ lies in $V_i - V_i'$ so that $\dim_Z L_Z(F|V_i) < n + \dim V_i - \dim V$; consequently, $\lambda(F|U'' - X'') > \dim(U'' - X'') - n$ so $\mu(F|U'' - X'') < n$. The inductive hypothesis can therefore be applied to the restriction $F|U'' - X''$, and it follows readily that its image $F(U'' - X'') = F(U') \cap (W_B' - Y') = F(V) \cap (W_B' - Y')$ is a holomorphic subvariety of $W_B' - Y'$. Moreover applying the inductive hypothesis to the further restriction of this mapping to the inverse image of arbitrarily small open neighborhoods of any point $A$ and recalling that $\dim V_i - \dim_Z L_Z(F|V_i) > \dim V - n$ at each point $Z \in V_i \cap (U'' - X'')$ shows that $\dim_A F(U'' - X'') > \dim V - n \geqq \dim Y'$ at each point $A \in F(U'' - X'')$. The theorem of Thullen, Remmert, and Stein, Theorem K7, then shows that the closure of $F(U'' - X'')$ in $W_B'$ is a holomorphic subvariety $W_1$ of $W_B'$. The image $F(U'') = F(K) \cap W_B'$ is a relatively closed subset of $W_B'$ since $K$ is compact, and consequently $W_1 \subseteq F(U'')$. Furthermore either $X'' \cap V_i$ is a proper subvariety of $V_i$ and hence the complement $(U'' - X'') \cap V_i$ is dense in $U'' \cap V_i$, or $V_i \cap X'' \subseteq X''$ and the image $F(V_i \cap U'') \subseteq Y'$, so that $F(U'') \subseteq W_1 \cup Y'$. Therefore $F(V) \cap W_B' = F(U'') = W_1 \cup Y'$ is a holomorphic subvariety of $W_B'$ as desired. Since

$F(V)$ is a holomorphic subvariety of $W$ it follows immediately from Theorem L10 by considering just the complement of the exceptional locus that dim $F(V)$ is also as desired, and the proof is thereby concluded.

There are many interesting and important classes of mappings to which the preceding theorems apply but to which the results of section L do not apply, in the sense that for these mappings $F: V \to W$ the full image $F(V)$ is a holomorphic subvariety of $W$ but at no point in the exceptional locus $E(F) \subset V$ is the image under $F$ of the germ of $V$ the germ of a holomorphic subvariety of $W$. The simplest and in many ways perhaps the basic example of this possibly surprising phenomenon is the **monoidal** or **quadratic transform**, a standard construction in algebraic geometry introduced into complex analysis by H. Hopf and consequently sometimes also called Hopf's $\sigma$-process. Consider the space $\mathbb{C}^n$ with coordinates $(z_1, \ldots, z_n)$ and the complex projective space $\mathbb{P}^{n-1}$ with homogeneous coordinates $[t_1, \ldots, t_n]$, and introduce the holomorphic subvariety $V \subset \mathbb{C}^n \times \mathbb{P}^{n-1}$ defined by

$$V = \{(Z, T) \in \mathbb{C}^n \times \mathbb{P}^{n-1} : z_i t_j - z_j t_i = 0, 1 \leq i, j \leq n\} \tag{1}$$

Since the defining equations for $V$ are homogeneous linear functions of the homogeneous coordinates in $\mathbb{P}^{n-1}$, it is clear that $V$ is a well-defined holomorphic subvariety of $\mathbb{C}^n \times \mathbb{P}^{n-1}$. Indeed in the coordinate neighborhood $U_k = \{T \in \mathbb{P}^{n-1} : t_k \neq 0\}$ with local coordinates $s_j = t_j/t_k, j \neq k$, it is easily verified that

$$V \cap (\mathbb{C}^n \times U_k) = \{(Z, S) \in \mathbb{C}^n \times U_k : z_j = s_j z_k, j \neq k\}$$

Thus $V$ is actually an $n$-dimensional complex submanifold of $\mathbb{C}^n \times \mathbb{P}^{n-1}$ and $s_1, \ldots, s_{k-1}, z_k, s_{k+1}, \ldots, s_n$ can be used as local coordinates on the piece $V \cap (\mathbb{C}^n \times U_k)$ of this manifold. The natural projection $\mathbb{C}^n \times \mathbb{P}^{n-1} \to \mathbb{C}^n$ restricts to a holomorphic mapping $\sigma: V \to \mathbb{C}^n$, and that is the mapping of interest here. For any point $A \in \mathbb{C}^n$ other than the origin, $\sigma^{-1}(A) = \{(A, T) \in \mathbb{C}^n \times \mathbb{P}^{n-1} : a_i t_j - a_j t_i = 0, 1 \leq i, j \leq n\}$; hence clearly $\sigma^{-1}(A)$ consists of those points $(A, T) \in \mathbb{C}^n \times \mathbb{P}^{n-1}$ for which the vectors $T$ and $A$ are linearly dependent. That of course means that $\sigma^{-1}(A)$ is just the single point of $\mathbb{C} \times \mathbb{P}^{n-1}$ described by the coordinates $(A, A)$. The restriction

$$\sigma | V - \sigma^{-1}(0) : V - \sigma^{-1}(0) \to \mathbb{C}^n - 0$$

is thus a one-to-one holomorphic mapping between these two $n$-dimensional complex manifolds—hence, is actually a biholomorphic mapping between them. On the other hand, for the origin $O \in \mathbb{C}^n$ the defining equations (1) impose no restrictions whatsoever on the coordinates $T \in \mathbb{P}^{n-1}$ for which $\sigma^{-1}(O) = O \times T \in V$, so that $\sigma^{-1}(O) = O \times \mathbb{P}^{n-1}$. The complex manifold $V$, the monoidal or quadratic transform of $\mathbb{C}^n$ at the origin, can thus be viewed as the result of replacing the origin in $\mathbb{C}^n$ by the entire projective space $\mathbb{P}^{n-1}$ in such a way that the resulting entity is still an $n$-dimensional connected complex manifold. This manifold $V$ is sometimes rather dramatically though appropriately described as the result of "blowing up the origin of $\mathbb{C}^n$ to $\mathbb{P}^{n-1}$." The image of $V$ under the mapping $\sigma: V \to \mathbb{C}^n$ is all of

$\mathbb{C}^n$—hence of course is a holomorphic subvariety of $\mathbb{C}^n$. The exceptional locus $E(\sigma)$ consists of the single level set $\sigma^{-1}(0) \cong \mathbb{P}^{n-1}$, while all the other level sets consist of just isolated points. On the piece $V \cap (\mathbb{C}^n \times U_k)$ of the manifold $V$, and in terms of the coordinates $s_1, \ldots, s_{k-1}, z_k, s_{k+1}, \ldots, s_n$ as introduced above, the exceptional locus is the submanifold defined by $z_k = 0$ and the mapping $\sigma$ has the form $\sigma(s_1, \ldots, s_{k-1}, z_k, s_{k+1}, \ldots, s_n) = (z_1, \ldots, z_n)$ where $z_k$ is the same on both sides of the equality while $z_j = s_j z_k$ for $j \neq k$. In the special case $n = 2$, $k = 2$, this is just the mapping $(s_1, z_2) \to (s_1 z_2, z_2)$ considered several times before, a simple example of a mapping that is not open at any point $(s_1, 0)$ and hence with the property that the image of the germ of $\mathbb{C}^2$ at any point $(s_1, 0)$ is not the germ of a holomorphic subvariety at the origin in $\mathbb{C}^2$; the same result holds in the general case. The only open subsets of $V$ with images under $\sigma$ that are open subsets of the origin in $\mathbb{C}^n$ are open neighborhoods of the entire exceptional locus $E(\sigma) \cong \mathbb{P}^{n-1}$.

When written out formally as above, the quadratic transform may seem less surprising or unnatural than it does sometimes initially appear. The geometry of the situation may possibly be made more intuitively clear by considering the analogous construction in a real three-dimensional case. Thus introduce the real submanifold

$$V = \{(t, x; [y_1, y_2]) \in \mathbb{R}^2 \times \mathbb{P}^1_{\mathbb{R}} : t y_2 - x y_1 = 0\}$$

In the open subset $U = \{(t, x; [y_1, y_2]) \in \mathbb{R}^2 \times \mathbb{P}^1_{\mathbb{R}} : y_1 \neq 0\}$, which can be identified with $\mathbb{R}^3$ by using as coordinate functions $t$, $x$, $y = y_2/y_1$, the submanifold $V \cap U$ is just the quadric surface

$$Q = \{(t, x, y) \in \mathbb{R}^3 : ty = x\}$$

and the mapping $\sigma | Q \colon Q \to \mathbb{R}^2$ is the restriction to $Q$ of the natural projection $(t, x, y) \to (t, x)$. For any fixed point $t \in \mathbb{R}^1$ the intersection of $Q$ with the plane consisting of those points $(t, x, y)$ for which the first coordinate is this fixed value is the straight line

$$L_t = \{(x, y) \in \mathbb{R}^2 : yt = x\}$$

and $\sigma(x, y) = x$. The line $L_t$ has slope $1/t$, so when $t \neq 0$ the mapping $\sigma\colon L_t \to \mathbb{R}^1$ is one-to-one, while when $t = 0$ the mapping $\sigma\colon L_0 \to \mathbb{R}^1$ takes the entire line to the origin in $\mathbb{R}^1$. The quadric surface $Q$ can thus be viewed as a family of lines with slope $1/t$ depending on the parameter $t$, so the projection $\sigma\colon Q \to \mathbb{R}^2$ is generally one-to-one except when the line has rotated to the vertical position. The cover illustration may help in visualizing this. The situation is really the same, but more difficult to visualize, in the complex case and for an arbitrary dimension.

The monoidal transform is a special case of an important general class of holomorphic mappings, and the consideration of this special case in detail first may have served to motivate and make more intuitively clear the subsequent brief discussion of the general class, defined as follows.

**4. DEFINITION.** *A* **holomorphic modification** *is a surjective proper holomorphic mapping $F: V \to W$ between two irreducible holomorphic varieties $V$ and $W$ such that if $E \subset V$ is the exceptional locus of $F$ and $C = F(E) \subset W$ is its image, then the restriction $F|V - E: V - E \to W - C$ is a biholomorphic mapping.*

The **exceptional locus** of a modification $F$ is as defined in section L, and in this particular case it is evidently the proper holomorphic subvariety of $V$ characterized by $E = \{Z \in V: \dim_Z L_Z(F) > 0\}$. The image $C = F(E)$ is then a holomorphic subvariety of $W$ by the proper mapping theorem; this subvariety is called the center of the modification. The requirement that $F$ be surjective is both natural and convenient, but since the image of $F$ is a holomorphic subvariety by the proper mapping theorem, it is a requirement that is evidently not terribly essential. The irreducibility of $W$ is a consequence of the surjectivity of $F$ and the irreducibility of $V$, but the latter requirement is quite essential. If $A$ and $B$ are any two points of $\mathbb{P}^1$, the restriction to the subvariety $V = (A \times \mathbb{P}^1) \cup (\mathbb{P}^1 \times B)$ of the natural projection mapping $\mathbb{P}^1 \times \mathbb{P}^1 \to A \times \mathbb{P}^1 = W$ is a holomorphic mapping $F: V \to W$ that satisfies all of the conditions of Definition 4 except for the irreducibility of $V$, but this is a rather trivial case that is not really of more than passing interest. The example of a monoidal transform is clearly a holomorphic modification and of a quite nontrivial sort. In a monodial transform both $V$ and $W$ are regular varieties, the center is a single point of $W$, and the exceptional locus is a subvariety of codimension one in $V$ and is biholomorphic to a complex projective space; actually these properties characterize a monoidal transform among all holomorphic modifications. There are, however, a great variety of holomorphic modifications other than monoidal transforms. The grandest and most amazing result about holomorphic modifications is without a doubt Hironaka's theorem that for any one of a broad class of analytic varieties called normal analytic varieties, which will be discussed later here, there exists a nonsingular holomorphic modification. It serves to reduce the study of general holomorphic varieties to that of regular varieties for some purposes. In the special case of two-dimensional varieties this result was long known and has been the principal tool in the detailed study of singularities of two-dimensional varieties. The proof in the general case is considerably more involved and the result more difficult to use in studying singularities. The theory of resolution of singularities is an extensive field in itself and will really not be treated at all here, but some acquaintance with holomorphic modifications is useful to anyone interested in complex analysis.

Before turning to the consideration of a common construction of holomorphic modifications, though, one further technical but useful comment about the definition should be inserted here. In that definition it is required that the restriction $F|V - E$ should be a biholomorphic mapping and moreover that the image $F(V - E)$ should be precisely the complement of $C = F(E)$. It is clear from the surjectivity of $F$ that the image $F(V - E)$ must cover $W - C$, but implicit in the definition is the further requirement that the image $F(V - E)$ must lie in $W - C$, or equivalently must be disjoint from $C$. Actually, *if the variety $W$ is irreducible at each of it points, then this further requirement is a consequence of the others, so with this additional hypothesis, in the last part of the definition it is sufficient merely* *to require that the restriction $F|V - E$ is a biholomorphic mapping from $V - E$ to its image.* To show that, suppose to the contrary that there is a point $A \in V - E$ such that $F(A) \in C$. Let $B = F^{-1}(A) \cap E$, noting that this is a nonempty subset of $V$, and choose disjoint open neighborhoods $U_A$ of the point $A$ and $U_B$ of the set $B$ in $V$. If $U_A$ is so small that $U_A \subseteq V - E$, then the restriction $F|U_A : U_A \to F(U_A)$ is a biholomorphic mapping. Since $W$ is irreducible at each of its points and $\dim W = \dim V$, it follows from Theorem G3 that $F(U_A)$ is an open neighborhood of the point $F(A)$ in $W$. After shrinking the neighborhood $U_B$ if necessary, it can further be supposed that $U_B \subseteq F^{-1}(F(U_A))$ and hence that $F(U_B) \subseteq F(U_A)$. Now the open subsets $V - F^{-1}(C)$ and $W - C$ are biholomorphic under $F$ and $F(U_A \cap (V - F^{-1}(C))) = F(U_A) \cap (W - C)$, so whenever $Z \in V - F^{-1}(C)$ and $F(Z) \in F(U_A)$, then $Z \in U_A$. On the other hand, since $V$ is irreducible, the subset $V - F^{-1}(C)$ is dense in $V$, as is evident from Corollary E20. There must consequently be some point $Z \in U_B \cap (V - F^{-1}(C))$, but then since $F(Z) \in F(U_B) \subseteq F(U_A)$, necessarily $Z \in U_A$ as well, contradicting the disjointness of $U_A$ and $U_B$. That contradiction serves to conclude the proof of the desired assertion. It should be observed also that the local irreducibility of $W$ at each of its points is really a necessary hypothesis. If $W$ is irreducible but is reducible at $A \in W$ with one branch regular, then simply take for $V$ the variety arising by performing a monoidal transform on that local branch of $W$ at $A$ to obtain a counterexample.

**5. THEOREM.** *If $F: V \to \mathbb{C}^{m+1}$ is a holomorphic mapping defined on an irreducible holomorphic variety $V$ and $Y = \{Z \in V: F(Z) = 0\}$, then $F$ induces a holomorphic mapping $F_0: V - Y \to \mathbb{P}^m$. If moreover* $\dim Y < \dim V - m$, *then there are an irreducible holomorphic variety $\tilde{V}$, a proper holomorphic mapping $P: \tilde{V} \to V$ onto all of $V$, and a holomorphic mapping $\tilde{F}: \tilde{V} \to \mathbb{P}^m$ such that with $X = P^{-1}(Y)$ the restriction*

$$P|\tilde{V} - X : \tilde{V} - X \to V - Y$$

*is a biholomorphic mapping and*

$$\tilde{F}|\tilde{V} - X = F_0 \circ P|\tilde{V} - X$$

*Proof.* The subset $Y$ is of course a holomorphic subvariety of $V$, the set of common zeros of the $m + 1$ coordinate functions of the mapping $F$. At any point $Z \in V - Y$ not all the coordinates of the image point $F(Z) \in \mathbb{C}^{m+1}$ are zero, so $F(Z)$ represents a point $F_0(Z) \in \mathbb{P}^m$. That determines a mapping $F_0: V - Y \to \mathbb{P}^m$, and since $F$ is the representation of $F_0$ in homogeneous coordinates and $F$ is holomorphic, it follows that $F_0$ is also holomorphic. The mapping $F_0$ is not defined at points $Z \in Y$, and indeed it may not even be possible to define it at such points so as to be continuous. To prove the second statement and principal point of the theorem introduce the graph of the mapping $F_0$, the holomorphic subvariety $V_0 = \{(Z, T) \in (V - Y) \times \mathbb{P}^m : T = F_0(Z)\}$. The restriction to $V_0$ of the natural projection $(V - Y) \times \mathbb{P}^m \to V - Y$ is a holomorphic mapping $V_0 \to V - Y$ that is inverse to the holomorphic mapping $V - Y \to V_0$ taking a point $Z \in V - Y$ to the point $(Z, F_0(Z)) \in V_0$; thus, $V_0$ and $V - Y$ are biholomorphic varieties. Since $V$ is irreducible, so also is $V - Y$, by Corollary E20; thus, $V_0$ is an irreducible variety with dim $V_0 = \dim V$. Now $V_0$ can be viewed as a holomorphic subvariety in the complement in $V \times \mathbb{P}^m$ of the subvariety $Y \times \mathbb{P}^m$, and since by hypothesis $\dim V_0 > \dim Y + m = \dim(Y \times \mathbb{P}^m)$ it follows from the theorem of Thullen, Remmert, and Stein, Theorem K7, that the closure of $V_0$ in $V \times \mathbb{P}^m$ is a holomorphic subvariety $\tilde{V}$ of $V \times \mathbb{P}^m$. This variety $\tilde{V}$ is also irreducible by Theorem E19, since the regular part of $V_0$ is a connected dense open subset of the regular part of $\tilde{V}$. The restriction to $\tilde{V}$ of the natural projection $V \times \mathbb{P}^m \to V$ is a holomorphic mapping $P: \tilde{V} \to V$ that is proper since $\mathbb{P}^m$ is compact, and is surjective since it is proper and its image contains the dense subset $V - Y \subseteq V$. Moreover, since $\tilde{V} - X = V_0$ where $X = P^{-1}(Y)$, the restriction $P|\tilde{V} - X: \tilde{V} - X \to V - Y$ is as already observed a biholomorphic mapping. On the other hand, the restriction to $\tilde{V}$ of the natural projection $V \times \mathbb{P}^m \to \mathbb{P}^m$ is a holomorphic mapping $\tilde{F}: \tilde{V} \to \mathbb{P}^m$, and the restriction of this mapping to $\tilde{V} - X = V_0$ takes a point $(Z, F(Z)) \in V_0$ to $\tilde{F}(Z) = F_0 \circ P(Z, F(Z))$ and hence coincides with $F_0 \circ P|\tilde{V} - X$. That suffices to conclude the proof.

**6. COROLLARY.** *If $F: M \to \mathbb{C}^{m+1}$ is a holomorphic mapping defined on a connected complex manifold $M$ and if $Y = \{Z \in M : F(Z) = 0\}$ is a subvariety for which* $\dim Y < \dim M - m$, *then there exists a holomorphic modification $P: \tilde{M} \to M$ with center a subvariety of $Y$ and a holomorphic mapping $\tilde{F}: \tilde{M} \to \mathbb{P}^m$ such that*

$$\tilde{F}|\tilde{M} - X = P \circ F_0|\tilde{M} - X$$

*where $X = P^{-1}(Y) \subseteq \tilde{M}$ and $F_0: M - Y \to \mathbb{P}^m$ is the holomorphic mapping naturally induced by $F|M - Y$.*

*Proof.* It follows immediately from the preceding theorem that in the circumstances of this corollary there are an irreducible holomorphic variety $\tilde{M}$, a holomorphic mapping $\tilde{F}: \tilde{M} \to \mathbb{P}^m$, and a surjective proper holomorphic mapping $P: \tilde{M} \to M$ such that $P|\tilde{M} - X : \tilde{M} - X \to M - Y$ is a biholomorphic mapping and $\tilde{F}|\tilde{M} - X = P \circ F_0|\tilde{M} - X$. It is thus only necessary to show that $P$ is a holomorphic modification with center a holomorphic subvariety of $Y$, and that merely amounts to showing that the restriction $P|\tilde{M} - E : \tilde{M} - E \to M - P(E)$ is a biholomorphic mapping where $E = \{Z \in \tilde{M} : \dim_Z L_Z(P) > 0\}$ is the exceptional locus, it being evident that the holomorphic subvariety $E$ is contained in $X$ and hence that $P(E) \subseteq Y$. There is of course something to show only when $E$ is a proper subvariety of $X$, as may indeed be the case. First at any point $Z \in X - E \subset \tilde{M}$ it is the case that $\dim_Z L_Z(F) = 0$, so that the germ of $P$ at $Z$ is a finite germ of a holomorphic mapping. Hence by Theorem E7 there is an open neighborhood $U_Z$ of $Z$ in $\tilde{M}$ such that the image $P(U_Z)$ is a holomorphic subvariety of an open neighborhood of $P(Z)$ in $M$ and the restriction $P|U_Z : U_Z \to P(U_Z)$ is a finite branched holomorphic covering of some order $\nu$. The intersection $U_Z \cap (\tilde{M} - X)$ is a dense open subset of $U_Z$ on which $P$ is biholomorphic, so $\dim U_Z = \dim M$; but $\dim P(U_Z) = \dim U_Z$ by Theorem G16, and since $M$ is a regular variety, it follows as in Theorem G3 that $P(U_Z)$ must be an open neighborhood of $P(Z)$ in $M$. A dense open subset of this neighborhood lies in the set $M - Y$ over which $P$ is biholomorphic and hence one-to-one, so that necessarily $\nu = 1$. But then $P|U_Z : U_Z \to P(U_Z)$ is a one-to-one holomorphic mapping from the variety $U_Z$ onto the regular variety $P(U_Z)$ and so by Corollary E10 must be a biholomorphic mapping. That shows altogether that the restriction of $P$ to $\tilde{M} - E$ is a locally biholomorphic mapping. Next if $Z_1$, $Z_2 \in X - E \subset \tilde{M}$ are two distinct points such that $P(Z_1) = P(Z_2)$, then as just demonstrated there are open neighborhoods $U_{Z_1}$, $U_{Z_2}$ of these points that are mapped biholomorphically by $P$ onto an open neighborhood $P(U_{Z_1}) = P(U_{Z_2})$ of $P(Z_1) = P(Z_2)$, and it can of course be supposed that $U_{Z_1} \cap U_{Z_2} = \emptyset$. That means that the mapping $P$ must be at least two-to-one over the neighborhood $P(U_{Z_1}) = P(U_{Z_2})$, which is impossible since a dense open subset of this neighborhood lies in the set $M - Y$ over which the mapping $P$ is one-to-one. Thus $P|\tilde{M} - E : \tilde{M} - E \to P(\tilde{M} - E)$ is a one-to-one as well as locally biholomorphic mapping and so is necessarily a biholomorphic mapping. Since $W$ is regular and hence irreducible at each of its points, nothing further is required to show that $P$ is a holomorphic modification, as noted in the discussion of Definition 4, so the proof is thereby concluded.

In Theorem 5 the assumption that $\dim Y < \dim V - m$ is not an unreasonable one; the subvariety $Y$ is the set of common zeros of the $m + 1$ coordinate functions of the mapping $F$ and so would generally be expected to be of codimension $m + 1$ in $V$ by repeated application of Theorem G7. In Corollary 6 the assumption that $M$ is a regular variety was used in two ways: first to show that $M$ is irreducible at each of its points, and second to show that any germ of a finite branched holomorphic covering of order one over $M$ is the germ of a biholomorphic mapping. These two properties do not hold for general holomorphic varieties, nor does the corollary as stated. However the corollary does hold with the proof as given whenever $M$ is a holomorphic variety with these two additional properties. A special class of holomorphic varieties having these properties, the class of normal varieties, will be discussed in some detail later. For the special case that $M = \mathbb{C}^{m+1}$ and $F: M \to \mathbb{C}^{m+1}$ is the identity mapping, it follows readily that the holomorphic modification of $M = \mathbb{C}^{m+1}$ arising from an application of Corollary 6 is just the monoidal transform, since the subvariety defined in equation (1) is clearly just the graph of the induced mapping $F_0: \mathbb{C}^{m+1} - 0 \to \mathbb{P}^m$. It is particularly easy to see in this case that $F_0$ cannot ever be extended to a continuous mapping from all of $\mathbb{C}^{m+1}$ to $\mathbb{P}^m$, showing that in general it is necessary to pass to a modification in order to obtain an extension. Indeed the restriction of $F_0$ to a complex line $\{(ta_1, \ldots, ta_{m+1}) : t \in \mathbb{C}\}$ outside the origin in $\mathbb{C}^{m+1}$ takes the entire line to a single point of $\mathbb{P}^m$, and distinct lines are mapped to distinct points. Another example of an application of this corollary will occur in the discussion of meromorphic functions in the next section. Finally it should be noted that although the variety $\tilde{V}$ in Theorem 5 was canonically constructed in terms of the given mapping $F$, no uniqueness was asserted; indeed, it is always possible to replace $\tilde{V}$ by the result of applying a monoidal transform at any regular point $A \in \tilde{V}$, and to extend the mappings $P$ and $\tilde{F}$ to take the projective space replacing $A$ to the point $P(A)$ or $\tilde{F}(A)$. The same is of course the case for Corollary 6 as well.

# O
# Meromorphic Functions in $\mathbb{C}^n$

The ring ${}_V\mathcal{M}_A$ of germs of meromorphic functions at a point $A$ on a holomorphic variety $V$ was defined in section B as the total quotient ring of the local ring ${}_V\mathcal{O}_A$ of the variety $V$ at $A$. It is thus the ring of all formal quotients $\mathbf{f}'/\mathbf{f}''$ of elements of ${}_V\mathcal{O}_A$ for which $\mathbf{f}''$ is not a zero-divisor, with the usual notion of equivalence and the usual ring operations. As the terminology suggests, this ring can be interpreted as the ring of germs of some globally defined meromorphic functions. The aim of the present discussion is to introduce these functions and to examine some of their basic properties. This endeavor is considerably easier on regular varieties than on arbitrary varieties, where there are some interesting subtleties at the singular points, so the present discussion will be limited to the case of regular varieties—primarily indeed just to the space $\mathbb{C}^n$ itself. Even in that case some care must be taken, since for example a quotient $f'/f''$ of holomorphic functions is not really at all well defined as a complex-valued function at points at which $f''$ vanishes. That suggests introducing the following definition, keeping as close as possible to the classical feeling that a meromorphic function should really after all be a function.

**1. DEFINITION.** *A* **representative** *of a meromorphic function in an open subset $D \subseteq \mathbb{C}^n$ is a pair $(D_f, f)$, where $D_f$ is a dense open subset of $D$ and $f$ is a complex-valued function in $D_f$ such that for each point $A \in D$ there are holomorphic functions $f'_A, f''_A$ in an open neighborhood $U_A$ of $A$ for which $f(Z) = f'_A(Z)/f''_A(Z)$ whenever $Z \in U_A \cap D_f$. Two such representatives $(D_f, f)$, $(D_g, g)$ of meromorphic functions in $D$ are called* **equivalent** *if there is a dense open set $U_{f,g} \subseteq D$ for which $f(Z) = g(Z)$ whenever $Z \in U_{f,g} \cap D_f \cap D_g$. This is an equivalence relation in the usual sense, and an equivalence class is called a* **meromorphic function** *in $D$. The set of all meromorphic functions in $D$ will be denoted by ${}_n\mathcal{M}_D$ or $\mathcal{M}_D$.*

Note particularly in the preceding definition that there must exist a local expression $f(Z) = f'_A(Z)/f''_A(Z)$ for a representative of a meromorphic function in $D$ at every point $A \in D$, not just at the points $A \in D_f$ Implicit in the existence of this local expression is the condition that $f(Z)$ be a well-defined complex number and hence that $f''_A(Z) \neq 0$ whenever $Z \in U_A \cap D_f$. As is both convenient and customary, a meromorphic function in D will be described by giving some representative $f$ and will be treated much as an ordinary complex-valued function in $D$, although it must be remembered that this representative function is not necessarily well defined at all points of $D$ and that there is an equivalence relation involved. Thus if $f$ and $g$ are any two (representatives of) meromorphic functions in $D$, then $f \pm g$ and $fg$ are well-defined (representatives in $D_f \cap D_g$ of) meromorphic functions in $D$, so that ${}_n\mathscr{M}_D$ has the natural structure of a ring. Any holomorphic function $f \in {}_n\mathcal{O}_D$ can be viewed as a representative of a meromorphic function in $D$, and it is clear from the identity theorem for holomorphic functions that no two distinct holomorphic functions are equivalent, in the sense that they represent the same meromorphic functions. Thus there is a well-defined inclusion ${}_n\mathcal{O}_D \subseteq {}_n\mathscr{M}_D$ of ${}_n\mathcal{O}_D$ as a subring of ${}_n\mathscr{M}_D$. In particular there is a natural inclusion $\mathbb{C} \subseteq {}_n\mathscr{M}_D$, so that ${}_n\mathscr{M}_D$ has the structure of an algebra as well. The zero element is the meromorphic function represented by the constant 0, and the multiplicative identity is the meromorphic function represented by the constant 1. There is an analogue of the identity theorem that holds for meromorphic functions as well.

**2. THEOREM (identity theorem for meromorphic functions).** *If $f$ and $g$ are meromorphic functions in a connected open set $D \subseteq \mathbb{C}^n$ and if there is a nonempty open subset $U \subseteq D$ such that $f(Z) = g(Z)$ whenever $Z \in U \cap D_f \cap D_g$, then actually $f(Z) = g(Z)$ for all points $Z \in D_f \cap D_g$.*

*Proof.* Let $E$ be the subset of $D$ consisting of those points $A \in D$ that have an open neighborhood $U_A$ for which $f(Z) = g(Z)$ whenever $Z \in U_A \cap D_f \cap D_g$. To prove the theorem it is evidently enough just to show that $E = D$. The set $E$ is nonempty since $E \supseteq U$ and is open as an immediate consequence of its definition; so since $D$ is connected, it is only necessary to show that $E$ is relatively closed in $D$. For a point $A \in \bar{E} \cap D$ there is a connected open neighborhood $U_A$ of $A$ in $D$ in which there are holomorphic functions $f'$, $f''$, $g'$, $g''$ such that $f(Z) = f'(Z)/f''(Z)$ whenever $Z \in U_A \cap D_f$ and $g(Z) = g'(Z)/g''(Z)$ whenever $Z \in U_A \cap D_g$. Since $A \in \bar{E}$ there must be some point $B \in U_A \cap E$, and since $B \in E$ there must be an open neighborhood $U_B$ of $B$ for which $f(Z) = g(Z)$ whenever $Z \in U_B \cap D_f \cap D_g$. But then $f'(Z)g''(Z) = f''(Z)g'(Z)$ for all points $Z$ in the nonempty open subset $U_A \cap U_B \cap D_f \cap D_g$ of $U_A$, and since $U_A$ is connected, the ordinary identity theorem for holomorphic functions, Theorem I-A3, shows that $f'(Z)g''(Z) = f''(Z)g'(Z)$ for points $Z \in U_A$. That means that $f(Z) = g(Z)$ whenever $Z \in U_A \cap D_f \cap D_g$, so that $A \in E$. Thus $E$ is relatively closed in $D$, and the proof is thereby concluded.

To return then to the consideration of the algebraic structure of the ring ${}_n\mathscr{M}_D$, if $f \in {}_n\mathscr{M}_D$ is a zero-divisor, then there must exist some $g \in {}_n\mathscr{M}_D$ not identically zero such that $fg = 0$. In terms of any local expressions $f = f'_A/f''_A$, $g = g'_A/g''_A$ near any point $A \in D$, the product $f'_A g'_A$ vanishes in a dense open subset of a neighborhood of $A$ and hence identically in that neighborhood. Since ${}_n\mathcal{O}_A$ is a local ring, either $f'_A$ or $g'_A$ vanishes identically near $A$ and hence either $f = 0$ or $g = 0$ near $A$. The identity theorem for meromorphic functions then shows that either $f = 0$ or $g = 0$ in any connected component of $D$; therefore, *${}_n\mathscr{M}_D$ has zero-divisors only when $D$ is not connected, and the zero-divisors are meromorphic functions that vanish identically in* at least one connected component of D. If $f \in {}_n\mathscr{M}_D$ is not a zero-divisor, then for any local expression $f = f_A'/f_A''$ near any point $A \in D$, the function $f_A'$ is not identically zero near $A$ and hence $1/f = f_A''/f_A'$ is also meromorphic; thus, *if* $f \in {}_n\mathscr{M}_D$ *is not a zero-divisor, then* $1/f \in {}_n\mathscr{M}_D$. In particular, ${}_n\mathscr{M}_D$ *is a field if D is connected.* In any case the ring ${}_n\mathscr{M}_D$ naturally contains the total quotient ring of ${}_n\mathcal{O}_D$, with the latter also being a field if $D$ is connected. The question whether ${}_n\mathscr{M}_D$ is actually equal to the total quotient ring of ${}_n\mathcal{O}_D$ is a much more difficult one; it really amounts to the question whether any meromorphic function $f \in {}_n\mathscr{M}_D$ can be expressed globally as a quotient $f = f'/f''$ of two holomorphic functions $f', f'' \in {}_n\mathcal{O}_D$, where by definition it is only assumed that there are local such expressions. The discussion of this matter will be left to Volume III.

If $f \in {}_n\mathscr{M}_D$ and $E$ is any open subset of $D$, then the restriction to $E$ of any representative of $f$ is the representative of a meromorphic function in $E$, and any two representatives of $f$ are equivalent in $E$ and hence represent the same meromorphic function in $E$. That leads to a well-defined restriction mapping $f \in {}_n\mathscr{M}_D \to f|E \in {}_n\mathscr{M}_E$, which is evidently a ring homomorphism. In terms of this restriction homomorphism, it is of course possible to introduce as usual the notion of the germ of a meromorphic function at any point $A \in D$. If $f \in {}_n\mathscr{M}_D$ has a local expression $f = f_A'/f_A''$ in an open subset $U_A \cap D_f$ and another local expression $f = \tilde{f}_A'/\tilde{f}_A''$ in another open subset $\tilde{U}_A \cap D_f$, where $U_A$ and $\tilde{U}_A$ are open neighborhoods of $A$, then $f_A'(Z)\tilde{f}_A''(Z) = \tilde{f}_A'(Z)f_A''(Z)$ whenever $Z \in U_A \cap \tilde{U}_A \cap D_f$, and by continuity the same holds whenever $Z \in U_A \cap \tilde{U}_A$. Therefore $\mathbf{f}_A'/\mathbf{f}_A'' = \tilde{\mathbf{f}}_A'/\tilde{\mathbf{f}}_A''$ in ${}_n\mathscr{M}_A$, and this common element can be identified with the germ of the meromorphic function $f$ near $A$. Thus the two different notions of the germ of a meromorphic function coincide and the notation ${}_n\mathscr{M}_A$ can be used for either.

As already noted, there are many different local expressions for a meromorphic function as quotients of holomorphic functions; among these are some expressions of a more canonical form, though still not unique. Since the local ring ${}_n\mathcal{O}_A$ is a unique factorization domain, as ensured by Theorem A7, any germ $\mathbf{f} \in {}_n\mathscr{M}_A$ other than zero can be expressed as a quotient $\mathbf{f} = \mathbf{f}'/\mathbf{f}''$ of coprime elements $\mathbf{f}'$, $\mathbf{f}'' \in {}_n\mathcal{O}_A$, and the latter are unique up to multiplication by a common unit in ${}_n\mathcal{O}_A$. Among these particular local expressions therefore either all possible germs $\mathbf{f}'$ belong to the maximal ideal or none do, and similarly for all possible germs $\mathbf{f}''$. The point $A$ is called a **regular point** of the germ $\mathbf{f} \in {}_n\mathscr{M}_A$ if $\mathbf{f}'' \notin {}_n\mathfrak{m}_A$, and is called a **singular point** or **pole** if $\mathbf{f}'' \in {}_n\mathfrak{m}_A$. That point is called a **zero** of $\mathbf{f}$ if $\mathbf{f}' \in {}_n\mathfrak{m}_A$, and a **point of indeterminacy** if both $\mathbf{f}' \in {}_n\mathfrak{m}_A$ and $\mathbf{f}'' \in {}_n\mathfrak{m}_A$. These notions are well defined but not mutually exclusive; a regular point may or may not be a zero, while a point of indeterminacy is simultaneously a zero and a pole. Correspondingly a point $A$ in an open subset $D \subseteq \mathbb{C}^n$ is called a regular point of a meromorphic function $f \in {}_n\mathscr{M}_D$ if it is a regular point of the germ $\mathbf{f}_A \in {}_n\mathscr{M}_D$, and so on.

What may at first blush seem rather confusing about this terminology is that a point may be both a zero and a pole of a meromorphic function. For functions of a single variable, any two germs in ${}_1\mathfrak{m}_A$ must necessarily have a common nonzero factor and so can never be coprime; the oddity just mentioned thus never arises in that case. However the meromorphic function $f(z_1, z_2) = z_1/z_2$ in $\mathbb{C}^2$ really does have what must be considered both a zero and a pole at the origin. This is actually not quite so different from the classical case of functions of one variable, though, since as shortly to be made more evident the zeros and poles should be considered as subvarieties of codimension one rather than as the points on these subvarieties. Before taking up that point, however, note that at a regular point a meromorphic function is actually a holomorphic function. Note further that at a pole $A$ that is not a point of indeterminacy, a meromorphic function $f$ has a local expression $f = f'/f''$ in which $f'(A) \neq 0$ but $f''(A) = 0$. Clearly then

$$\lim_{Z \to A, Z \in D_f} |f(Z)| = \infty$$

Note further that at a point of indeterminacy $A$, a meromorphic function $f$ has a local expression $f = f'/f''$ in which $f'(A) = f''(A) = 0$, and it can be supposed that $D_f \cap U$ is the complement of the subvariety $V_\infty = \{Z \in U : f''(A) = 0\}$ for some open neighborhood $U$ of the point $A$. For any complex constant $c$ introduce the subvariety $V_c = \{Z \in U : f'(Z) - cf''(Z) = 0\}$ which passes through $A$ and is by Theorem G5 of pure dimension $n - 1$. The intersection $V_c \cap V_\infty$ is a proper subvariety of $V_c$ near $A$, since otherwise, as follows readily from Theorem G5, the germs $\mathbf{f}' - c\mathbf{f}''$ and $\mathbf{f}''$ in ${}_n\mathcal{O}_A$ must have a common factor and that contradicts the condition that $\mathbf{f}'$ and $\mathbf{f}''$ are coprime. Now $n - 1 > 0$, it having already been noted that there can be points of indeterminacy only when $n > 1$, so that any open neighborhood of $A$ must contain some point $Z \in V_c - V_c \cap V_\infty = V_c \cap D_f$ at which $f(Z) = c$. Thus $f$ takes arbitrary values in every open neighborhood of $A$ so can really have no determined limiting value at $A$, justifying the terminology to some extent. To discuss these properties further it is convenient to have the following auxiliary observations at hand.

**3. THEOREM.** *If the two nonzero nonunits* $\mathbf{f}, \mathbf{g} \in {}_n\mathcal{O}$ *generate an ideal* $\mathfrak{A} \subset {}_n\mathcal{O}$, then $\mathbf{f}, \mathbf{g}$ *are coprime precisely when* $\dim \operatorname{loc} \mathfrak{A} = n - 2$.

*Proof.* First note as a consequence of Corollary G8 that $n - 2 \leqq \dim \operatorname{loc} \mathfrak{A} \leqq n - 1$. Thus the theorem can be rephrased as the assertion that $\mathbf{f}, \mathbf{g}$ have a common nonunit factor precisely when $\dim \operatorname{loc} \mathfrak{A} = n - 1$. If $\mathbf{f}, \mathbf{g}$ have a common nonunit factor $\mathbf{h}$, then since both $\mathbf{f}$ and $\mathbf{g}$ lie in the ideal ${}_n\mathcal{O}\mathbf{h}$ generated by $\mathbf{h}$, necessarily $\mathfrak{A} \subseteq {}_n\mathcal{O}\mathbf{h}$. Hence $\operatorname{loc} \mathfrak{A} \supseteq \operatorname{loc} {}_n\mathcal{O}\mathbf{h}$, and since an application of Theorem G5 shows that $\dim \operatorname{loc} {}_n\mathcal{O}\mathbf{h} = n - 1$, it follows from Theorem G3 that $\dim \operatorname{loc} \mathfrak{A} = n - 1$. On the other hand, if $\dim \operatorname{loc} \mathfrak{A} = n - 1$ and $\mathbf{V}$ is any component of $\operatorname{loc} \mathfrak{A}$ of pure dimension $n - 1$, then from Theorem G5 it follows that $\operatorname{id} \mathbf{V}$ is principal and hence that $\operatorname{id} \mathbf{V} = {}_n\mathcal{O}\mathbf{h}$ for some nonunit $\mathbf{h} \in {}_n\mathcal{O}$. Now $\mathbf{V} \subseteq \operatorname{loc} \mathfrak{A}$ so that $\mathfrak{A} \subseteq \operatorname{id} \operatorname{loc} \mathfrak{A} \subseteq \operatorname{id} \mathbf{V} = {}_n\mathcal{O}\mathbf{h}$, and therefore both $\mathbf{f}$ and $\mathbf{g}$ belong to ${}_n\mathcal{O}\mathbf{h}$ and so must have $\mathbf{h}$ as a common factor. That suffices for the proof of the theorem.

**4. COROLLARY.** *If $f, g$ are holomorphic functions in an open neighborhood of a point $A \in \mathbb{C}^n$ and if the germs $\mathbf{f}_A, \mathbf{g}_A \in {}_n\mathcal{O}_A$ are coprime, then the germs $\mathbf{f}_Z, \mathbf{g}_Z \in {}_n\mathcal{O}_Z$ are coprime for all points $Z$ in some open neighborhood of $A$.*

*Proof.* If $f$, $g$ are holomorphic in an open neighborhood $U$ of $A$ in $\mathbb{C}^n$ and if $V = \{Z \in U : f(Z) = g(Z) = 0\}$, then the germ of the subvariety $V$ at any point

$Z \in U$ is the locus of the ideal in ${}_n\mathcal{O}_Z$ generated by the germs of the functions $f, g$ at that point $Z$. If $\mathbf{f}_A$, $\mathbf{g}_A$ are coprime, it follows from the preceding theorem that $\dim \mathbf{V}_A = n - 2$. All those irreducible components of $V$ passing through $A$ must be of dimension $n - 2$, although there may be other components of $V$ in $U$ having dimension $n - 1$. But by choosing a subneighborhood $U'$ of $A$ in $U$ sufficiently small to avoid any of these other components, it is clear that $\dim \mathbf{V}_Z = n - 2$ whenever $Z \in U'$, and then by the preceding theorem $\mathbf{f}_Z$, $\mathbf{g}_Z$ are coprime in $U'$ as desired, concluding the proof.

A simple application of the preceding corollary yields the following observation.

**5. THEOREM.** *If $f$ is a meromorphic function in an open subset $D \subseteq \mathbb{C}^n$ and is not identically zero, then the set of zero points and the set of singular points of $f$ are each holomorphic subvarieties of $D$ of pure dimension $n - 1$ if nonempty, while the set of indeterminacy points of $f$ is a holomorphic subvariety of $D$ of pure dimension $n - 2$ if nonempty.*

*Proof.* For any point $A \in D$ choose holomorphic functions $f'$, $f''$ in some open neighborhood $U_A$ of $A$ such that $f(Z) = f'(Z)/f''(Z)$ whenever $Z \in U_A \cap D_f$ and the germs $\mathbf{f}'_A$, $\mathbf{f}''_A$ in ${}_n\mathcal{O}_A$ are coprime. It follows from the preceding corollary that if $U_A$ is sufficiently small, then the germs $\mathbf{f}'_Z$, $\mathbf{f}''_Z$ in ${}_n\mathcal{O}_Z$ will also be coprime whenever $Z \in U_A$. The set of zeros of $f$ in $U_A$ can then be described as the set $\{Z \in U_A : f'(Z) = 0\}$ and the set of poles of $f$ in $U_A$ can be described correspondingly as the set $\{Z \in U_A : f''(Z) = 0\}$, and by Theorem G5 each is a holomorphic subvariety of $U_A$ of pure dimension $n - 1$ if nonempty. If the function $f'$ vanishes on any irreducible component of the latter subvariety at any point $Z \in U_A$, then both $\mathbf{f}'_Z$ and $\mathbf{f}''_Z$ are in the ideal of the germ of that component at $Z$. This ideal is by Theorem G5 a principal ideal, so the generator is a common nonunit divisor of $\mathbf{f}'_Z$ and $\mathbf{f}''_Z$, but that is impossible since these two germs are coprime. The subvariety $\{Z \in U_A : f'(Z) = f''(Z) = 0\}$, which is the set of points of indeterminacy of $f$ in $U_A$, is then by Corollary G8 of pure dimension $n - 2$ if nonempty. The desired theorem being a local assertion, the proof is thereby completed.

In view of this observation the set of zeros of a meromorphic function $f$ in an open subset $D \subseteq \mathbb{C}^n$ is called the **zero variety** of $f$ and will be denoted by $Z(f)$. Correspondingly the set of singular points is called the **singular variety** or **polar variety** of $f$ and will be denoted by $P(f)$. The intersection of these two subvarieties is the **indeterminacy variety** $I(f) = Z(f) \cap P(f)$. In the original definition of a meromorphic function it is always possible to take as $D_f$ the complement of the polar variety $P(f)$; that is the maximal possible set that can be taken as $D_f$ for some representative of $f$ and thereby determines a unique maximal representative of the meromorphic function $f$. For some purposes it is convenient to use this canonical representative, although when considering sums or products of distinct meromorphic functions, for instance, it is usually more convenient not to use these canonical representatives.

The notions of the divisor and the local divisor of a holomorphic function were introduced in section G and can be extended to meromorphic functions quite readily. The definitions of a divisor and of a local divisor in general are of course just as in section G. If $\mathbf{f} \in {}_n\mathcal{M}_A$ is the germ of a meromorphic function at a point $A \in \mathbb{C}^n$ and it is expressed as a quotient $\mathbf{f} = \mathbf{f}'/\mathbf{f}''$ of coprime elements $\mathbf{f}', \mathbf{f}'' \in {}_n\mathcal{O}_A$, then these two germs are uniquely determined up to a common unit factor in ${}_n\mathcal{O}_A$, so their local divisors $\mathfrak{d}_A(\mathbf{f}'), \mathfrak{d}_A(\mathbf{f}'')$ are actually uniquely determined. The **local divisor** of the germ $\mathbf{f}$ is then defined to be the divisor $\mathfrak{d}_A(\mathbf{f}) = \mathfrak{d}_A(\mathbf{f}') - \mathfrak{d}_A(\mathbf{f}'') \in \mathcal{D}_A$. It is clear that the local divisor $\mathfrak{d}_A(\mathbf{f})$ is positive precisely when $\mathbf{f} \in {}_n\mathcal{O}_A \subseteq {}_n\mathcal{M}_A$ and that the mapping $\mathbf{f} \in {}_n\mathcal{M}_A \rightarrow \mathfrak{d}_A(\mathbf{f}) \in \mathcal{D}_A$ is a homomorphism from the multiplicative group of nonzero elements in ${}_n\mathcal{M}_A$ onto the local group of divisors at $A$, with the kernel being the multiplicative subgroup in ${}_n\mathcal{M}_A$ consisting of the units of ${}_n\mathcal{O}_A$. If $f \in {}_n\mathcal{M}_D$ is a meromorphic function in an open subset $D \subseteq \mathbb{C}^n$ and is not identically zero in any connected component of $D$, then the zero variety $Z(f)$ and the polar variety $P(f)$ are proper holomorphic subvarieties in each component of $D$ and so can be written as unions $Z(f) = \bigcup_j V_j'$ and $P(f) = \bigcup_k V_k''$ of their irreducible components, each of which by Theorem 5 is a subvariety of $D$ of codimension one. *There is a uniquely determined divisor* $\mathfrak{d}(f) = \sum_j \nu_j' V_j' - \sum_k \nu_k'' V_k'' \in \mathcal{D}_D$ *such that the germ of this divisor at any point* $A \in D$ *is the local divisor of the germ of the meromorphic function* $f$ *at that point*; this will be called the **divisor** of the function $f$ in $D$. To demonstrate this assertion, note first that at any point $A \in D$ there are coprime germs $\mathbf{f}_A', \mathbf{f}_A'' \in {}_n\mathcal{O}_A$ such that $\mathbf{f}_A = \mathbf{f}_A'/\mathbf{f}_A''$. In a sufficiently small open neighborhood $U_A$ of $A$, these germs will have by Corollary 4 representatives $f_A', f_A'' \in {}_n\mathcal{O}_{U_A}$ that are coprime at all points of $U_A$. The divisor $\mathfrak{d}_{U_A} = \mathfrak{d}(f_A') - \mathfrak{d}(f_A'') \in \mathcal{D}_{U_A}$ then has the desired property in this set, in the sense that the germ of this divisor at any point of $U_A$ is the local divisor of the germ of the meromorphic function $f$ at that point. The divisor $\mathfrak{d}_{U_A}$ is of course formed from the irreducible components of the intersections $V_j' \cap U_A$ and $V_k'' \cap U_A$, and just as earlier in demonstrating the corresponding assertion for holomorphic functions, so also here it is now only necessary to show that the integers appearing as the coefficients of the various irreducible components of $V_j' \cap U_A$ for a fixed $j$ are all equal, and similarly for the components of $V_k'' \cap U_A$. That follows immediately from the observation that at all points $A \in V_j' \cap \mathfrak{R}(V)$, where $V = Z(f) \cup P(f) = \bigcup_j V_j' \cup \bigcup_k V_k''$, this divisor must have the form $\mathfrak{d}_{U_A} = \nu_j' V_j'$ for some integer $\nu_j'$, and $\nu_j'$ is independent of the choice of $A$ since $V_j' \cap \mathfrak{R}(V)$ is connected.

The notion of the divisor of a meromorphic function is very useful in the study of global properties of these functions, as will become apparent in the discussion of such properties in Volume III. For the present this notion may at least serve to clarify some of the elementary properties of meromorphic functions. For emphasis it is worth repeating that the divisor of a meromorphic function $f \in {}_n\mathcal{M}_D$ is defined only when $f$ is not identically zero in any connected component of $D$—that is, when $f$ is not a zero-divisor in ${}_n\mathcal{M}_D$. The mapping $f \in {}_n\mathcal{M}_D \rightarrow \mathfrak{d}(f) \in \mathcal{D}_D$ is clearly a homomorphism from the multiplicative group of non–zero-divisors in ${}_n\mathcal{M}_D$ to the group of divisors in $D$. It is not necessarily surjective, as noted earlier for the special case of holomorphic functions, but the more detailed discussion of that must be delayed to Volume III. The kernel of this homomorphism is the multiplicative group ${}_n\mathcal{O}_D^*$ of nowhere-vanishing holomorphic functions in $D$. If $f \in {}_n\mathcal{M}_D$ and if $\mathfrak{d}(f) = \sum_j \nu_j' V_j' - \sum_k \nu_k'' V_k'' \in \mathcal{D}_D$ where $\nu_j' > 0$ and $\nu_k'' > 0$, then it is really both more natural and more useful to view the irreducible subvarieties $V_j'$ themselves rather than the points on them as the zeros of the function $f$, and correspondingly to view the irreducible subvarieties $V_k''$ rather than points on them as the **poles** of the function $f$. The integer $v_j'$ is the **order** of the zero $V_j'$ of the function $f$ and correspondingly the integer $v_k''$ is the **order** of the pole $V_k''$ of $f$. This of course reduces to the classical form for functions of a single variable. In any case the zeros and poles are quite distinct as holomorphic subvarieties; they may well intersect and the subvariety $\bigcup_{j,k} V_j' \cap V_k''$ is the indeterminacy locus of $f$ as before.

In addition to the identity theorem as already demonstrated, some of the other function-theoretic properties of meromorphic functions are analogous to those already established for holomorphic functions; among these are the extension theorems to be established next, the analogues of the first results demonstrated in section K. Although probably not really necessary, nonetheless a few words of caution will be inserted here to clarify what is meant by the extension of meromorphic functions. A meromorphic function is really well defined as such even at its singular points; it does not extend as a holomorphic function at its singular points, but as a meromorphic function it is already well defined at such points, so there is really no nontrivial notion of extension involved at such points. The function $f(z) = 1/z$ is well defined either as a holomorphic or as a meromorphic function in the subset $D = \{z \in \mathbb{C} : z \neq 0\}$; it extends as a meromorphic function but not as a holomorphic function to the entire complex plane $\mathbb{C}$. On the other hand, the function $f(z) = \sin(1/z)$ is also well defined either as a holomorphic or as a meromorphic function in the same subset $D \subseteq \mathbb{C}$, but it does not extend to the entire complex plane $\mathbb{C}$ either as a holomorphic or even as a meromorphic function. With this in mind, the extension theorems are as follows.

**6. THEOREM (Levi's theorem).** *If $W$ is a holomorphic subvariety of an open subset $D \subseteq \mathbb{C}^n$ with* $\dim W \leqq n - 2$ *and if $f$ is a meromorphic function in $D - W$, then there is a unique meromorphic function $\tilde{f}$ in $D$ such that $\tilde{f}|D - W = f$.*

*Proof.* The divisor of the meromorphic function $f$ is an expression $\mathfrak{d}(f) = \sum_j v_j V_j$ where $v_j \in \mathbb{Z}$ and $V_j$ are irreducible holomorphic subvarieties of $D - W$ of dimension $n - 1$; the union $V = \bigcup_j V_j$ is also a holomorphic subvariety of $D - W$. The theorem of Thullen, Remmert, and Stein, Theorem K7, shows first that each subvariety $V_j \subset D - W$ extends to a holomorphic subvariety of all of $D$, the extension being $\tilde{V}_j = D \cap \bar{V}_j$ and hence also an irreducible subvariety of dimension $n - 1$, and second that the entire subvariety $V$ also extends to the holomorphic subvariety $\tilde{V} = D \cap \bar{V}$. Then $\tilde{\mathfrak{d}} = \sum_j v_j \tilde{V}_j$ is a well-defined divisor in $D$ such that $\tilde{\mathfrak{d}}|(D - W) = \mathfrak{d}$. The second observation in the preceding sentence is needed to ensure that only finitely many of the subvarieties $\tilde{V}_j$ meet any sufficiently small open neighborhood of a point of $W$. Now for any point $A \in W$ the germ of the divisor $\mathfrak{d}$ at $A$ is as noted before the divisor of some germ $\mathbf{f}_A \in {}_n\mathcal{M}_A$. In a sufficiently small open neighborhood $U_A$ of $A$ there will be a meromorphic function $f_A$ representing this germ $\mathbf{f}_A$, and as also noted before the germ of the divisor $\mathfrak{d}$ at any point $Z \in U_A$ is also the local divisor of the germ of the function $f_A$ at that point $Z$. The germs of the meromorphic functions $f$ and $f_A$ thus have the same local divisor at any point $Z \in U_A \cap (D - W)$, so that the quotient $u_A = f/f_A$ is a holomorphic and nowhere vanishing function throughout

$U_A \cap (D - W)$. It follows then from Theorem K1 that $u_A$ extends to a holomorphic function throughout $U_A$ and hence that $f = u_A f_A$ extends to a meromorphic function throughout $U_A$. That is true for any point $A \in W$, so any representative of the meromorphic function $f$ in $D - W$ can also be viewed as the representative of a meromorphic function in all of $D$. This latter function is the desired extension $\tilde{f}$, and the proof is thereby concluded.

If $W$ is a holomorphic subvariety of dimension $n - 1$ in an open subset $D \subseteq \mathbb{C}^n$, it is not the case that every meromorphic function in $D - W$ extends to a meromorphic function in all of $D$; but just as in the case of holomorphic functions, at least the following is true.

**7. THEOREM.** *If $W$ is an irreducible holomorphic subvariety of dimension $n - 1$ in an open subset $D \subseteq \mathbb{C}^n$, if $f$ is a meromorphic function in $D - W$, and if there is a point $A \in W$ such that $f$ extends to a meromorphic function in an open neighborhood of $A$, then $f$ extends to a meromorphic function in all of $D$.*

*Proof.* Note first that it is sufficient just to prove the theorem in the special case that $W$ is a regular holomorphic subvariety of $D$. Indeed if $W$ is an arbitrary irreducible holomorphic subvariety of dimension $n - 1$ in $D$, then its singular locus $\mathfrak{S}(W)$ is a holomorphic subvariety of $D$ of dimension at most $n - 2$, while its regular locus $\mathfrak{R}(W)$ is a connected submanifold of $D - \mathfrak{S}(W)$. If $f$ extends to a meromorphic function in an open neighborhood of some point $A \in W$, that neighborhood must contain some point $B \in \mathfrak{R}(W)$, since $\mathfrak{R}(W)$ is dense in $W$, and hence can also be viewed as an open neighborhood of $B$. The special case of the theorem then shows that $f$ extends to a meromorphic function in $D - \mathfrak{S}(W)$, and it then follows from the preceding theorem that $f$ extends further to a meromorphic function in all of $D$.

To consider then the special case that $W$ is a regular holomorphic subvariety of $D$, let $W_0$ be the subset of $W$ consisting of those points $A \in W$ such that $f$ extends to a meromorphic function in an open neighborhood of $A$. To complete the proof it is of course only necessary to show that $W_0 = W$. The submanifold $W$ is connected, since it is irreducible, the subset $W_0$ is an open subset of $W$ by its definition, and $W_0$ is nonempty by hypothesis; thus it is finally sufficient just to show that $W_0$ is closed in $W$. Consider then a point $A \in W \cap \overline{W_0}$, and choose an open neighborhood $U_A$ of $A$ so that the intersection $U_A \cap W$ is a connected and hence irreducible holomorphic submanifold of $U_A$. The divisor of the meromorphic function $f$ in the subset $U_A - U_A \cap W$ has the form $\mathfrak{d}_A = \sum_j \nu_j V_j$ where $V_j$ are irreducible subvarieties of dimension $n - 1$ in $U_A - U_A \cap W$ and the union $V = \bigcup_j V_j$ is also a holomorphic subvariety of $U_A - U_A \cap W$. Since $A \in \overline{W_0}$, there must be some point $B \in U_A \cap W_0$ so that the function $f$ extends to a meromorphic function in an open neighborhood of $B$. It is clear that the subvarieties $V_j$ and $V$ then also extend to holomorphic subvarieties in that neighborhood of $B$ as well. The theorem of Thullen, Remmert, and Stein, Theorem K7, can then be applied to show that the subvarieties $V_j$ and $V$ and hence the divisor $\mathfrak{d}$ extend to the full neighborhood $U_A$. Just as in the proof of Theorem 6, that leads to an extension of $f$ to a meromorphic function in $U_A$, showing therefore that $A \in W_0$ and with that concluding the proof of the theorem.

Any holomorphic mapping $F: D \to E$ between open subsets $D \subseteq \mathbb{C}^m$ and $E \subseteq \mathbb{C}^n$ induces a ring homomorphism $F^*: {}_n\mathcal{O}_E \to {}_m\mathcal{O}_D$ by composition. Such a mapping does not necessarily induce a well-defined ring homomorphism $F^*: {}_n\mathcal{M}_E \to {}_m\mathcal{M}_D$ though. For instance, if the image $F(D)$ is contained in a proper holomorphic subvariety $V \subseteq E$ and if $V$ is contained in the singular variety of a meromorphic function $f \in {}_n\mathcal{M}_E$, then the composition $f \circ F$ is not a well-defined meromorphic function on $D$, while if $V$ is contained in the indeterminacy variety of $f$, then $f \circ F$ cannot even be viewed as a well-defined meromorphic function on $D$ by modifying the definitions to include among the meromorphic functions that which is identically equal to $\infty$ in some sense. Thus some care must be taken when considering homomorphisms of rings of meromorphic functions induced by holomorphic mappings. If $D$ and $E$ are connected open subsets and if $F: D \to E$ is a surjective holomorphic mapping, then such difficulties do not arise. The composition $f \circ F$ with any meromorphic function $f \in {}_n\mathcal{M}_E$ is clearly a well-defined meromorphic function $F^*(f) \in {}_m\mathcal{M}_D$, and the resulting mapping $F^*: {}_n\mathcal{M}_E \to {}_m\mathcal{M}_D$ is a ring homomorphism. In particular whenever $F: D \to E$ is a biholomorphic mapping, then $F^*: {}_n\mathcal{M}_E \to {}_m\mathcal{M}_D$ is a well-defined ring isomorphism. That means of course that it is possible to introduce in a natural way the notion of a meromorphic function on an arbitrary complex manifold. The procedure is quite obvious and the details straightforward and not terribly interesting, so nothing more will be added here in general.

It may be helpful though to consider at least one example in a bit more detail; a useful and interesting though not typical example is that provided by the complex projective space $\mathbb{P}^n$. This is a compact connected complex manifold, so an application of the maximum modulus theorem shows that the only holomorphic functions on $\mathbb{P}^n$ are just the constant functions. There are, however, nontrivial meromorphic functions on $\mathbb{P}^n$ whenever $n > 0$. To see this, note that whenever $P$ and $Q$ are homogeneous polynomials of the same degree $d$ in homogeneous coordinates $[z_0, \ldots, z_n]$ on $\mathbb{P}^n$ and $Q \neq 0$, then the quotient $f(Z) = P(Z)/Q(Z)$ has the same value at all homogeneous coordinates of the same point in $\mathbb{P}^n$, since $P(tZ) = t^d P(Z)$ and $Q(tZ) = t^d Q(Z)$. This leads to a well-defined meromorphic function $f$ on $\mathbb{P}^n$, the function that can also be represented as a quotient of holomorphic functions $P(1, z_1, \ldots, z_n)/Q(1, z_1, \ldots, z_n)$ in the coordinate neighborhood $U_0$ and correspondingly in the other coordinate neighborhoods. Meromorphic functions of this special form are called rational functions; what is particularly interesting is that all the meromorphic functions on $\mathbb{P}^n$ are of this form. This assertion is quite familiar in the case $n = 1$ but a bit more complicated for general $n$.

**8. THEOREM (theorem of Weierstrass and Hurwitz).** *The meromorphic functions on $\mathbb{P}^n$ are precisely the rational functions.*

*Proof.* In view of what has already been observed, it only remains to show that every meromorphic function on $\mathbb{P}^n$ is a rational function. If $f$ is a meromorphic function on $\mathbb{P}^n$ and $H: (\mathbb{C}^{n+1})^* \to \mathbb{P}^n$ is the natural mapping that associates to any point $Z = (z_0, \ldots, z_n) \in \mathbb{C}^{n+1}$ other than the origin the point $H(Z) \in \mathbb{P}^n$ having homogeneous coordinates $[z_0, \ldots, z_n]$, then the composition $\tilde{f} = f \circ H$ is a well-defined meromorphic function on $(\mathbb{C}^{n+1})^*$, and it follows from Levi's theorem, Theorem 6, that $\tilde{f}$ extends to a meromorphic function at the origin as well. The extended function will also be denoted by $\tilde{f}$. In an open polydisc $\Delta$ about the origin in $\mathbb{C}^{n+1}$ this function $\tilde{f}$ has the expression $\tilde{f}(Z) = f'(Z)/f''(Z)$ for some holomorphic functions $f', f'' \in {}_{n+1}\mathcal{O}_\Delta$, and these functions can as usual be written as sums $f' = \sum_\nu f'_\nu, f'' = \sum_\nu f''_\nu$ where $f'_\nu, f''_\nu$ are homogeneous polynomials of degree $\nu$. Now by construction it is clear that $\tilde{f}(tZ) = \tilde{f}(Z)$ whenever $t \in \mathbb{C}^*$ and $Z \in \mathbb{C}^{n+1}$, and therefore $f'(tZ) = \tilde{f}(Z)f''(tZ)$ whenever $|t| \leq 1$ and $Z \in \Delta \cap D_{\tilde{f}}$. When the latter identity is written out in terms of the homogeneous polynomials summing to the functions $f', f''$, it takes the form

$$\sum_\nu f'_\nu(Z)t^\nu = \sum_\nu \tilde{f}(Z)f''_\nu(Z)t^\nu$$

and since this holds identically in $t$ for $|t| \leq 1$ and for any fixed point $Z \in \Delta \cap D_f$, it follows that $f'_\nu(Z) = \tilde{f}(Z)f''_\nu(Z)$ whenever $Z \in \Delta \cap D_{\tilde{f}}$. There must be at least one integer $\nu_o$ for which $f''_{\nu_o} \neq 0$, and then $\tilde{f}(Z) = f'_{\nu_o}(Z)/f''_{\nu_o}(Z)$ so that $f$ is a rational function as desired and the proof is thereby concluded.

The preceding theorem can be used to complete one point of the discussion of complex projective spaces from section M. It was demonstrated there that $\mathbb{P}^n$ can be viewed as a complex homogeneous space, by exhibiting a transitive group of biholomorphic mappings $F: \mathbb{P}^n \to \mathbb{P}^n$, the projective transformations described explicitly in equation M(2). It will now be demonstrated here that there are no other biholomorphic mappings $F: \mathbb{P}^n \to \mathbb{P}^n$ than these projective transformations. This is the natural generalization of a familiar result from the theory of holomorphic functions of one variable, the assertion that any biholomorphic mapping $F: \mathbb{P}^1 \to \mathbb{P}^1$ of the Riemann sphere to itself is necessarily a linear fractional or Moebius transformation. First though it is probably worth inserting one general observation about arbitrary holomorphic mappings $F: \mathbb{P}^m \to \mathbb{P}^n$. In terms of inhomogeneous coordinates $(z_1, \ldots, z_m)$ in $\mathbb{P}^m$, the mapping $F$ can clearly be described by $n$ coordinate functions $f_1, \ldots, f_n$, which are meromorphic functions on $\mathbb{P}^m$ and hence by Theorem 8 are rational functions. It is thus possible to write $f_j = p_j/p_0$, where $p_0, p_1, \ldots, p_n$ are some polynomial functions in the inhomogeneous coordinates $z_1, \ldots, z_m$. It is further possible to find homogeneous polynomials $P_j$ all of the same degree $r$ in the homogeneous coordinates $t_0, t_1, \ldots, t_m$ in $\mathbb{P}^m$ so that $p_j(z_1, \ldots, z_m) = P_j(1, z_1, \ldots, z_m)$. These homogeneous polynomials define a holomorphic mapping $P: \mathbb{C}^{m+1} \to \mathbb{C}^{n+1}$ that induces the given holomorphic mapping $F: \mathbb{P}^m \to \mathbb{P}^n$ in the natural manner. Thus an arbitrary holomorphic mapping $F: \mathbb{P}^m \to \mathbb{P}^n$ can be described in a reasonably simple and purely algebraic manner. What is to be demonstrated is that if $m = n$ and $F$ is a biholomorphic mapping, then the polynomials $P_j$ can be taken to be homogeneous linear polynomials, so that $F$ is a projective transformation. It is convenient for that purpose to have available the following auxiliary result.

**9. LEMMA.** *If* $\mathbf{Q} \in {}_n\mathcal{O}_O$ *is the germ of a homogeneous polynomial and can be factored nontrivially as a product* $\mathbf{Q} = \mathbf{q}'\mathbf{q}''$ *in* ${}_n\mathcal{O}_O$, *then it can also be factored nontrivially as a product of homogeneous polynomials.*

*Proof.* Suppose that $\mathbf{Q} \in {}_n\mathcal{O}_O$ is the germ of a homogeneous polynomial of degree $r > 0$ and that $\mathbf{Q} = \mathbf{q}'\mathbf{q}''$ for some nonunits $\mathbf{q}', \mathbf{q}'' \in {}_n\mathcal{O}_O$. The power series expansion of the germ $\mathbf{q}'$ can be written $\mathbf{q}' = \sum_{\nu=0}^{\infty} Q'_\nu$, where $Q'_\nu$ is a homogeneous polynomial of degree $\nu$, and similarly for $\mathbf{q}''$, so that $Q = \sum_{\nu,\mu} Q'_\nu Q''_\mu$. For any fixed point $Z$ sufficiently near the origin that these series converge at $Z$ and for any complex number $t$ with $|t| < 1$, it follows from the homogeneity that

$$t^r Q(Z) = Q(tZ) = \sum_{\nu,\mu} Q'(tZ)Q''_\mu(tZ) = \sum_{\nu,\mu} t^{\nu+\mu} Q'_\nu(Z)Q'_\mu(Z)$$

This is an identity between two convergent power series in $t$ for $|t| < 1$, so the coefficients must be the same and consequently

$$\sum_{\nu+\mu=\lambda} Q'_\nu(Z)Q''_\mu(Z) = \begin{cases} Q(Z) & \text{if } \lambda = r \\ 0 & \text{if } \lambda \neq r \end{cases}$$

Furthermore the last identity holds for all points $Z$ sufficiently near the origin and hence of course for all values of $Z$. In particular, if $\nu_0$, $\mu_0$ are the least indices for which $Q'_{\nu_0} \neq 0$, $Q''_{\mu_0} \neq 0$, then

$$Q'_{\nu_0}(Z)Q''_{\mu_0}(Z) = \begin{cases} Q & \text{if } \nu_0 + \mu_0 = r \\ 0 & \text{if } \nu_0 + \mu_0 \neq r \end{cases}$$

So since ${}_n\mathcal{O}_O$ is an integral domain, it must actually be the case that $\nu_0 + \mu_0 = r$ and $Q = Q'_{\nu_0}Q''_{\mu_0}$. The hypothesis that $\mathbf{q}'$, $\mathbf{q}''$ are nonunits implies that $\nu_0 > 0$, $\mu_0 > 0$, so this yields a nontrivial factorization of $Q$ into homogeneous polynomials as desired and therewith concludes the proof.

It is perhaps worth pointing out as an immediate consequence of the preceding lemma that a homogeneous polynomial is irreducible in the ring of homogeneous polynomials precisely when its germ in ${}_n\mathcal{O}_O$ is irreducible, and that the decomposition of a homogeneous polynomial $Q$ into a product of irreducible homogeneous polynomials corresponds precisely to the decomposition of the germ $\mathbf{Q} \in {}_{n+1}\mathcal{O}_O$ as a product of irreducible elements so that there is also a unique decomposition of any homogeneous polynomial as a product of irreducible homogeneous polynomials. This will be applied in deriving the following consequence of Theorem 8.

**10. THEOREM.** *Every biholomorphic mapping* $F: \mathbb{P}^n \to \mathbb{P}^n$ *is a projective transformation.*

*Proof.* Suppose that $F: \mathbb{P}^n \to \mathbb{P}^n$ is a biholomorphic mapping, and choose homogeneous polynomials $P_j$ of some degree $r$ in the homogeneous coordinates of $\mathbb{P}^n$ such that the mapping $P: \mathbb{C}^{n+1} \to \mathbb{C}^{n+1}$ defined by these polynomials induces the mapping $F$. Thus in terms of the standard inhomogeneous coordinates $z_j = t_j/t_0$, the mapping

$F$ is described by the meromorphic coordinate functions

$$f_j(z_1, \ldots, z_n) = \frac{P_j(t_0, t_1, \ldots, t_n)}{P_0(t_0, t_1, \ldots, t_n)} \tag{1}$$

and of course there is a corresponding description of the mapping $F$ in terms of the inhomogeneous coordinates $t_0/t_k, \ldots, t_{k-1}/t_k, t_{k+1}/t_k, \ldots, t_n/t_k$ in each of the other $n$ coordinate neighborhoods of $\mathbb{P}^n$, $1 \leq k \leq n$. If $n = 1$, the mapping is that described by a rational function, and as is well known can be a biholomorphic mapping only when that function is a linear fractional and hence projective mapping. If $n > 1$, the mapping $F$ is biholomorphic, so its Jacobian matrix at any point $T \in \mathbb{P}^n$, when $F$ is expressed as a holomorphic mapping from a coordinate neighborhood about $T$ to a coordinate neighborhood about $F(T)$, must be a nonsingular matrix. The proof of the theorem will be accomplished by expressing this condition in terms of the Jacobian matrix of the mapping $P$ and using the result appropriately. First at any point $T = [t_0, t_1, \ldots, t_n] \in \mathbb{P}^n$ at which $t_0 P_0(T) \neq 0$, the mapping $F$ can be expressed in terms of the standard inhomogeneous coordinates by the coordinate functions (1). Note that $t_k = z_k t_0$ so that

$$\frac{\partial f_j}{\partial z_k}(Z) = t_0 P_0(T)^{-2}\left[P_0(T)\frac{\partial P_j}{\partial t_k}(T) - P_j(T)\frac{\partial P_0}{\partial t_k}(T)\right] \tag{2}$$

for any indices $1 \leq j, k \leq n$. Note also that $P_j(xt_0, xt_1, \ldots, xt_n) = x^r P_j(t_0, t_1, \ldots, t_n)$ since $P_j$ is homogeneous of degree $r$, so that upon differentiating with respect to $x$ and setting $x = 1$, it follows that

$$\sum_{i=0}^{n} t_i \frac{\partial P_j}{\partial t_i}(T) = rP_j(T) \tag{3}$$

If the $(n + 1) \times (n + 1)$ Jacobian matrix $\partial P_j/\partial t_i$ is singular at the point $T$, there are constants $c_i$ not all zero such that

$$\sum_{i=0}^{n} c_i \frac{\partial P_j}{\partial t_i}(T) = 0 \tag{4}$$

for $0 \leq j \leq n$. Since $t_0 \neq 0$, setting $\tilde{c}_0 = c_0/t_0$ and $\tilde{c}_k = -c_k + c_0 t_k/t_0$ for $1 \leq k \leq n$ and using (3) lead to

$$\begin{aligned} 0 &= \tilde{c}_0 t_0 \frac{\partial P_j}{\partial t_0}(T) + \sum_{k=1}^{n} (\tilde{c}_0 t_k - \tilde{c}_k)\frac{\partial P_j}{\partial t_k}(T) \\ &= \tilde{c}_0 \sum_{i=0}^{n} t_i \frac{\partial P_j}{\partial t_i}(T) - \sum_{k=1}^{n} \tilde{c}_k \frac{\partial P_j}{\partial t_k}(T) \\ &= \tilde{c}_0 r P_j(T) - \sum_{k=1}^{n} \tilde{c}_k \frac{\partial P_j}{\partial t_k}(T) \end{aligned}$$

Using this formula for $j = 0$ and for another value of the index $j$ in the range $1 \leq j \leq n$ yields

$$0 = \tilde{c}_0 r[P_0(T)P_j(T) - P_j(T)P_0(T)]$$

$$= \sum_{k=1}^{n} \tilde{c}_k \left[ P_0(T)\frac{\partial P_j}{\partial t_k}(T) - P_j(T)\frac{\partial P_0}{\partial t_k}(T) \right]$$

and since $P_0(T) \neq 0$ as well, it then follows from (2) that

$$0 = \sum_{k=1}^{n} \tilde{c}_k t_0^{-1} P_0(T)^2 \frac{\partial f_j}{\partial z_k}(Z)$$

Since the matrix $\partial f_j/\partial z_k$ is nonsingular, it must be the case that $\tilde{c}_k = 0$ for $1 \leq k \leq n$ and hence that $c_k = c_0 t_k/t_0$ for $1 \leq k \leq n$ and of course also trivially for $k = 0$. But then from (4) and (3) it follows that

$$0 = c_0 t_0^{-1} \sum_{i=0}^{n} t_i \frac{\partial P_j}{\partial t_i}(T) = c_0 t_0^{-1} r P_j(T)$$

for all indices $j$, and since $P_0(T) \neq 0$ that implies that $c_0 = 0$ and hence that $c_k = c_0 t_k/t_0 = 0$ for all indices $k$, contradicting the assumption that not all these constants $c_k$ are zero. The matrix $\partial P_j/\partial t_k$ is therefore nonsingular at any point $T \in \mathbb{C}^{n+1}$ at which $t_0 P_0(T) \neq 0$. The same argument in the other homogeneous coordinate neighborhoods shows that this matrix is also nonsingular at any point $T \in \mathbb{C}^{n+1}$ at which $t_k P_k(T) \neq 0$ for any index $k$, so that it can be singular only at points $T$ of the proper subvariety

$$V = \{T \in \mathbb{C}^{n+1} : t_0 P_0(T) = t_1 P_1(T) = \cdots = t_n P_n(T) = 0\}$$

If the polynomials $P_j$ are all homogeneous of degree $r = 1$, then the matrix $\partial P_j/\partial t_i$ is a constant matrix, which must be nonsingular, and the mapping $F$ is a projective transformation as desired. If the polynomials $P_j$ are all homogeneous of some degree $r > 1$, then $Q = \det\{\partial P_j/\partial t_k\}$ is a homogeneous polynomial of degree $(n + 1)(r - 1) > 0$, and $Q(T) = 0$ implies that $T \in V$. Write $Q$ as a product $Q = \prod_i Q_i$ of irreducible homogeneous polynomials, noting by Lemma 9 that the germs $\mathbf{Q}_i \in {}_{n+1}\mathcal{O}_O$ are irreducible and hence generate prime ideals $\mathfrak{p}_i = {}_{n+1}\mathcal{O}_O \mathbf{Q}_i$. Since loc $\mathfrak{p}_i \subseteq \mathbf{V}$, it follows from Hilbert's zero-theorem in the form of Corollary E3 that $\mathbf{Q}_i$ must divide each of the germs $\mathbf{t}_k \mathbf{P}_k$ for $0 \leq k \leq n$. If $n > 1$ it cannot be the case that $\mathbf{Q}_i = \mathbf{t}_j$ for some index $j$; for then all of the polynomials $\mathbf{P}_k$ for $k \neq j$ would be divisible by $t_j$, and the mapping $F$ would take the entire $(n - 1)$-dimensional subvariety $\{T \in \mathbb{P}^n : t_j = 0\}$ to the point $\{T \in \mathbb{P}^m : t_0 = \cdots = t_{j-1} = t_{j+1} = \cdots = t_n = 0\}$, thus contradicting the assumption that $F$ is a biholomorphic mapping. Therefore $\mathbf{Q}_i$ must actually divide each of the germs $\mathbf{P}_k$, so that as a consequence of Lemma 9 as noted in the discussion after that lemma, it is possible to write $P_k = P'_k Q_i$ for $0 \leqq k \leqq n$, where $P'_k$ is a homogeneous polynomial of degree strictly less than the degree of $P_k$. Now the original mapping can be described by the polynomials $P'_k$ rather than $P_k$, and if the polynomials $P'_k$ are of degree strictly greater than one, the process can be repeated. Eventually then the mapping $F$ can be described by homogeneous linear polynomials, so it is as a projective transformation as desired and the proof is thereby concluded.

There is another way in which projective spaces naturally occur in the consideration of meromorphic functions. As is no doubt quite familiar, a meromorphic function $f$ in an open subset $D \subseteq \mathbb{C}^1$ can be viewed as a holomorphic mapping $f: D \to \mathbb{P}^1$. What that really means of course is that for any meromorphic function $f$ on $D \subseteq \mathbb{C}^1$, there is a holomorphic mapping $\tilde{f}: D \to \mathbb{P}^1$ so that $f(z)$ is the inhomogeneous coordinate of the point $\tilde{f}(z) \in \mathbb{P}^1$ whenever $z \in D_f$. For the case of meromorphic functions of several variables, there is a similar but slightly more complicated statement, as follows.

**11. THEOREM.** *If $f$ is a meromorphic function defined in an open subset $D \subseteq \mathbb{C}^n$, then there are a holomorphic modification $P: \tilde{D} \to D$ with exceptional locus $E$ for which $P(E) \subseteq D - D_f$ and a holomorphic mapping $\tilde{f}: \tilde{D} \to \mathbb{P}^1$ such that $f(P(Z))$ is the inhomogeneous coordinate of the point $\tilde{f}(Z)$ whenever $Z \in \tilde{D} - E$.*

*Proof.* In an open neighborhood $U$ of any point $A \in D$ the function $f$ can be expressed as a quotient $f = f'/f''$ of two holomorphic functions $f', f'' \in {}_n\mathcal{O}_U$, and if the germs $\mathbf{f}'_A, \mathbf{f}''_A \in {}_n\mathcal{O}_A$ are taken to be coprime, then by Corollary 4 the germs $\mathbf{f}'_Z, \mathbf{f}''_Z \in {}_n\mathcal{O}_Z$ are coprime at any point $Z \in U$ provided that $U$ is sufficiently small. The mapping $F: U \to \mathbb{C}^2$ with coordinate functions $f', f''$ then has by Theorem 3 the property that $\dim\{Z \in U : F(Z) = 0\} = n - 2$, so by Corollary N6 there exist a holomorphic modification $P: \tilde{U} \to U$ and a holomorphic mapping $\tilde{f}: \tilde{U} \to \mathbb{P}^1$ such that $f(P(Z))$ is the inhomogeneous coordinate of the point $\tilde{f}(Z)$ whenever $f''(Z) \neq 0$. Thus the desired theorem holds locally near any point of $D$. To complete the proof it is sufficient to observe that the modification $\tilde{U}$, which is actually canonically associated to the mapping $F$, is independent of the choice of the holomorphic functions $f'$ and $f''$. Indeed these functions are unique up to multiplication by a common nowhere-vanishing holomorphic function $u$ in $U$. On the other hand, $\tilde{U}$ is just the point set closure in $U \times \mathbb{P}^1$ of the subvariety

$$U_0 = \{(Z, T) \in U \times \mathbb{P}^1 : f''(Z) \neq 0 \text{ and } T = [f'(Z), f''(Z)]\}$$

But $[f'(Z), f''(Z)] = [u(Z)f'(Z), u(Z)f''(Z)] \in \mathbb{P}^1$ whenever $u(Z) \neq 0$, so this locus is really independent of the representation of $f$. That serves to conclude the proof.

It is worth noting that, as discussed at the end of section N, the holomorphic mapping $F: \mathbb{C}^2 \to \mathbb{C}^2$ given by $F(Z) = (z_1, z_2)$ does not lead to a holomorphic mapping $F: \mathbb{C}^2 \to \mathbb{P}^1$ without some modification, so the meromorphic function $f(Z) = z_1/z_2$ cannot be viewed as a holomorphic mapping $\mathbb{C}^2 \to \mathbb{P}^1$ without some modification.

# P

# Meromorphic Functions on Varieties

It is possible to define a meromorphic function in an open subset of an arbitrary holomorphic variety in much the same way as in an open subset of $\mathbb{C}^n$, but even when considering the most elementary properties of meromorphic functions, there are marked differences between the cases of functions on general varieties and in $\mathbb{C}^n$. As for the definition itself, the goal here as before is to keep as close as possible to the classical feeling that a meromorphic function is a function.

1. **DEFINITION.** *A* **representative** *of a meromorphic function in an open subset $U$ of a holomorphic variety $V$ is a pair $(U_f, f)$, where $U_f$ is a dense open subset of $U$ and $f$ is a complex-valued function in $U_f$ with the property that for any point $A \in U$ there are holomorphic functions $f'_A$ and $f''_A$ in an open neighborhood $U_A$ of $A$ in $V$ for which $f(Z) = f'_A(Z)/f''_A(Z)$ whenever $Z \in U_A \cap U_f$. Two such representatives $(U_f, f)$ and $(U_g, g)$ of meromorphic functions in $U$ are called* **equivalent** *if there is a dense open subset $U_{f,g} \subseteq U$ for which $f(Z) = g(Z)$ whenever $Z \in U_{f,g} \cap U_f \cap U_g$. This is an equivalence relation in the usual sense, and an equivalence class is called a* **meromorphic function** *in $U$. The set of all meromorphic functions in $U$ will be denoted by ${}_V\mathcal{M}_U$ or $\mathcal{M}_U$.*

Note particularly that there must exist a local expression $f(Z) = f'_A(Z)/f''_A(Z)$ for a representative of a meromorphic function in $U$ at every point $A \in U$, not just at the points $A \in U_f$. Note also that implicit in this definition is the condition that $f''_A(Z) \neq 0$ whenever $Z \in U_A \cap U_f$. Since $U_A \cap U_f$ is dense in $U_A$, it follows that $f''_A$ cannot vanish identically on any irreducible component of $U_A$. The definition does not explicitly make any mention of global or local irreducible branches of $V$, only of the topology of $V$; but it is often helpful to consider the significance of this and other definitions and results for the separate global or local branches of $V$. Note finally that the restriction of any meromorphic function in an open subset $U$ of $V$ to $U \cap \mathfrak{R}(V)$ is a meromorphic function on the complex manifold $U \cap \mathfrak{R}(V)$, as discussed in section O. It is indeed quite clear that the restriction of a representative of a meromorphic function in $U$ to $U \cap \mathfrak{R}(V)$ is a representative of a meromorphic function in $U \cap \mathfrak{R}(V)$, and that the restrictions of equivalent representatives are equivalent. Of course, for any meromorphic function in $U$ it is always possible to choose a representative $(U_f, f)$ for which $U_f \subseteq U \cap \mathfrak{R}(V)$, so the given meromorphic function is uniquely determined by its restriction to $U \cap \mathfrak{R}(V)$. This provides at times a useful tool for handling the meromorphic functions on $V$, although it is entirely obvious but should nonetheless be pointed out quite explicitly that not every meromorphic function on the complex manifold $U \cap \mathfrak{R}(V)$ is necessarily the restriction of a meromorphic function on $V$. As is both customary and convenient, a meromorphic function in $U$ will be described by giving some representative $f$ and will be treated much as an ordinary complex-valued function in $V$, although it must be remembered that $f$ is not necessarily well defined at all points of $U$ and that there is an equivalence relation involved. Note that the subset $U$ is itself a holomorphic variety, and that there is a natural identification ${}_U\mathcal{M}_U \cong {}_V\mathcal{M}_U$. Thus for many purposes it is enough just to consider the set ${}_V\mathcal{M}_V = \mathcal{M}_V$.

The set $\mathcal{M}_V$ has a natural ring structure, where addition and multiplication of meromorphic functions are defined by the pointwise addition and multiplication of representatives in any common dense open subset of $V$ in which these representatives are defined. Any holomorphic function in $V$ can be viewed as a representative of a meromorphic function in $V$, and it is clear from the identity theorem for holomorphic functions on varieties, Theorem H1, that distinct holomorphic functions represent distinct meromorphic functions. Thus there is a natural inclusion $\mathcal{O}_V \subseteq \mathcal{M}_V$ of $\mathcal{O}_V$ as a subring of $\mathcal{M}_V$ and consequently of course a natural inclusion $\mathbb{C} \subseteq \mathcal{M}_V$ so that $\mathcal{M}_V$ has the natural structure of a complex algebra. The zero element in the ring $\mathcal{M}_V$ is the constant 0, and the unit element the constant 1.

**2. THEOREM (identity theorem for meromorphic functions).** *If $f$ and $g$ are meromorphic functions on an irreducible holomorphic variety $V$ and if there is a nonempty open subset $U \subseteq V$ such that $f(Z) = g(Z)$ whenever $Z \in U \cap U_f \cap U_g$, then $f = g$ in $\mathcal{M}_V$.*

*Proof.* If $(U_f, f)$ and $(U_g, g)$ are representatives of the meromorphic functions under consideration, then as noted it can always be supposed that $U_f = U_g \subseteq \mathfrak{R}(V)$, and it is enough just to consider the restriction of these meromorphic functions to the complex manifold $\mathfrak{R}(V)$. However on this complex manifold the identity theorem is an evident consequence of the identity theorem for meromorphic functions in $\mathbb{C}^n$, Theorem O2, since $\mathfrak{R}(V)$ is a connected complex manifold for an irreducible variety $V$ by Theorem E9; the details probably need not be given here. Then the representatives $(U_f, f)$ and $(U_g, g)$ coincide and so determine the same meromorphic function on $V$; the proof is thereby concluded.

To proceed further in parallel to the discussion of meromorphic functions in $\mathbb{C}^n$ in section O, note that if $f \in \mathcal{M}_V$ is a zero-divisor, then there must be a nonzero element $g \in \mathcal{M}_V$ such that $fg = 0$. By choosing representatives $(U_f, f)$ and $(U_g, g)$ for which $U_f = U_g \subseteq \mathfrak{R}(V)$, it must be the case that $f(Z)g(Z) = 0$ for all points $Z \in U_f = U_g$. Referring back to the properties of meromorphic functions in $\mathbb{C}^n$ shows that either $f = 0$ or $g = 0$ on any connected component of $\mathfrak{R}(V)$. By Theorem E9 again, it follows that *$\mathcal{M}_V$ has zero-divisors only when $V$ is reducible, and the zero-divisors are meromorphic functions that vanish identically in at least one irreducible component of $V$.* Further, *if $f \in \mathcal{M}_V$ is not a zero-divisor, then $1/f \in \mathcal{M}_V$*, so in particular $\mathcal{M}_V$ is *a field if V is irreducible.* The ring $\mathcal{M}_V$ in any case contains the ring of quotients of the subring $\mathcal{O}_V \subseteq \mathcal{M}_V$, but again any global discussion of the precise relations between these rings will be left to Volume III. For the local questions, however, it is of course possible to introduce the ring ${}_V\mathcal{M}_A$ of germs of meromorphic functions at a point $A$ of a holomorphic variety $V$, and to identify this with the ring of quotients of the subring $\mathcal{O}_V \subseteq \mathcal{M}_V$. The details are just as in the case of meromorphic functions in $\mathbb{C}^n$, so need not be repeated here.

From this point on the discussion of meromorphic functions on general holomorphic varieties must pursue a rather different and necessarily slightly more complicated path than that traversed in the earlier discussion of meromorphic functions in $\mathbb{C}^n$. The local rings ${}_V\mathcal{O}_A$ are not necessarily unique factorization domains, so the rather simple and algebraic discussion of zeros, singularities, and points of indeterminacy in section O cannot be used, but it can be replaced by a geometric discussion, which may well shed a different light on the properties of meromorphic functions in $\mathbb{C}^n$.

If $(U_f, f)$ is a representative of a meromorphic function on a holomorphic variety $V$, then introduce the subset

$$\Gamma_f^0 = \left\{(Z, T) \in U_f \times \mathbb{P}^1 : \frac{t_1}{t_0} = f(Z)\right\} \subseteq V \times \mathbb{P}^1$$

and let $\Gamma_f$ be the point set closure of $\Gamma_f^0$ in $V \times \mathbb{P}^1$. Note that if $(U_g, g)$ is any other representative of the same meromorphic function and $(U_h, h)$ is the representative defined by $U_h = U_f \cap U_g$, $h(Z) = f(Z) = g(Z)$ whenever $Z \in U_h$, then clearly $\Gamma_h^0$ is a dense subset both of $\Gamma_f^0$ and of $\Gamma_g^0$, so that $\Gamma_h = \Gamma_f = \Gamma_g$. The set $\Gamma_f$ is thus really independent of the representative chosen. This subset $\Gamma_f \subseteq V \times \mathbb{P}^1$ will be called the **graph** of the meromorphic function $f$ on $V$.

**3. LEMMA.** *The graph $\Gamma_f$ of the meromorphic function $f$ on the holomorphic variety $V$ is a holomorphic subvariety of $V \times \mathbb{P}^1$.*

*Proof.* For any point $A \in V$ there are holomorphic functions $f_A'$ and $f_A''$ in an open neighborhood $U_A$ of $A$ in $V$ such that $f(Z) = f_A'(Z)/f_A''(Z)$ whenever $Z \in U_A \cap U_f$. Since the graph $\Gamma_f$ is as noted above independent of the representative chosen for the meromorphic function $f$, it can be supposed without loss of generality that $U_A \cap U_f = \{Z \in U_A : f_A''(Z) \neq 0\}$. In the open subset $U_A \times \mathbb{P}^1 \subseteq V \times \mathbb{P}^1$ introduce the holomorphic subvarieties

$$\Gamma_A = \{(Z, T) \in U_A \times \mathbb{P}^1 : t_0 f_A'(Z) - t_1 f_A''(Z) = 0\}$$

$$\Gamma_A^* = \{(Z, T) \in U_A \times \mathbb{P}^1 : t_0 f_A'(Z) - t_1 f_A''(Z) = f_A''(Z) = 0\}$$

and note that $\Gamma_A^* \subseteq \Gamma_A$, while from the definition of the graph of $f$ and in terms of the particular choice of representative of $f$ made here,

$$\Gamma_f^0 \cap (U_A \times \mathbb{P}^1) = \left\{(Z, T) \in U_A \times \mathbb{P}^1 : f_A''(Z) \neq 0, \frac{f_A'(Z)}{f_A''(Z)} = \frac{t_1}{t_0}\right\}$$
$$= \Gamma_A - \Gamma_A^*$$

Therefore $\Gamma_f \cap (U_A \times \mathbb{P}^1)$ is just the closure of $\Gamma_A - \Gamma_A^*$ in $U_A \times \mathbb{P}^1$ and so by Corollary E20 is a holomorphic subvariety of $U_A \times \mathbb{P}^1$. That is true for any point $A \in V$, and therefore suffices to conclude the proof of the lemma.

Note that whenever $(Z, T) \in \Gamma_f^0$, then $t_1/t_0$ is the value at $Z$ of the representative of the meromorphic function $f$. In extension of this observation it is quite reasonable to view $t_1/t_0$ as a value of the meromorphic function $f$ at the point $Z$ whenever $(Z, T) \in \Gamma_f$, thus providing in some sense a notion of the value of a meromorphic function at all points of $V$ independent of any choice of representative. Of course if $t_0 = 0$ this value is in some sense $\infty$; alternatively the value can be considered to lie in $\mathbb{P}^1$. In these terms a point $Z \in V$ will be called a **zero** of the meromorphic function $f$ if $(Z, (1, 0)) \in \Gamma_f$, and will be called a **pole** of $f$ if $(Z, (0, 1)) \in \Gamma_f$. On the other hand, $Z$ will be called a **point of indeterminacy** of $f$ if $(Z, T_1) \in \Gamma_f$ and $(Z, T_2) \in \Gamma_f$ for two distinct points $T_1, T_2 \in \mathbb{P}^1$. The preceding lemma can then be used to establish that these various sets are holomorphic subvarieties of $V$ as follows.

**4. THEOREM.** *If $f$ is a meromorphic function on a holomorphic variety $V$, then the set of zeros of $f$ and the set of poles of $f$ are both holomorphic subvarieties of $V$. If $V$ is of pure dimension $n$ and $f$ is not identically zero on any irreducible component of $V$, these subvarieties are either empty or of pure dimension $n - 1$. If, further, $V$ is irreducible at each of its points, then the set of points of indeterminacy of $f$ is also a holomorphic subvariety of $V$, and if $Z \in V$ is any point of indeterminacy of $f$, then $(Z, T) \in \Gamma_f$ for all $T \in \mathbb{P}^1$.*

*Proof.* The natural projection mapping $V \times \mathbb{P}^1 \rightarrow V$, which is proper since $\mathbb{P}^1$ is compact, induces a proper holomorphic mapping $\pi: \Gamma_f \rightarrow V$. The intersection $\Gamma_f \cap (V \times (1, 0)) \subseteq V \times \mathbb{P}^1$ is a holomorphic subvariety of $\Gamma_f$, and by Remmert's proper mapping theorem, Theorem N1, the image of this subvariety under $\pi$ is a holomorphic subvariety of $V$. That image is just the set of zeros of $f$, which is thus a holomorphic subvariety of $V$ as desired. Similarly the set of poles of $f$ is a holomorphic subvariety of $V$.

If now $V$ is of pure dimension $n$, note that the restriction $\pi|\Gamma_f^0 : \Gamma_f^0 \rightarrow U_f$ is a one-to-one mapping and is hence a finite branched holomorphic covering of order one, and that it follows from Theorem G16 that $\Gamma_f^0$ is also of pure dimension $n$; the closure $\Gamma_f$ of $\Gamma_f^0$ is then of pure dimension $n$ as well. For all points $(Z, T)$ in a sufficiently small open neighborhood $U_A$ of any point $(A, (1, 0))$ in $V \times \mathbb{P}^1$, the homogeneous coordinate $t_0$ will be nonzero, so that $t = t_1/t_0$ can be viewed as a holomorphic function in $U_A$. But $U_A \cap \Gamma_f \cap (V \times (1, 0))$ is just the set of zeros of this holomorphic function $t$ in the subset $U_A \cap \Gamma_f$ of $\Gamma_f$ and so by Theorem G7 is a holomorphic subvariety of $U_A \cap \Gamma_f$ of pure dimension $n-1$. That holds for any point $A$, so that $\Gamma_f \cap (V \times (1, 0))$ itself is a holomorphic subvariety of $\Gamma_f$ of pure dimension $n-1$. The image of this subvariety under $\pi$, the set of zeros of the function $f$, is then a subvariety of $V$ of pure dimension $n-1$ by Theorem G16. Similarly the set of poles of $V$ is of pure dimension $n-1$ as well.

Finally suppose that $V$ is also an irreducible subvariety at each of its points. If $Z \in V$ is a point of indeterminacy of $f$, then $\pi^{-1}(Z) = \Gamma_f \cap (Z \times \mathbb{P}^1)$ can be viewed as a holomorphic subvariety of the compact manifold $\mathbb{P}^1$, so it is either a finite subset of $\mathbb{P}^1$ or all of $\mathbb{P}^1$. If $\pi^{-1}(Z) = \{(Z, T_1), \ldots, (Z, T_r)\}$ where $1 < r < \infty$, then the germ of the mapping $\pi$ at each point $(Z, T_j)$ is finite, so by Theorem E7 an open neighborhood $U_j$ of $(Z, T_j)$ can be written as a union $U_j = U_j' \cup U_j'' \cup \cdots$ of irreducible components for each of which the restriction $\pi | U_j' : U_j' \to \pi(U_j')$ is a finite branched holomorphic covering. Since $\Gamma_f$ is of pure dimension $n$, each subvariety $U_j'$ is of dimension $n$, so by Theorem G16 the subset $\pi(U_j')$ must be an open neighborhood of $Z$ in $V$; thus, $\pi | U_j : U_j \to \pi(U_j)$ will be a finite branched holomorphic covering of an open neighborhood of $Z$ in $V$. That means of course that over an open neighborhood of $Z$ the mapping $\pi$ is a finite branched holomorphic covering of order $\geqq r > 1$; but that is impossible, since the restriction of $\pi$ to the dense open subset $\Gamma_f^0 \subseteq \Gamma_f$ is a finite branched holomorphic covering of order one. Consequently $\pi^{-1}(Z) = Z \times \mathbb{P}^1$ for each point of indeterminacy $Z$ of $f$, so that the set of points of indeterminacy of $f$ is just the image $\pi(X)$ of the subset $X = \{(Z, T) \in \Gamma_f : \dim_{(Z,T)} L_{(Z,T)}(\pi) \geqq 2\}$, where $L_{(Z,T)}(\pi)$ is the level set of the mapping $\pi$ at the point $(Z, T)$. Now $X$ is a holomorphic subvariety of $\Gamma_f$ by Theorem L7, so the image $\pi(X)$ is a holomorphic subvariety of $V$ by Remmert's proper mapping theorem, Theorem N1. That suffices to conclude the proof.

In view of the preceding results and in parallel with the terminology introduced earlier in the discussion of meromorphic functions in $\mathbb{C}^n$, the set of zeros of a meromorphic function $f$ on a holomorphic variety $V$ will be called the **zero variety** of $f$ and will be denoted by $Z(f)$, while the set of poles will be called the **pole variety** or the **singular variety** of $f$ and will be denoted by $P(f)$. The set of points of indeterminacy will be called the **indeterminacy set** of $f$, or the **indeterminacy variety** in those cases in which it is known to be a holomorphic subvariety of $V$, and will be denoted by $I(f)$.

**Examples.** It will not have passed without observation that in comparison with the corresponding statements for meromorphic functions on $\mathbb{C}^n$, there are some gaps or weaknesses in the results obtained so far about meromorphic functions on arbitrary varieties, primarily in connection with points of indeterminacy. That reflects the nature of meromorphic functions on general holomorphic varieties though. For a meromorphic function $f$ on an open subset of $\mathbb{C}^n$, it was noted that $I(f) = Z(f) \cap P(f)$, and it is clear that *for a meromorphic function $f$ on an arbitrary holomorphic variety, it is the case that $Z(f) \cap P(f) \subseteq I(f)$; but this may be a proper inclusion.* To see that, consider the holomorphic subvariety $V = V_1 \cup V_2 \subseteq \mathbb{C}^2$, where $V_1 = \{(z_1, z_2) : z_1 = 0\}$ and $V_2 = \{(z_1, z_2) : z_2 = 0\}$, and the function $f$ on

$V - V_1 \cap V_2$ defined by

$$f(Z) = \begin{cases} -1 & \text{if} \quad Z \in V_1, Z \notin V_2 \\ +1 & \text{if} \quad Z \in V_2, Z \notin V_1 \end{cases}$$

This function $f$ is a well-defined holomorphic function on $V$ outside the origin $O = V_1 \cap V_2$, and since it can be expressed for instance as

$$f(Z) = \frac{z_1 + z_2}{z_1 - z_2} \quad \text{whenever} \quad Z \in V, (z_1, z_2) \neq (0, 0)$$

it is indeed a meromorphic function on $V$. The graph $\Gamma_f \subseteq V \times \mathbb{P}^1$ is clearly just the union $\Gamma_f = [V_1 \times (-1)] \cup [V_2 \times (+1)]$, so the projection $\pi: \Gamma_f \to V$ is two-to-one over the origin $O$, which must therefore be a point of indeterminacy. In this case $Z(f) = P(f) = \emptyset$, while $I(f) = O \in V$. As a somewhat related observation, *the set of indeterminacy of a meromorphic function $f$ on a general holomorphic variety $V$ cannot necessarily be described as* $\{Z \in V : \mathbf{f}'(Z) = \mathbf{f}''(Z) = 0$ *whenever* $\mathbf{f}', \mathbf{f}'' \in {}_V\mathcal{O}_Z$ and $\mathbf{f} = \mathbf{f}'/\mathbf{f}'' \in {}_V\mathcal{M}_Z\}$. Indeed consider again thc by now familiar holomorphic subvariety $V = \{(z_1, z_2) : z_1^3 = z_2^2\} \subseteq \mathbb{C}^2$. Recall that this is the image of the proper holomorphic mapping $F: \mathbb{C}^1 \to \mathbb{C}^2$ given by $F(t) = (t^2, t^3)$, and that $F: \mathbb{C}^1 \to V$ is one-to-one but not biholomorphic since $V$ has a singularity at the origin $O = F(O)$. The function $F^{-1}(Z) = t$ can be expressed outside the origin as the quotient $t = z_2/z_1$ and hence is a meromorphic function on $V$. It is actually a continuous function on $V$, since $F$ is a homeomorphism, so its graph is evidently $\Gamma_f = \{(Z, T) \in V \times \mathbb{P}^1 : t_1/t_2 = t\}$; thus, $Z(f) = (0, 0)$, $P(f) = \emptyset$, and $I(f) = \emptyset$. However in any representation $t = f'/f''$ where $f'$ and $f''$ are holomorphic functions near the origin on $V$, it must be the case that $f'(0) = f''(0) = 0$; for if $f''(0) \neq 0$ then $t = F^{-1}(Z)$ would be holomorphic near the origin, contradicting the fact that $F$ is not biholomorphic there, hence $f''(0) = 0$ and the continuity of $t$ implies that $f'(0) = 0$ as well. Next, to see that *the set of indeterminacy is not necessarily a holomorphic subvariety*, consider the holomorphic subvariety $V = V_1 \cup V_2 \subseteq \mathbb{C}^3$ where $V_1 = \{(z_1, z_2, z_3) : z_1 = 0\}$, $V_2 = \{(z_1, z_2, z_3) : z_2 = 0\}$, and the function $f$ on $V - V_1 \cap V_2$ defined by

$$f(Z) = \begin{cases} -z_3 & \text{if} \quad Z \in V_1, Z \notin V_2 \\ +z_3 & \text{if} \quad Z \in V_2, Z \notin V_1 \end{cases}$$

This function $f$ is a well-defined holomorphic function on $V - V_1 \cap V_2$, and since it can be expressed as

$$f(Z) = \frac{z_3(z_1 + z_2)}{(z_1 - z_2)} \quad \text{whenever} \quad Z \in V, (z_1, z_2) \neq (0, 0)$$

it is a meromorphic function on $V$. The graph is evidently the union $\Gamma_f = \Gamma_f^1 \cup \Gamma_f^2$ where $\Gamma_f^1 = \{(Z, T) \in V_1 \times \mathbb{P}^1 : t_1/t_0 = -z_3\}$ and $\Gamma_f^2 = \{(Z, T) \in V_2 \times \mathbb{P}^1 : t_1/t_0 = z_3\}$, so that $I(f) = \{(0, 0, z_3) : z_3 \neq 0\}$ and is thus not a holomorphic subvariety of $V$ at the origin. Finally to see that *even when the set of indeterminacy is a holomorphic subvariety of an everywhere locally irreducible variety, it need not be of codimension two,* consider another familiar example—the subvariety

$$V = \{(Z, W) \in \mathbb{C}^n \times \mathbb{C}^n : w_i z_j = w_j z_i \text{ for } 1 \leqq i, j \leqq n\}$$

As noted in section H this is a subvariety of $\mathbb{C}^{2n}$ of dimension $n+1$, and it is nonsingular outside the origin but singular though irreducible at the origin. The function $f(Z, W) = w_1/z_1$ is a well-defined meromorphic function on $V$, and in view of the defining equations for $V$ can also be represented as $f(Z, W) = w_j/z_j$ for any index $j$, $1 \leqq j \leqq n$. At any point outside the origin at least one of the coordinates is nonzero, so that point is at least not a point of indeterminacy of $f$. On the other hand, the origin is in both the zero variety $w_1 = 0$ and the pole variety $z_1 = 0$ and so must be a point of indeterminacy. Therefore $I(f) = 0$, and this is a subvariety of codimension $n+1$.

Somewhat more generally, if $f$ is a meromorphic function on a holomorphic variety $V$, a point $Z \in V$ at which $f$ is not holomorphic will be called a **singular point** or **singularity** of $f$, and the set of all the singular points of $f$ will be denoted by $S(f)$. Of course if $Z \notin S(f)$, then $f$ is holomorphic and hence can be represented by a continuous function near $Z$, so that $Z$ is neither a pole nor a point of indeterminacy of $f$; that is to say, $P(f) \cup I(f) \subseteq S(f)$. Although this is an equality for a meromorphic function in $\mathbb{C}^n$ or more generally on a complex manifold, it can be a strict inclusion in general. As already observed in the examples considered earlier in this section, on the variety $V = \{(z_1, z_2) \in \mathbb{C}^2 : z_1^3 = z_2^2\}$ the meromorphic function $z_2/z_1$ can be represented by a continuous function on $V$, so that $P(f) = I(f) = \emptyset$, but nonetheless this function cannot be holomorphic at the origin, so $S(f) = 0$. Thus in general *a meromorphic function can have a singularity that is neither a pole nor a point of indeterminacy.* Such singularities cannot be handled by the geometric construction used to describe the poles and points of indeterminacy, but can quite easily be treated algebraically. If $f$ is a meromorphic function on a holomorphic variety $V$, then to every point $Z \in V$ associate the **ideal of denominators** of $f$ at that point, defined as $\mathfrak{d}_Z(f) = \{\mathbf{g} \in {}_V\mathcal{O}_Z : \mathbf{g}\mathbf{f}_Z \in {}_V\mathcal{O}_Z\}$; it is obviously an ideal in the local ring ${}_V\mathcal{O}_Z$. Since the function $f$ is clearly holomorphic at $Z$, in the sense that $\mathbf{f}_Z \in {}_V\mathcal{O}_Z$, precisely when $1 \in \mathfrak{d}_Z(f)$, it follows that

$$S(f) = \{Z \in V : \mathfrak{d}_Z(f) \neq {}_V\mathcal{O}_Z\} \tag{1}$$

It is worth observing in passing that the elements in $\mathfrak{d}_Z(f)$ that are not zero-divisors in ${}_V\mathcal{O}_Z$ are indeed precisely those germs $\mathbf{f}'' \in {}_V\mathcal{O}_Z$ for which there is a representation $\mathbf{f}_Z = \mathbf{f}'/\mathbf{f}''$ in ${}_V\mathcal{M}_Z$, whence the terminology. More interesting though is the following observation.

5. LEMMA. *For any meromorphic function $f$ on a variety $V$, the family of ideals $\{\mathfrak{d}_Z(f)\}$ is finitely generated in an open neighborhood of any point of $V$.*

*Proof.* In some open neighborhood $V_A$ of any point $A$ in $V$ there will exist holomorphic functions $f', f'' \in {}_V\mathcal{O}_{V_A}$ such that the given meromorphic function can be represented as $f'(Z)/f''(Z)$ on the subset $\{Z \in V_A : f''(Z) \neq 0\}$. At any point $Z \in V_A$ the germ $\mathbf{f}_Z \in {}_V\mathcal{O}_Z$ can be represented as the quotient $\mathbf{f}_Z = \mathbf{f}'_Z/\mathbf{f}''_Z$ of the germs $\mathbf{f}'_Z$, $\mathbf{f}''_Z \in {}_V\mathcal{O}_Z$. A germ $\mathbf{g} \in {}_V\mathcal{O}_Z$ clearly lies in the ideal $\mathfrak{d}_Z(f)$ precisely when there exists some germ $\mathbf{h} \in {}_V\mathcal{O}_Z$ for which $\mathbf{g}\mathbf{f}'_Z = \mathbf{f}''_Z\mathbf{h}$ in ${}_V\mathcal{O}_Z$. Equivalently, if $\mathfrak{B}'_Z = {}_V\mathcal{O}_Z\mathbf{f}'_Z$ is the ideal in ${}_V\mathcal{O}_Z$ generated by $\mathbf{f}'_Z$ and $\mathfrak{B}''_Z = {}_V\mathcal{O}_Z\mathbf{f}''_Z$ that generated by $\mathbf{f}''_Z$, then $\mathbf{g} \in \mathfrak{d}_Z(f)$ precisely when $\mathbf{g}\mathfrak{B}'_Z \subseteq \mathfrak{B}''_Z$, so that $\mathfrak{d}_Z(f) = \mathfrak{B}''_Z : \mathfrak{B}'_Z$. It then follows from Corollary F10 to Oka's theorem that the family of ideals $\{\mathfrak{d}_Z(f)\}$ is finitely generated at least over a subneighborhood of $V_A$, but that suffices to conclude the proof of the desired result.

**6. THEOREM.** *The set $S(f)$ of singular points of a meromorphic function $f$ on a holomorphic variety $V$ is a holomorphic subvariety of $V$.*

*Proof.* It follows from the preceding lemma that on some open neighborhood $V_A$ of any point $A \in V$ there are holomorphic functions $f_1, \ldots, f_r \in {}_V\mathcal{O}_{V_A}$ that generate the ideal $\mathfrak{d}_Z(f)$ of denominators of $f$ at any point $Z \in V_A$. Then upon recalling (1) it is clear that $S(f) \cap V_A = \{Z \in V_A : f_1(Z) = \cdots = f_r(Z) = 0\}$; hence, $S(f)$ is a holomorphic subvariety of $V$ as desired and the proof is thereby completed.

In view of the preceding result, the set $S(f)$ can also be called the **singular variety** of the function $f$. It should be noted, as is evident from the last of the examples considered earlier in this section, that $S(f)$ is not necessarily a subvariety of pure codimension one in $V$, even when $V$ is irreducible. When the local ring ${}_V\mathcal{O}_Z$ is not a unique factorization domain, then in a representation of a meromorphic function as a quotient $\mathbf{f} = \mathbf{f}'/\mathbf{f}''$ of coprime germs it is not necessarily the case that all the zeros of $\mathbf{f}''$ are singularities of $\mathbf{f}$. There is at least one positive result vaguely in this direction though.

**7. THEOREM.** *If $f$ is a meromorphic function on a variety $V$ and $g$ is a holomorphic function on $V$ that vanishes on the subvariety $S(f)$, then for any compact subset $K \subseteq V$ there is an integer $\nu$ for which $fg^\nu$ is holomorphic in an open neighborhood of $K$.*

*Proof.* It is clear from (1) that the germ of the subvariety $S(f)$ at any point $Z \in V$ can be described as $\mathbf{S}(f)_Z = \text{loc}\ \mathfrak{d}_Z(f)$. If $g$ is holomorphic on $V$ and vanishes on $S(f)$, then it follows from Hilbert's zero-theorem on $V$, Corollary E3, that $\mathbf{g}_Z^{\nu_Z} \in \mathfrak{d}_Z(f)$ for some integer $\nu_Z \geqq 0$. That means of course that $g^{\nu_Z}f$ is holomorphic at $Z$ and hence in an open neighborhood of $Z$. A finite number of such neighborhoods will cover any compact subset $K \subseteq V$, and if $\nu$ is the maximum of the finitely many integers $\nu_Z$ for these neighborhoods, it follows that $g^\nu f$ is holomorphic in an open neighborhood of $K$, thereby concluding the proof.

Before turning to a discussion of other properties of meromorphic functions on arbitrary varieties, it is convenient to establish the following generally useful auxiliary result, which essentially means that as far as meromorphic functions are concerned, it is often possible to limit the consideration to irreducible holomorphic varieties.

**8. THEOREM.** *If $V = \bigcup_j V_j$ is a holomorphic variety with irreducible components $V_j$ and if $f_j$ is a meromorphic function on $V_j$ for each $j$, then there is a meromorphic function $f$ on $V$ such that $f|V_j = f_j$ for each $j$.*

*Proof.* It is clearly enough just to show that if $f_j$ is a meromorphic function on $V_j$ for some fixed $j$, then the function $f$ on $V$ defined by $f|V_j = f_j$ and $f = 0$ otherwise is meromorphic at each point $A \in V$. For any fixed point $A \in V$, an open neighborhood of $A$ in $V$ can always be represented by a holomorphic subvariety of an open subset $U_A$ in some space $\mathbb{C}^n$. If the neighborhood $U_A$ is sufficiently small, it will meet only finitely many of the irreducible components $V_k$, and since all components are holomorphic subvarieties of $V$, there will exist holomorphic functions $h_k$ in $U_A$ such that

$$\bigcup_{\ell \neq k} V_\ell \cap U_A = \{Z \in V \cap U_A : h_k(Z) = 0\}$$

for each $k$. Consider only those indices $k$ for which $V_k$ meets $U_A$, and introduce the meromorphic function $g_j = h_j/\sum_k h_k$ in $U_A$. The restriction $g_j|V \cap U_A$ is a well-defined meromorphic function on $V \cap U_A$—indeed is even a well-defined complex-valued function on $V \cap U_A - \bigcup_{k \neq \ell} V_k \cap V_\ell \cap U_A$ such that $g_j(Z) = 1$ if $Z \in V_j$ and $g_j(Z) = 0$ otherwise. Now if $f_j$ is a meromorphic function on $V_j$, then after shrinking the neighborhood $U_A$ if necessary, there will exist a meromorphic function $\tilde{f}_j$ in $U_A$ such that $\tilde{f}_j|U_A \cap V_j = f_j|U_A \cap V_j$. The function $g_j\tilde{f}_j$ is then also meromorphic in $U_A$, so its restriction $g_j\tilde{f}_j|U_A \cap V$ is meromorphic on $U_A \cap V$. But this function is $f_j$ on $U_A \cap V_j$ and zero otherwise, and as noted that suffices to conclude the proof.

The preceding result certainly does not hold for holomorphic functions, since it does not even hold for continuous functions. It indicates that in some ways meromorphic functions are more flexible as well as being more plentiful than holomorphic functions. It also means as already noted that for some purposes it is possible to consider only irreducible subvarieties when analyzing meromorphic functions. Thus the often convenient machinery developed earlier in the discussion of finite branched holomorphic coverings can be brought into play. That is the case, for instance, when considering extensions of meromorphic functions, as in the following further extension of Riemann's theorem on removable singularities.

**9. THEOREM.** *If $V$ is a holomorphic variety, $W$ is a holomorphic subvariety of $V$, and $f$ is a uniformly bounded holomorphic function in $V - W$, then there exists a meromorphic function $\tilde{f}$ on $V$ such that $\tilde{f}|(V - W) = f$.*

*Proof.* First in view of the preceding theorem it is only necessary to prove the present theorem in the special case that $V$ is irreducible. If $W = V$, the desired result holds quite trivially, since any meromorphic function $\tilde{f}$ on $V$ will serve; thus, it can be assumed not only that $V$ is irreducible but also that $W$ is a proper holomorphic subvariety of $V$. The complement $V - W$ is then a dense open subset of $V$ by Corollary E20, so that it is indeed just necessary to show that the given function $f$ on $V - W$ is a representative of a meromorphic function on $V$. That is, of course, a purely local question so it is enough to prove it in an open neighborhood $V_A$ of any point $A$ in $V$. Now since $V$ is pure dimensional—say, of dimension $n$—some open neighborhood $V_A$ of $A$ can be represented as a finite branched holomorphic covering over an open subset $U \subseteq \mathbb{C}^n$, and if the neighborhood $V_A$ is sufficiently small, it can be assumed that there is a holomorphic function $h$ on $V_A$ that separates the sheets of this covering. It then follows from Corollary C7 that there is a nontrivial holomorphic function $d_h \in {}_n\mathcal{O}_V$ such that $d_h \cdot f$ extends to a holomorphic function in all of $V_A$. But that means that $f$ itself represents a meromorphic function in $V_A$ as desired, and that serves to conclude the proof.

It is evident from the proof of the preceding theorem that the function $\tilde{f}$ is not always uniquely determined, although it is so provided that the subvariety $W$ contains no irreducible components of $V$. It should be emphasized that unlike the corresponding result for regular varieties, this theorem asserts only the existence of a meromorphic rather than a holomorphic extension $\tilde{f}$. The example considered in section H shows that nothing more could be expected to hold in general; but as will be discussed in detail in the next section, there is a class of holomorphic varieties with singularities for which the extension $\tilde{f}$ will automatically be holomorphic. Before turning to that, though, one more extension theorem for meromorphic functions should be demonstrated here. This too uses the standard techniques established for finite branched holomorphic coverings but requires a slight extension to hold for meromorphic functions.

Consider then a finite branched holomorphic covering $\pi\colon V \to U$ of order $\nu$ over a connected open subset $U \subseteq \mathbb{C}^n$, and suppose that $f$ is a meromorphic function on the holomorphic variety $V$. The singular locus $S(f)$ is a proper holomorphic subvariety of $V$ by Theorem 6, so its image $Y = \pi(S(f))$ is a proper holomorphic subvariety of $U$ by Remmert's proper mapping theorem. The function $f$ is then holomorphic on $V - \pi^{-1}(Y)$, so applying Theorem C5 to the finite branched holomorphic covering $\pi\colon V - \pi^{-1}(Y) \to U - Y$ shows that there is canonically associated to $f$ a monic polynomial $P_f(X) \in {}_n\mathcal{O}_{U-Y}$ of degree $\nu$ such that $P_f(f) = 0$. In a sufficiently small open neighborhood $U_A$ of any point $A \in Y \subset U$, there will exist a holomorphic function $h_A$ that vanishes on $Y \cap U_A$ but is not identically zero in $U_A$. The composite function $\tilde{h}_A = h_A \circ \pi$ is then a nontrivial holomorphic function in $\pi^{-1}(U_A)$ that vanishes on $\pi^{-1}(U_A) \cap S(f)$, and after shrinking $U_A$ and replacing $h_A$ by some sufficiently large positive power $h_A^N$ if necessary, the product $\tilde{h}_A f$ will be holomorphic in $\pi^{-1}(U_A)$ by Theorem 7. Now to this holomorphic function $\tilde{h}_A f$ there is also canonically associated a monic polynomial $P_{\tilde{h}_A f}(X)$ of degree $\nu$ such that $P_{\tilde{h}_A f}(\tilde{h}_A f) = 0$, and upon recalling from section C the construction of this polynomial, it is evident that $P_{\tilde{h}_A f}(\tilde{h}_A X) = \tilde{h}_A^{\nu} P_f(X)$ over $U_A - Y \cap U_A$. But that obviously implies that the coefficients of $P_f(X)$ extend to meromorphic functions in $U_A$, and since that is the case for every point $A$, the canonical polynomial $P_f(X)$ can be viewed as an element $P_f(X) \in {}_n\mathcal{M}_U[X]$. This can then be used to show that Levi's theorem, Theorem O6, holds on arbitrary holomorphic varieties.

**10. THEOREM (Levi's theorem).** *If $V$ is a holomorphic variety of pure dimension $n$, $W$ is a holomorphic subvariety of $V$ with* $\dim W \leq n - 2$, *and $f$ is a meromorphic function on $V - W$, then there is a unique meromorphic function $\tilde{f}$ on $V$ such that $\tilde{f}|(V - W) = f$.*

*Proof.* Since $V - W$ is a dense open subset of $V$, the uniqueness of the function $\tilde{f}$ follows immediately from the identity theorem for meromorphic functions, Theorem 2, and it is enough just to prove the desired result locally. It can therefore be supposed that there is a finite branched holomorphic covering $\pi: V \to U$ over some open subset $U \subseteq \mathbb{C}^n$. The image $Y = \pi(W)$ is a holomorphic subvariety of $U$ by Remmert's proper mapping theorem, and it follows readily upon recalling Theorem G16 that $\dim Y \leqq n - 2$. Now the function $f$ is meromorphic on $V - \pi^{-1}(Y)$, and for the finite branched holomorphic covering $\pi: V - \pi^{-1}(Y) \to U - Y$ there is as already noted a canonical monic polynomial $P_f(X) \in {}_n\mathcal{M}_{U-Y}[X]$ for which $P_f(f) = 0$. The coefficients of this polynomial extend to meromorphic functions in $U$ by Levi's theorem, Theorem O6, so that actually $P_f(X) \in {}_n\mathcal{M}_U[X]$. In an open neighborhood $U_A$ of any point $A \in U$ there will exist a nontrivial holomorphic function $h_A$ for which all the coefficients of the polynomial $h_A P_f(X)$ are actually holomorphic functions in $U_A$. If $P_f(X) = X^\nu + a_1 X^{\nu-1} + \cdots + a_{\nu-1} X + a_\nu$, then

$$\begin{aligned} 0 &= h_A^\nu P_f(f) \\ &= (h_A f)^\nu + (h_A a_1)(h_A f)^{\nu-1} + \cdots + (h_A^{\nu-1} a_{\nu-1})(h_A f) + (h_A^\nu a_\nu) \end{aligned}$$

The values of the function $h_A f$ whenever defined in $\pi^{-1}(U_A)$ are thus the roots of a monic polynomial with coefficients holomorphic in $U_A$ and so are uniformly bounded. It then follows easily that $h_A f$ is meromorphic in $\pi^{-1}(U_A)$, since $h_A f$ is certainly holomorphic outside the subvariety $W \cup S(f)$ in $\pi^{-1}(U_A)$. But then $f$ is itself meromorphic in $U_A$, and since that is the case for any point $A$, it follows that $f$ is meromorphic on $V$ as desired. That serves to conclude the proof.

**11. COROLLARY.** *If $V$ is a holomorphic variety of pure dimension $n$, $W$ is a holomorphic subvariety with $\dim W \leqq n - 2$, and $f$ is a holomorphic function on $V - W$, then there exists a unique meromorphic function $\tilde{f}$ on $V$ such that $\tilde{f}|(V - W) = f$. Moreover for any compact subset $K \subseteq V$ the function $f$ is uniformly bounded on $K \cap (V - W)$.*

*Proof.* The first assertion of the corollary is an immediate consequence of Levi's theorem. As for the second assertion, an open neighborhood $V_A$ of any point $A \in V$ can be represented as a finite branched holomorphic covering $\pi: V_A \to U_A$ over an open subset $U_A \subseteq \mathbb{C}^n$, and to the function $f$ on the restriction $\pi: V_A - V_A \cap \pi^{-1}(Y_A) \to U_A - Y_A$ where $Y_A = \pi(W \cap V_A)$ there is associated as usual a monic polynomial $P_{A,f}(X) \in {}_n\mathcal{O}_{U_A - Y_A}[X]$ such that $P_{A,f}(f) = 0$. Since $\dim Y_A \leqq n-2$, the coefficients of this polynomial extend across $Y_A$ as holomorphic functions, so that actually $P_{A,f}(X) \in {}_n\mathcal{O}_{U_A}[X]$. After restricting $V_A$ if necessary, the roots of this polynomial will be uniformly bounded for $Z \in U_A$; hence, the values of the function $f$ will be uniformly bounded on $V_A \cap (V - W)$. That clearly suffices to conclude the proof.

It should possibly be mentioned here that Theorem O7 also holds on arbitrary holomorphic varieties, but it is more convenient to delay proving this result until the machinery of normalization is in place.

# Q

# Normal Varieties

It has been pointed out several times in the discussion so far that on some holomorphic varieties $V$ there are subvarieties $W \subset V$ and uniformly bounded holomorphic functions on $V - W$ that do not extend to holomorphic functions at points of $W$. The present section will be devoted to a more detailed discussion of such functions, which are quite important in the investigation of holomorphic varieties. Of course the extended form of Riemann's removable singularities theorem holds on complex manifolds, so such functions can arise only when $W \subseteq \mathfrak{S}(V)$. That suggests the following notion.

**1. DEFINITION.** *A* **weakly holomorphic function** *in an open subset $U$ of a holomorphic variety $V$ is a holomorphic function on $U \cap \mathfrak{R}(V)$ that is locally bounded in $U$. The set of all weakly holomorphic functions in $U$ will be denoted by ${}_V\hat{\mathcal{O}}_U$.*

It should perhaps be remarked for emphasis that, according to the usage adopted earlier in the discussion here, to say that a function $f$ defined in $U \cap \mathfrak{R}(V)$ is locally bounded in $U$ means that for every point $A \in U$ there is an open neighborhood $U_A$ of $A$ in $U$ such that $f$ is uniformly bounded in $U_A \cap \mathfrak{R}(V)$. The point is of course that this must hold at every point $A \in U$, not just at points $A \in U \cap \mathfrak{R}(V)$. It is clear that ${}_V\hat{\mathcal{O}}_U$ is a ring under pointwise addition and multiplication of functions, and that ${}_V\hat{\mathcal{O}}_U \supseteq {}_V\mathcal{O}_U$ so that ${}_V\hat{\mathcal{O}}_U$ has the natural structure of a module over ${}_V\mathcal{O}_U$—hence, in particular the natural structure of an algebra over the complex numbers. The extended form of Riemann's removable singularities theorem, Theorem I-D2, shows that ${}_V\hat{\mathcal{O}}_U = {}_V\mathcal{O}_U$ whenever $V$ is a regular variety. It is also apparent from that theorem that whenever $W$ is a holomorphic subvariety of any holomorphic variety $V$ and $W$ does not contain any irreducible component of $V$, then any function that is holomorphic on $V - W$ and locally bounded in $V$ can be identified with a weakly holomorphic function on $V$ by extending it to all points in $W \cap \mathfrak{R}(V)$, with that extension being uniquely determined. It should also be noted that by Theorem P9 any weakly holomorphic function on a holomorphic variety $V$ represents a meromorphic function on $V$, so that there is a natural inclusion ${}_V\hat{\mathcal{O}}_U \subseteq {}_V\mathcal{M}_U$. *The weakly holomorphic functions in $U$ can then be defined equivalently as the locally bounded meromorphic functions in $U$*, meaning of course the meromorphic functions in $U$

that have representatives that are locally bounded in $U$. Finally it should be noted that the restriction of a weakly holomorphic function in $U$ to an open subset $U' \subseteq U$ is a weakly holomorphic function in $U'$. It is evident then that it is possible to introduce in a natural way the notion of the germ of a weakly holomorphic function at a point $A \in V$. The set of all such germs form a ring that will be denoted by ${}_V\hat{\mathcal{O}}_A$, and ${}_V\mathcal{O}_A \subseteq {}_V\hat{\mathcal{O}}_A \subset {}_V\mathcal{M}_A$ at each point $A \in V$.

Since any weakly holomorphic function is meromorphic, it follows immediately from the identity theorem for meromorphic functions, Theorem P2, that the same result holds for weakly holomorphic functions. Thus *if $f$ and $g$ are weakly holomorphic functions on an irreducible holomorphic variety $V$ and if there is a nonempty open subset $U \subseteq V$ on which these functions agree, then $f = g$ in ${}_V\mathcal{O}_V$.* Moreover since the weakly holomorphic functions can be characterized as locally bounded meromorphic functions, Theorem P8 extends immediately to hold for them as well. Thus, *if $V = \bigcup_j V_j$ is a holomorphic variety with irreducible components $V_j$ and if $f_j$ is a weakly holomorphic function on $V_j$ for each $j$, then there is a weakly holomorphic function $f$ on $V$ such that $f|V_j = f_j$ for each $j$. Equivalently whenever $V = \bigcup_j V_j$ is a holomorphic variety with irreducible components $V_j$, there is a natural isomorphism*

$$ {}_V\hat{\mathcal{O}}_V \cong \sum_j {}_{V_j}\hat{\mathcal{O}}_{V_j} \tag{1} $$

Furthermore Corollary P11 to Levi's theorem can be restated as the assertion that *if $V$ is a holomorphic variety of pure dimension $n$ and $W$ is a holomorphic subvariety of $V$ with* $\dim W \leqq n - 2$, *then any holomorphic function on $V - W$ extends uniquely to a weakly holomorphic function on $V$.*

Although weakly holomorphic functions are meromorphic functions, they are quite evidently meromorphic functions of a rather special sort. One reflection of their special nature is the result that they can all be taken to be meromorphic functions represented as quotients of holomorphic functions, all having the same denominator. Indeed somewhat more can be asserted, as follows.

**2. THEOREM.** *For every point $A$ on a holomorphic variety $V$ there are an open neighborhood $V_A$ of $A$ in $V$ and a holomorphic function $d \in {}_V\mathcal{O}_{V_A}$ such that at each point $Z \in V_A$ the germ $\mathbf{d}_Z \in {}_V\mathcal{O}_Z$ is not a zero divisor in ${}_V\mathcal{O}_Z$ and $\mathbf{d}_Z \cdot {}_V\hat{\mathcal{O}}_Z \subseteq {}_V\mathcal{O}_Z$.*

*Proof.* First if $V$ is of pure dimension $n$ at the point $A$, then some open neighborhood $V_A$ of $A$ in $V$ can be represented as a finite branched holomorphic covering $\pi: V_A \to U$ over a connected open subset $U \subseteq \mathbb{C}^n$, and if $V_A$ is sufficiently small, there will exist a holomorphic function $f \in {}_V\mathcal{O}_{V_A}$ that separates the sheets of this covering. To this function $f$ there is associated as in Theorem C6 a canonical monic polynomial $P_f(X) \in {}_n\mathcal{O}_U[X]$ such that $P_f(f) = 0$, and the discriminant of this polynomial is a nonzero element $d \in {}_n\mathcal{O}_U$. When viewed as a function on the variety $V_A$, the germ $\mathbf{d}_Z$ of this function is therefore not a zero-divisor in ${}_V\mathcal{O}_Z$ at any point $Z \in V_A$. The proof of the theorem in this special case can be concluded merely by showing that $\mathbf{d}_Z \cdot {}_V\hat{\mathcal{O}}_Z \subseteq {}_V\mathcal{O}_Z$ at each point $Z \in V_A$. Consider therefore a point $Z \in V_A$ and a germ

$\mathbf{g} \in {}_V\hat{\mathcal{O}}_Z$, and choose an open subneighborhood $V_Z \subseteq V_A$ in which there is a weakly holomorphic function $g$ representing the germ $\mathbf{g}$. For a suitable such neighborhood the restriction $\pi\colon V_Z \to U'$ will be a finite branched holomorphic covering of an open subset $U' \subseteq U$, and the restriction of the function $f$ to $V_Z$ will also separate the sheets of this covering. In this restriction there is also associated as in Theorem C6 a canonical polynomial $P_f'(X) \in {}_n\mathcal{O}_{U'}[X]$, and upon recalling the construction of that polynomial, it is evident that it is a divisor of the restriction of the polynomial $P_f(X)$ to $U'$. Consequently the discriminants of these polynomials are related correspondingly, so the discriminant $d' \in {}_n\mathcal{O}_{U'}$ of $P_f'(X)$ can be written $d' = dd''$ for some function $d'' \in {}_n\mathcal{O}_{U'}$. Now $g$ is a bounded holomorphic function outside a proper holomorphic subvariety of $V_Z$, and it follows immediately from Corollary C7 that $d'g$ extends to a holomorphic function in all of $V_Z$. But then $\mathbf{d}_Z\mathbf{g} = \mathbf{d}_Z''\mathbf{d}_Z'\mathbf{g} \in {}_V\mathcal{O}_Z$ as desired, thus demonstrating this special case of the theorem.

Next if $V$ is not pure dimensional at $A$, some open neighborhood $V_A$ of $A$ in $V$ can be written as a union $V_A = \bigcup_j V_j$ of distinct pure-dimensional varieties $V_j$, and all of these varieties $V_j$ and hence $V_A$ as well can be represented as holomorphic subvarieties of an open neighborhood $U_A$ of the point representing $A$ in some space $\mathbb{C}^n$. After shrinking the neighborhood $U_A$ if necessary, there will exist by the first part of the proof holomorphic functions $d_j$ on the subvarieties $V_j$ such that at each point $Z \in V_j$ the germ $\mathbf{d}_{jZ} \in {}_{V_j}\mathcal{O}_Z$ is not a zero-divisor and $\mathbf{d}_{jZ}\,{}_{V_j}\hat{\mathcal{O}}_Z \subseteq {}_{V_j}\mathcal{O}_Z$. Moreover these functions $d_j$ can be viewed as the restrictions to $V_j$ of some holomorphic functions $\tilde{d}_j \in {}_n\mathcal{O}_{U_A}$ and hence can of course be considered as holomorphic functions on all of $V_A$ if desired. After shrinking the neighborhood even further if necessary, there will exist holomorphic functions $\tilde{h}_j \in {}_n\mathcal{O}_{U_A}$ such that $\tilde{h}_j$ vanishes identically on all the pieces $V_k$ for $k \neq j$ but $\tilde{h}_j$ does not vanish identically on any irreducible component of $V_j$. The proof will be concluded by showing that the function $d = \sum_j(\tilde{d}_j\tilde{h}_j | V_A) \in {}_V\mathcal{O}_{V_A}$ has all the desired properties. First, since $d|V_j = d_j(\tilde{h}_j|V_j)$, it is clear that $d$ does not vanish identically on any irreducible component of $V_A$ and hence the germ $\mathbf{d}_Z \in {}_V\mathcal{O}_Z$ is not a zero-divisor at any point $Z \in V_A$. Next if $\mathbf{g} \in {}_V\hat{\mathcal{O}}_Z$ at some point $Z \in V_A$, then of course $\mathbf{g}|V_j \in {}_{V_j}\hat{\mathcal{O}}_Z$ whenever $Z \in V_j$ so that $\mathbf{d}_{jZ}(\mathbf{g}|V_j) \in {}_{V_j}\mathcal{O}_Z$; thus, there will be some germ $\tilde{\mathbf{g}}_j \in {}_n\mathcal{O}_Z$ for which $\mathbf{d}_{jZ}(\mathbf{g}|V_j) = \tilde{\mathbf{g}}_j|V_j$. The germ $\tilde{\mathbf{g}} = \sum_j \tilde{\mathbf{h}}_j\tilde{\mathbf{g}}_j \in {}_n\mathcal{O}_Z$ then has the property that $\tilde{\mathbf{g}}|V_j = (\tilde{\mathbf{h}}_j\tilde{\mathbf{g}}_j|V_j) = \mathbf{d}_{jZ}(\tilde{\mathbf{h}}_j\mathbf{g}|V_j) = \mathbf{d}_Z\mathbf{g}|V_j$; hence, $\mathbf{d}_Z\mathbf{g} \in {}_V\mathcal{O}_Z$ as desired and the proof is thereby completed.

**3. DEFINITION.** *A* **universal denominator** *at a point $A$ of a holomorphic variety $V$ is any germ $\mathbf{d} \in {}_V\mathcal{O}_A$ for which $\mathbf{d}\;{}_V\hat{\mathcal{O}}_A \subseteq {}_V\mathcal{O}_A$. The set of all universal denominators at $A$ will be denoted by $\mathfrak{d}_A$.*

The zero element of the ring ${}_V\mathcal{O}_A$ is of course a universal denominator for $V$ at $A$, albeit a rather trivial one. The preceding theorem shows among other things that there always exists a nontrivial universal denominator—indeed, an element $\mathbf{d} \in \mathfrak{d}_A$ that is not a zero-divisor in ${}_V\mathcal{O}_A$. Any weakly holomorphic germ $\mathbf{g} \in {}_V\hat{\mathcal{O}}_A$ can then be represented as a meromorphic germ of the form $\mathbf{g} = \mathbf{f}/\mathbf{d}$ for some $\mathbf{f} \in {}_V\mathcal{O}_A$, whence the terminology. The set $\mathfrak{d}_A$ of all universal denominators for $V$ at $A$ then clearly form a nonzero ideal in the local ring ${}_V\mathcal{O}_A$; this ideal is called the **conductor** of the local ring ${}_V\mathcal{O}_A$. These universal denominators are very useful in the further study of weakly holomorphic functions, as is nicely illustrated in a purely algebraic characterization of the ring ${}_V\hat{\mathcal{O}}_A$ of germs of weakly holomorphic functions.

**4. THEOREM.** *At any point A of a holomorphic variety V, the ring ${}_V\hat{\mathcal{O}}_A$ of germs of weakly holomorphic functions is a finitely generated module over the local ring ${}_V\mathcal{O}_A$.*

*Proof.* If $\mathbf{d} \in {}_V\mathcal{O}_A$ is a universal denominator at $A$ and is not a zero-divisor in the local ring ${}_V\mathcal{O}_A$, then the mapping that associates to any germ $\mathbf{f} \in {}_V\hat{\mathcal{O}}_A$ the germ $\mathbf{df} \in {}_V\mathcal{O}_A$ is an isomorphism of ${}_V\mathcal{O}_A$-modules. The image ${}_V\mathfrak{A}_A = \mathbf{d}\,{}_V\hat{\mathcal{O}}_A \subseteq {}_V\mathcal{O}_A$ is clearly an ideal in ${}_V\mathcal{O}_A$ and hence is a finitely generated ${}_V\mathcal{O}_A$-module since the local ring ${}_V\mathcal{O}_A$ is Noetherian; and since ${}_V\hat{\mathcal{O}}_A$ and ${}_V\mathfrak{A}_A$ are isomorphic as ${}_V\mathcal{O}_A$-modules it follows that ${}_V\hat{\mathcal{O}}_A$ is also finitely generated, thereby concluding the proof.

**5. THEOREM.** *The ring ${}_V\hat{\mathcal{O}}_A$ of germs of weakly holomorphic functions at a point A of a holomorphic variety V can be characterized as the integral closure of ${}_V\mathcal{O}_A$ in its total quotient ring ${}_V\mathcal{M}_A$.*

*Proof.* Since ${}_V\hat{\mathcal{O}}_A$ is a finitely generated ${}_V\mathcal{O}_A$-module by the preceding theorem, it follows immediately that any element $\mathbf{f} \in {}_V\hat{\mathcal{O}}_A \subseteq {}_V\mathcal{M}_A$ is integral over ${}_V\mathcal{O}_A$, a well-known result from algebra. On the other hand, if $\mathbf{f} \in {}_V\mathcal{M}_A$ is integral over ${}_V\mathcal{O}_A$, then by definition it must be the root of some monic polynomial $P(X) = X^\nu + \mathbf{a}_1 X^{\nu-1} + \cdots + \mathbf{a}_\nu \in {}_V\mathcal{O}_A[X]$. In some open neighborhood $U$ of the point $A$ in $V$, the germs $\mathbf{f}, \mathbf{a}_j$ will have representatives $f \in {}_V\mathcal{M}_U$, $a_j \in {}_V\mathcal{O}_U$, and

$$f(Z)^\nu + a_1(Z)f(Z)^{\nu-1} + \cdots + a_\nu(Z) = 0$$

whenever $Z \in U$ is a point at which $f(Z)$ is well defined. The coefficients $a_j$ are uniformly bounded in $U$ if $U$ is sufficiently small; hence, the meromorphic function $f$ must also be uniformly bounded in $U$. But that means that $f$ is weakly holomorphic in $U$, so that $\mathbf{f} \in {}_V\hat{\mathcal{O}}_A$ and the proof is thereby concluded.

**6. COROLLARY.** *In the local ring ${}_V\mathcal{O}_A$ of a holomorphic variety V at a point A, let $\mathfrak{A}$ be an ideal such that $\mathbf{h}\mathfrak{A} = 0$ for some $\mathbf{h} \in {}_V\mathcal{O}_A$ only when $\mathbf{h} = 0$. Then any germ $\mathbf{f} \in {}_V\mathcal{M}_A$ for which $\mathbf{f}\mathfrak{A} \subseteq \mathfrak{A}$ must actually be the germ of a weakly holomorphic function on V.*

*Proof.* If $\mathbf{f} \in {}_V\mathcal{M}_A$ is the germ of a meromorphic function for which $\mathbf{f}\mathfrak{A} \subseteq \mathfrak{A}$ and if $\mathbf{h}_j \in {}_V\mathcal{O}_A$ are a finite number of generators of the ideal $\mathfrak{A}$, then there must be some germs $\mathbf{a}_{jk} \in {}_V\mathcal{O}_A$ for which $\mathbf{f}\mathbf{h}_j = \sum_k \mathbf{a}_{jk}\mathbf{h}_k$, or equivalently for which

$$\sum_k (\mathbf{a}_{jk} - \mathbf{f}\delta_k^j)\mathbf{h}_k = 0 \quad \text{for all } j$$

where $\delta_k^j$ is the Kronecker symbol. It then follows from Cramer's rule that $\mathbf{a}\mathbf{h}_k = 0$ for all $k$ where $\mathbf{a} \in {}_V\mathcal{M}_A$ is the determinant of the matrix $\{\mathbf{a}_{jk} - \mathbf{f}\delta_k^j\}$. If $\mathbf{a} \neq 0$, then there must be an element $\mathbf{b} \in {}_V\mathcal{O}_A$ such that $\mathbf{ab} \in {}_V\mathcal{O}_A$ and $\mathbf{ab} \neq 0$. But then $\mathbf{abh}_k = 0$ for all $k$ so that $\mathbf{ab}\mathfrak{A} = 0$, an impossibility by the hypothesis of the theorem. Therefore $\mathbf{a} = 0$, and when that determinant is written out, it clearly has the form of a monic polynomial in f with coefficients in ${}_V\mathcal{O}_A$ so that f must be integral over ${}_V\mathcal{O}_A$. It then follows from the preceding theorem that $\mathbf{f} \in {}_V\hat{\mathcal{O}}_A$ as desired, thereby concluding the proof.

An obviously interesting category of holomorphic varieties are those for which all weakly holomorphic functions are actually holomorphic. The function theory of such varieties is in many ways particularly simple, and as will eventually be seen, these varieties play an important role in the further study of weakly holomorphic functions on general varieties. The traditional terminology used in describing such varieties is the following.

**7. DEFINITION.** *A holomorphic variety $V$ is said to be* **normal at a point** $A \in V$ if ${}_V\hat{\mathcal{O}}_A = {}_V\mathcal{O}_A$. *That variety is said simply to be* **normal** *if it is normal at each of its points.*

The condition that a holomorphic variety be normal at one of its points clearly involves only the germ of the variety at that point. By Theorem 5 this condition can be restated purely algebraically in terms of the local ring of the variety at that point. Indeed, it is clear from that theorem that *a holomorphic variety $V$ is normal at a point $A \in V$ precisely when its local ring ${}_V\mathcal{O}_A$ is integrally closed in its total quotient ring ${}_V\mathcal{M}_A$.* The most immediately obvious properties of normal varieties as distinct from arbitrary holomorphic varieties are that the extension theorems that hold generally only for weakly holomorphic functions actually hold for ordinary holomorphic functions. The first of these is the extended form of Riemann's removable singularities theorem.

**8. THEOREM.** *If $V$ is a normal holomorphic variety, $W$ is a holomorphic subvariety of $V$, and $f$ is a holomorphic function on $V - W$ that is locally bounded in $V$, then there is a holomorphic function $\tilde{f}$ on $V$ such that $\tilde{f}|(V - W) = f$. If $W$ does not contain any irreducible components of $V$, the extension $\tilde{f}$ is unique.*

*Proof.* It follows from Theorem P9 that $f$ extends to a meromorphic function $\tilde{f}$ on all of $V$. This extension is locally bounded on $V$ and so must be weakly holomorphic, and since $V$ is normal, it must indeed be holomorphic on $V$. The identity theorem for holomorphic functions implies that $\tilde{f}$ is determined uniquely by $f$ so long as $V - W$ is dense in $V$, and that is the case whenever $W$ does not contain an irreducible component of $V$, thereby concluding the proof.

Since the extended form of Riemann's removable singularities theorem holds on normal varieties, so also do many of the consequences of that theorem, results derived and used so far only for regular varieties, as well as versions of these results that have separate implications about the singularities of normal varieties. An example of the latter is the following simple but important observation.

**9. THEOREM.** *A normal holomorphic variety is irreducible at each of its points.*

*Proof.* Suppose to the contrary that there is a holomorphic variety $V$ for which some connected open neighborhood $V_A$ of a point $A \in V$ can be written as a union

$V_A = V' \cup V''$ of proper holomorphic subvarieties. The function that is zero on $V' - V' \cap V''$ and one on $V'' - V' \cap V''$ is then holomorphic and uniformly bounded on $V_A - V' \cap V''$, so by Theorem 8 extends to a holomorphic function on $V_A$. But that is impossible, since this function cannot even extend to a continuous function on $V_A$, and that contradiction serves to complete the proof.

Note that this does not assert that a normal variety is irreducible—only that it is locally irreducible. A reducible normal variety must therefore be a disjoint union of its irreducible components, each of which is also normal. This last result also makes the preceding assertion perhaps more believable. Another application of the removable singularities theorem refers back to properties of finite branched holomorphic coverings as discussed in section C. A frequently used result was Theorem C4, the assertion that a finite branched holomorphic covering over an open subset of $\mathbb{C}^n$ is biholomorphic at every point of branching order one, and the proof of that result extends immediately to the following.

**10. THEOREM.** *If $W$ is a normal holomorphic variety, then any finite branched holomorphic covering $\pi: V \to W$ is locally biholomorphic at each point $A \in V$ for which $o_\pi(A) = 1$.*

*Proof.* Any finite branched holomorphic covering $\pi: V \to W$ has a regular part $\pi | V_o : V_o \to W_o$ for which $W_o$ is the complement of a proper holomorphic subvariety of $W$. Since $W$ is irreducible at each of its points by Theorem 9, it then follows from Theorem E13 that $W_o$ is locally connected in $W$. As noted in the topological discussions of finite branched coverings at the beginning of section C, it further follows that $\pi$ is a local homeomorphism at each point $A \in V$ for which $o_A(\pi) = 1$. Then if $V_A$ is a sufficiently small open neighborhood of $A$ in $V$, the restriction $\pi | V_A : V_A \to U_{\pi(A)}$ will be a homeomorphism between $V_A$ and an open neighborhood $U_{\pi(A)}$ of the image $\pi(A)$ in $W$, and moreover $V_A$ itself can be represented by a holomorphic subvariety of an open subset of some space $\mathbb{C}^n$. The coordinate functions of the inverse mapping $(\pi | V_A)^{-1} : U_{\pi(A)} \to V_A$ are then continuous functions in $U_{\pi(A)}$ and are holomorphic in $U_{\pi(A)} \cap W_o$, the complement of a holomorphic subvariety of $U_{\pi(A)}$. But these functions are then holomorphic in $U_{\pi(A)}$ by Theorem 8, so that $(\pi | V_A)^{-1}$ is also holomorphic and hence $\pi | V_A$ is biholomorphic as desired. That concludes the proof.

**11. COROLLARY.** *If $W$ is a normal holomorphic variety, then any finite branched holomorphic covering $\pi: V \to W$ of order one is a biholomorphic mapping.*

*Proof.* This is an immediate consequence of the preceding theorem.

The interest of the preceding two assertions will become more evident in the discussion of normalization in the next section, but it may also be indicated by the following application.

**12. THEOREM.** *If $V$ is an irreducible normal holomorphic variety with* $\dim V = n$, *then* $\dim \mathfrak{S}(V) \leq n - 2$.

*Proof.* The desired result is really of a local nature, so it can be supposed that $V$ is represented as a finite branched holomorphic covering $\pi: V \to U$ over an open subset $U \subseteq \mathbb{C}^n$. Since $V$ is irreducible at each of its points by Theorem 9, it then follows from Corollary G18 that the branch locus $B_\pi \subset V$ and its image $\pi(B_\pi) \subset U$ are of pure dimension $n - 1$. The singular locus of the subvariety $\pi(B_\pi)$ must then be a holomorphic subvariety of $U$ of dimension at most $n - 2$, and by Theorem G16 its inverse image $\pi^{-1}(\mathfrak{S}(\pi(B_\pi)))$ must be a holomorphic subvariety of $V$ of dimension at most $n - 2$. Now the variety $V$ is regular at all points outside $B_\pi$, since the mapping $\pi$ is locally biholomorphic at all such points by Theorem C4. The theorem will be proved by demonstrating that $V$ is also regular at all points $Z \in B_\pi$ for which $\pi(Z) \in \mathfrak{R}(\pi(B_\pi))$; for then $\mathfrak{S}(V) \subseteq \pi^{-1}(\mathfrak{S}(\pi(B_\pi)))$ and the desired result follows.

Consider then a point $A \in \mathfrak{R}(\pi(B_\pi))$. After shrinking the set $U$ to a suitable open neighborhood of $A$ and choosing appropriate coordinates in $U$, it can be supposed that $U$ is an open polydisc $\Delta(0; 1) \subseteq \mathbb{C}^n$, the point $A$ being the origin $O \in \Delta(0; 1)$, and that $\pi(B_\pi)$ is the linear subvariety $L = \{Z \in \Delta(0; 1) : z_n = 0\}$. Thus $V$ is locally represented as a finite branched holomorphic covering $\pi: V \to \Delta(0; 1)$ of some order $\nu$, branched only over the linear subvariety $L \subset \Delta(0; 1)$. The topological situation is then particularly simple, since the complement $\Delta(0; 1) - L$ is the Cartesian product of a punctured disc and a polydisc in $\mathbb{C}^{n-1}$ and consequently has fundamental group $\pi_1(\Delta(0; 1) - L) \cong Z$. There is thus a uniquely determined connected topological covering of any order $\nu$ over $\Delta(0; 1) - L$, and all can obviously be extended to finite branched coverings over $\Delta(0; 1)$ with a branch point of order $\nu$ over each point of $L$. The mapping $\tau: \Delta(0; 1) \to \Delta(0; 1)$ defined by $\tau(t_1, \ldots, t_{n-1}, t_n) = (t_1, \ldots, t_{n-1}, t_n^\nu)$ is such a branched covering, so there must be a homeomorphism $\sigma: \Delta(0; 1) \to V$ for which $\pi\sigma = \tau$. The mapping $\tau$ is of course also a holomorphic mapping, from which it follows immediately that $\sigma$ is a holomorphic mapping outside the branch locus. But the coordinate functions of the mapping $\sigma$ are continuous functions, so by the extended form of Riemann's removable singularities theorem, the mapping $\sigma$ must be everywhere holomorphic. Then however $\sigma$ is a finite branched holomorphic covering of order 1, so by Corollary 11 must be a biholomorphic mapping; therefore $V$ is biholomorphic to the polydisc $\Delta(0; 1)$ and is hence a regular variety, and the proof is thereby concluded.

**13. COROLLARY.** *A one-dimensional holomorphic variety is normal precisely when it is regular.*

*Proof.* That a regular variety is normal was already noted, while that conversely a normal one-dimensional variety is necessarily regular follows immediately from Theorem 12, thereby concluding the proof.

There are special cases in which the converse of Theorem 12 holds. The preceding corollary can be viewed as one such case, albeit one in which the converse is quite trivial; a more interesting case is the following.

**14. THEOREM.** *If $V$ is a holomorphic subvariety of an open subset of $\mathbb{C}^{n+1}$ and is of pure dimension $n$ then $V$ is normal precisely when* $\dim \mathfrak{S}(V) \leq n - 2$.

*Proof.* In view of Theorem 12 it is only necessary to show that any holomorphic subvariety $V$ of an open subset of $\mathbb{C}^{n+1}$ with the properties that $V$ is of pure dimension $n$ and $\dim \mathfrak{S}(V) \leqq n-2$ must be a normal variety. For any point $A \in V$ there is by Theorem G4 a one-dimensional linear subvariety $L \subseteq \mathbb{C}^{n+1}$ such that $A$ is an isolated point of the intersection $L \cap V$; after changing coordinates in $\mathbb{C}^{n+1}$ so that $L$ is the axis of the variable $z_{n+1}$, it follows as in Theorem E5 that the natural projection $\pi: \mathbb{C}^{n+1} = \mathbb{C}^n \times \mathbb{C} \to \mathbb{C}^n$ exhibits an open neighborhood $V_A$ of $A$ in $V$ as a finite branched holomorphic covering $\pi: V_A \to U$ of some order $\nu$ over a connected open subset $U \subseteq \mathbb{C}^n$. The coordinate function $z_{n+1}$ separates the sheets of this covering, so as in Theorem C6 there is a unique monic polynomial $P(X) \in {}_n\mathcal{O}_U[X]$ of degree $\nu$ such that $P(z_{n+1}) = 0$ on $V_A$. Of course here $P$ is considered as a holomorphic function in $U \times \mathbb{C}$ that is a polynomial in $z_{n+1}$, and the preceding condition just means that $P(Z) = 0$ whenever $Z = (Z', z_{n+1}) \in V_A \subset U \times \mathbb{C}$. Indeed as in Theorem D2 the subvariety $V_A$ can be described as $V_A = \{Z \in U \times \mathbb{C} : P(Z) = 0\}$, and the germ of the function $P$ generates the ideal of the subvariety $V_A$ at any point of $U \times \mathbb{C}$. The discriminant of the polynomial $P$ is a holomorphic function $d \in {}_n\mathcal{O}_U$ that is not identically zero, and the image under $\pi$ of the branch locus of this covering is the holomorphic subvariety $W = \{Z' \in U : d(Z') = 0\}$. The image of the singular locus $\mathfrak{S}(V_A)$ under the mapping $\pi$ is then a holomorphic subvariety $W' = \pi(\mathfrak{S}(V_A)) \subset W \subset U$, and in view of the hypotheses and Theorem G16 it is clear that $\dim W' \leqq n-2$.

Now suppose that $f$ is a weakly holomorphic function on $V_A$. This function is then holomorphic outside the subvariety $\pi^{-1}(W') \supseteq \mathfrak{S}(V_A)$, so by Theorem C6 there is a unique polynomial $Q(X) = Q(Z'; X) \in {}_n\mathcal{O}_{U-W'}[X]$ of degree $\nu - 1$ such that $d(Z')f(Z) = Q(Z'; z_{n+1})$ whenever $Z = (Z'; z_{n+1}) \in V_A - \pi^{-1}(W')$. The coefficients of the polynomial $Q(X)$ are themselves polynomials in the values of $f(Z)$ and $z_{n+1}$ at the points $Z = (Z', z_{n+1}) \in V_A$ over any point $Z' \in U - W'$ and hence as in the proof of Corollary C7 extend to holomorphic functions in $U$, so that $Q(X)$ can be viewed as a polynomial $Q(X) \in {}_n\mathcal{O}_U[X]$. The quotient $Q'(X) = Q(X)/d \in {}_n\mathcal{M}_U[X]$ is then a polynomial with coefficients that are meromorphic in $U$ and have poles at most in the subvariety $W \subset U$, and $f(Z) = Q'(Z'; z_{n+1})$ whenever $Z = (Z', z_{n+1}) \in V_A$ and $Z' \in U - W$. The proof of the theorem will be concluded by showing that the coefficients of this polynomial extend to holomorphic functions at all points of $W - W'$. These coefficients being holomorphic outside the subvariety $W' \subset U \subseteq \mathbb{C}^n$ with $\dim W' \leqq n-2$ must necessarily extend to holomorphic functions in all of $U$, and since $f = Q(Z'; z_{n+1}) | V_A$ it follows that $f$ is actually holomorphic on $V_A$ and hence that $V$ is normal at $A$ as desired.

Consider then a point $B' \in W - W' \subset U$, and let $B_i = (B', b^i_{n+1})$ be the distinct points of $V_A$ for which $\pi(B_i) = B'$. The restriction of the mapping $\pi$ to a suitable open neighborhood $V_i$ of $B_i$ in $V_A$ is also a finite branched holomorphic covering $\pi: V_i \to U'$ over an open neighborhood $U'$ of $B'$, which neighborhood can be supposed the same for all $i$, and if $\nu_i$ is the order of this covering, then $\nu = \sum_i \nu_i$. For each of these coverings there is as in Theorem C6 a monic polynomial $P_i(x) \in {}_n\mathcal{O}_{U'}[X]$ of degree $\nu_i$ so that $P_i(z_{n+1}) = 0$ on $V_i$, and it follows from the uniqueness of the original polynomial $P(X)$ that $P(X) = \prod_i P_i(X)$ over $U'$. Since $B_i \notin \mathfrak{S}(V_A)$ the function $f$ must be holomorphic near $B_i$, so after shrinking the neighborhoods $V_i$ if necessary, there will be holomorphic functions $F_i$ near $B_i$ in $\mathbb{C}^{n+1}$ such that $F_i|V_i = f|V_i$. The polynomials $P_j(z_{n+1})$ are nonzero on $V_i$ if $j \neq i$, provided that the neighborhoods $V_i$ are chosen to be disjoint, so that $F_i \cdot \prod_{j\neq i} P_j(z_{n+1})^{-1}$ is also holomorphic near $B_i$ in $\mathbb{C}^{n+1}$. By the Weierstrass division theorem, which is applicable since $P_i(z_{n+1})$ is evidently a Weierstrass polynomial in terms of local coordinates centered at $B_i$, after shrinking the neighborhoods $V_i$ further if necessary, it is possible to write $F_i \cdot \prod_{j\neq i} P_j(z_{n+1})^{-1} = P_i(z_{n+1})G_i + Q_i(z_{n+1})$ where $G_i$ is holomorphic near $B_i$ in $\mathbb{C}^{n+1}$ and $Q_i(X) \in {}_n\mathcal{O}_{U'}[X]$ is a polynomial of degree $v_i - 1$. Since $P_i(z_{n+1}) = 0$ on $V_i$, it follows that

$$f(Z) = F_i(Z) = Q_i(Z'; z_{n+1}) \cdot \prod_{j\neq i} P_j(Z'; z_{n+1})$$

whenever $Z = (Z', z_{n+1}) \in V_i$. Now

$$Q_{B'}(X) = \sum_i \left[ Q_i(X) \cdot \prod_{j\neq i} P_j(X) \right] \in {}_n\mathcal{O}_{U'}[X]$$

is a polynomial of degree $v - 1$ for which it is clear that $Q_{B'}(Z'; z_{n+1}) = f_i(Z)$ whenever $Z = (Z'; z_{n+1}) \in V_A$ and $Z' \in U'$. But the polynomial $Q$ satisfies the same condition provided that $Z' \notin W$, and from the uniqueness of $G$ it follows that $Q = Q_{B'}$ in $U' - U' \cap W$. Thus $Q_{B'}$ provides a holomorphic extension of $Q$ to $U'$; hence, the coefficients of $Q$ are holomorphic in $W - W'$ as desired and the proof of the theorem is thereby concluded.

**Examples.** In the proof of the preceding theorem the hypothesis that $V$ is of pure codimension one was used in showing that holomorphic functions on $V$ have unique representations as polynomial functions in one of the coordinates in the ambient space. The uniqueness of that representation was crucial in showing that weakly holomorphic functions on $V$ are necessarily holomorphic. If $V$ is of pure codimension $k$ and there are $k$ holomorphic functions in the ambient space that generate the ideal of the subvariety at each of its points, then there is correspondingly a unique representation of the holomorphic functions on $V$ as polynomials in $k$ of the coordinates in the ambient space, and the proof of the preceding theorem can be extended to this case as well. But there are simpler results of greater generality, so the further pursuit of this method of proof is somewhat pointless. The preceding theorem is often useful, though, and well worth seeing here. It should be noted that the condition that the singular locus be of codimension at least two is not by itself enough to ensure that a holomorphic variety is normal. For example, if $V_1$ and $V_2$ are linear subvarieties of pure dimension $n$ in $\mathbb{C}^{n+2}$ and if $\dim(V_1 \cap V_2) \leqq n - 2$, then $V = V_1 \cup V_2$ has singular locus $\mathfrak{S}(V) = V_1 \cap V_2$ of codimension at least two, but $V_1 \cup V_2$ is not irreducible at each of its points and so is not normal. Theorem G10 shows that such a construction cannot be carried out for subvarieties of pure dimension $n$ in $\mathbb{C}^{n+1}$, as is of course also evident from Theorem 12. There are also holomorphic varieties that are irreducible at all of their points and have singularities of codimension two but are not normal. To construct an example of such a variety consider the holomorphic mapping $F: \mathbb{C}^2 \to \mathbb{C}^5$ defined by

$$F(t_1, t_2) = (t_1^2, t_1^3, t_2^2, t_2^3, t_1 t_2)$$

and let $W = F(\mathbb{C}^2) \subset \mathbb{C}^5$ be its image. The mapping $F$ is clearly proper, so that $W$ is a holomorphic subvariety of $\mathbb{C}^5$, and the mapping $F: \mathbb{C}^2 \to W$ is moreover clearly one-to-one. Furthermore the Jacobian matrix of the mapping $F$ has the form

$$J_F(t_1, t_2) = \begin{pmatrix} 2t_1 & 3t_1^2 & 0 & 0 & t_2 \\ 0 & 0 & 2t_2 & 3t_2^2 & t_1 \end{pmatrix}$$

so is evidently of rank two except when $t_1 = t_2 = 0$; hence, $W$ is nonsingular except at the origin $F(0, 0)$. If $W$ were normal at the origin, then by Corollary 11 the mapping $F$ would be biholomorphic. But it is evident that the coordinate functions of the mapping $F$ do not generate the maximal ideal ${}_2\mathfrak{m} \subset {}_2\mathcal{O}$, since it is impossible to write $t_1 = \mathbf{f}_1 \cdot t_1^2 + \mathbf{f}_2 \cdot t_1^3 + \mathbf{f}_3 \cdot t_2^2 + \mathbf{f}_4 \cdot t_2^3 + \mathbf{f}_5 \cdot t_1 t_2$ for any germs $\mathbf{f}_j \in {}_2\mathcal{O}$, and it then follows from Corollary I19 that $F$ is not biholomorphic. The origin is thus the only singular point of $W$, so that the singular locus of $W$ is of codimension two, but $W$ is not normal there. The complement of the origin in $W$ is homeomorphic to the complement of the origin in $\mathbb{C}^2$ under $F$ and hence is connected, so $W$ is also irreducible at the origin. There are obviously a plethora of examples of this sort, with singular loci of arbitrary codimension.

The preceding observations can now be used to complete Theorem 8 by showing that the other holomorphic extension theorems established in $\mathbb{C}^n$ also hold on normal varieties.

**15. THEOREM.** *Suppose that $V$ is an irreducible normal holomorphic variety, that $W$ is a proper holomorphic subvariety of $V$, and that $f$ is a holomorphic function on $V - W$.*

(i) *If* $\dim W \leq \dim V - 2$, *then $f$ extends to a holomorphic function on all of $V$.*
(ii) *If $W$ is irreducible,* $\dim W = \dim V - 1$, *and $f$ extends to a holomorphic function in $V$ at some point of $W$, then $f$ extends to a holomorphic function on all of $V$.*

*Proof.* (i) It follows from Levi's theorem, Theorem P10, that $f$ extends to a meromorphic function on all of $V$, and moreover by Corollary P11 this extension is locally bounded on $V$ and hence is weakly holomorphic. The normality of $V$ then implies that this function is actually holomorphic on all of $V$ as desired.

(ii) Since $V$ and $W$ are irreducible, then $\mathfrak{R}(V)$ is a connected complex manifold and $W \cap \mathfrak{R}(V)$ is an irreducible holomorphic subvariety of $\mathfrak{R}(V)$, as is evident upon recalling Corollary E20. The given function $f$ is holomorphic in $\mathfrak{R}(V) - W \cap \mathfrak{R}(V)$, and if it extends to a holomorphic function at some point $A \in W$, that extension is holomorphic in an open neighborhood of $A$ and hence at some point of $W \cap \mathfrak{R}(V)$. It then follows from Theorem K2, noting as mentioned before that this theorem holds for complex manifolds as well as for open subsets of $\mathbb{C}^n$, that $f$ extends to a holomorphic function on $\mathfrak{R}(V)$. Now if $V$ is normal, then $\dim \mathfrak{S}(V) \leqq \dim V - 2$ by Theorem 12, and since $f$ has been extended to a holomorphic function in the complement of $\mathfrak{S}(V)$, it follows from the first part of the present theorem that $f$ extends further to a holomorphic function on all of $V$. That suffices to conclude the proof.

To return now to general holomorphic varieties, the set of points at which any variety fails to be normal is obviously a thin subset, being contained within the singular locus, but rather less obviously is even a holomorphic subvariety. This section will be concluded by giving a particularly elegant and simple proof due to H. Grauert and R. Remmert of that assertion. For any given holomorphic variety $V$, to every point $A \in V$ there are by Theorem 2 an open neighborhood $V_A$ of $A$ in $V$ and a holomorphic function $d \in {}_V\mathcal{O}_{V_A}$ that is a universal denominator at each point $Z \in V_A$ but is not a zero-divisor at any point $Z \in V_A$. The zero locus of this function $d$ is a holomorphic subvariety $W \subset V_A$, and the family of ideals $\mathfrak{A}_Z = \text{id } \mathbf{W}_Z \subseteq {}_V\mathcal{O}_Z$ is a finitely generated family of ideals over a neighborhood of each point of $V_A$ by Cartan's theorem, Theorem F6. Now to each point $Z \in V_A$ associate the ${}_V\mathcal{O}_Z$-submodule $\mathcal{S}_Z \subseteq {}_V\mathcal{M}_Z$ defined as

$$\mathcal{S}_Z = \{\mathbf{f} \in {}_V\mathcal{M}_Z : \mathbf{f} \cdot \mathfrak{A}_Z \subseteq \mathfrak{A}_Z\} \tag{2}$$

It follows from Corollary 6 that $\mathcal{S}_Z \subseteq {}_V\hat{\mathcal{O}}_Z$, since $\mathbf{d}_Z \in \mathfrak{A}_Z$ and $\mathbf{d}_Z$ is not a divisor of zero in ${}_V\mathcal{O}_Z$. Thus $\mathcal{S}_Z$ is necessarily a finitely generated submodule of ${}_V\mathcal{M}_Z$, since ${}_V\hat{\mathcal{O}}_Z$ is so by Theorem 4. In these terms there is the following criterion for the normality of $V$.

**16. LEMMA.** *The variety $V$ is normal at a point $Z \in V_A$ precisely when $\mathcal{S}_Z \subseteq {}_V\mathcal{O}_Z$.*

*Proof.* On the one hand, since $\mathcal{S}_Z \subseteq {}_V\hat{\mathcal{O}}_Z$ as already noted, it is obvious that $\mathcal{S}_Z \subseteq {}_V\mathcal{O}_Z$ if $V$ is normal at $Z$. On the other hand, if $V$ is not normal at $Z$, there must be some germ $\mathbf{g} \in {}_V\hat{\mathcal{O}}_Z$ such that $\mathbf{g} \notin {}_V\mathcal{O}_Z$. Now since $\mathfrak{A}_Z = \text{id loc } {}_V\mathcal{O}_Z \cdot \mathbf{d}_Z$, it follows readily from Hilbert's zero-theorem on $V$ and the Noetherian property of the local ring ${}_V\mathcal{O}_Z$ as in Lemma G13 that $\mathfrak{A}_Z^\mu \subseteq {}_V\mathcal{O}_Z \cdot \mathbf{d}_Z$ for some integer $\mu \geqq 1$, and since $d$ is a universal denominator at $Z$ so that $\mathbf{d}_Z \cdot \mathbf{g} \in {}_V\mathcal{O}_Z$, it further follows that $\mathfrak{A}_Z^\mu \cdot \mathbf{g} \subseteq {}_V\mathcal{O}_Z$. If $\nu$ is the least integer for which $\mathfrak{A}_Z^\nu \cdot \mathbf{g} \subseteq {}_V\mathcal{O}_Z$, there must be some germ $\mathbf{f} \in \mathfrak{A}_Z^{\nu-1}\mathbf{g} \subseteq {}_V\hat{\mathcal{O}}_Z$ such that $\mathbf{f} \notin {}_V\mathcal{O}_Z$, and for this germ $\mathfrak{A}_Z \cdot \mathbf{f} \subseteq {}_V\mathcal{O}_Z$. Of course if $\nu = 1$, the germ $\mathbf{f} = \mathbf{g}$ itself satisfies $\mathfrak{A}_Z\mathbf{f} \subseteq {}_V\mathcal{O}_Z$. If $\mathbf{h} \in \mathfrak{A}_Z$, then $\mathbf{hf} \in {}_V\mathcal{O}_Z$, and since $\mathbf{h}$ vanishes on $W$ while $\mathbf{f}$ can be represented by a uniformly bounded function in a dense open subset of a neighborhood of $Z$, the product $\mathbf{hf}$ also vanishes on $W$; that is, $\mathbf{hf} \in \mathfrak{A}_Z$ as well. Thus $\mathfrak{A}_Z \cdot \mathbf{f} \subseteq \mathfrak{A}_Z$, so that $\mathbf{f} \in \mathcal{S}_Z$, and since $\mathbf{f} \notin {}_V\mathcal{O}_Z$ that shows that $\mathcal{S}_Z \nsubseteq {}_V\mathcal{O}_Z$ and thereby concludes the proof.

That the set of points at which a variety is not normal form a holomorphic subvariety then follows readily from this criterion for normality, with the family of modules $\{\mathcal{S}_Z\}$ being locally finitely generated since the family of ideals $\{\mathfrak{A}_Z\}$ is.

Some care must still be taken, though, since the discussion of finitely generated families of modules in section F was limited to families of submodules of free modules.

**17. THEOREM.** *The set of points at which a holomorphic variety $V$ fails to be a normal variety form a holomorphic subvariety in the singular locus of $V$.*

*Proof.* For any point $A \in V$ choose an open neighborhood $V_A$ and a holomorphic function $d \in {}_V\mathcal{O}_{V_A}$ such that $d$ is a universal denominator at each point of $V_A$ but is not a zero-divisor at any point of $V_A$, and introduce the family of submodules $\mathcal{S}_Z \subseteq {}_V\mathcal{M}_Z$ defined as in (2). Since $\mathcal{S}_Z \subseteq {}_V\hat{\mathcal{O}}_Z$ as already noted and $d$ is a universal denominator at $Z$, it follows that $\mathbf{d}_Z \cdot \mathcal{S}_Z = \mathfrak{B}_Z$ is an ideal in the local ring ${}_V\mathcal{O}_Z$ at any point $Z \in V_A$. To describe this ideal more directly, note that if $\mathbf{f} \in \mathfrak{B}_Z$, then $\mathbf{f} = \mathbf{d}_Z\mathbf{g}$ for some germ $\mathbf{g} \in \mathcal{S}_Z \subseteq {}_V\hat{\mathcal{O}}_Z$ and hence by (2) necessarily $\mathbf{f} \cdot \mathfrak{A}_Z = \mathbf{d}_Z\mathbf{g} \cdot \mathfrak{A}_Z \subseteq \mathbf{d}_Z \cdot \mathfrak{A}_Z$. On the other hand, if $\mathbf{f} \in {}_V\mathcal{O}_Z$ and $\mathbf{f} \cdot \mathfrak{A}_Z \subseteq \mathbf{d}_Z \cdot \mathfrak{A}_Z$, then $(\mathbf{f}/\mathbf{d}_Z) \in {}_V\mathcal{M}_Z$ has the property that $(\mathbf{f}/\mathbf{d}_Z) \cdot \mathfrak{A}_Z \subseteq \mathfrak{A}_Z$ so that $(\mathbf{f}/\mathbf{d}_Z) \in \mathcal{S}_Z$ and hence $\mathbf{f} \in \mathbf{d}_Z \cdot \mathcal{S}_Z = \mathfrak{B}_Z$. Thus $\mathfrak{B}_Z$ can be described as $\mathfrak{B}_Z = \{\mathbf{f} \in {}_V\mathcal{O}_Z : \mathbf{f} \cdot \mathfrak{A}_Z \subseteq \mathbf{d}_Z \cdot \mathfrak{A}_Z\}$, or equivalently as $\mathfrak{B}_Z = (\mathbf{d}_Z \cdot \mathfrak{A}_Z) : \mathfrak{A}_Z$. The family of ideals $\{\mathfrak{A}_Z\}$ and hence of course also the family of ideals $\{\mathbf{d}_Z \cdot \mathfrak{A}_Z\}$ are finitely generated over an open neighborhood of each point of $V_A$, and it follows from Corollary F10 that the family of ideals $\mathfrak{B}_Z$ is also finitely generated over an open neighborhood of each point of $V_A$. Note further that if $\mathcal{S}_Z \subseteq {}_V\mathcal{O}_Z$, then $\mathfrak{B}_Z = \mathbf{d}_Z \cdot \mathcal{S}_Z \subseteq \mathbf{d}_Z \cdot {}_V\mathcal{O}_Z$, and conversely that if $\mathfrak{B}_Z \subseteq \mathbf{d}_Z \cdot {}_V\mathcal{O}_Z$, then for any germ $\mathbf{f} \in \mathcal{S}_Z \subseteq {}_V\hat{\mathcal{O}}_Z$ the germ $\mathbf{d}_Z\mathbf{f} \in \mathfrak{B}_Z \subseteq \mathbf{d}_Z \cdot {}_V\mathcal{O}_Z$ can be written as $\mathbf{d}_Z\mathbf{f} = \mathbf{d}_Z\mathbf{g}$ for some $\mathbf{g} \in {}_V\mathcal{O}_Z$, so since $\mathbf{d}_Z$ is not a divisor of zero, $\mathbf{f} = \mathbf{g} \in {}_V\mathcal{O}_Z$. Thus Lemma 16 can be rephrased equivalently as the assertion that $V$ is normal at a point $Z \in V_A$ precisely when $\mathfrak{B}_Z \subseteq \mathbf{d}_Z \cdot {}_V\mathcal{O}_Z$. Now the family of ideals $\{\mathfrak{C}_Z\}$ defined as $\mathfrak{C}_Z = (\mathbf{d}_Z \cdot {}_V\mathcal{O}_Z) : \mathfrak{B}_Z$ is also finitely generated over an open neighborhood of each point of $V_A$ by Corollary F10, and $\mathfrak{B}_Z \subseteq \mathbf{d}_Z \cdot {}_V\mathcal{O}_Z$ precisely when $1 \in \mathfrak{C}_Z$ and hence precisely when $\mathfrak{C}_Z = {}_V\mathcal{O}_Z$. So if $f_1, \ldots, f_n \in {}_V\mathcal{O}_U$ are holomorphic functions in an open subset $U \subseteq V_A$ such that the germs of these functions generate the ideal $\mathfrak{C}_Z \subseteq {}_V\mathcal{O}_Z$ at each point $Z \in U$, then $\{Z \in U : V \text{ is not normal at } Z\} = \{Z \in U : f_1(Z) = \cdots = f_n(Z) = 0\}$, so is a holomorphic subvariety of $U$. That suffices to conclude the proof of the theorem.

# R

# Normalization of Varieties

The next topic to be considered here is the structure of the ring ${}_V\hat{\mathcal{O}}_A$ of germs of weakly holomorphic functions at a point $A$ on a holomorphic variety $V$. Of course if $A \in \mathfrak{R}(V)$, then ${}_V\hat{\mathcal{O}}_A = {}_V\mathcal{O}_A$ and that is as much as need be said about the structure of the ring ${}_V\hat{\mathcal{O}}_A$ on its own right. Thus this question is actually interesting only at a singular point $A$ of the variety $V$. As defined in the preceding section, a weakly holomorphic function is really just a function on the regular locus of $V$, but it is quite easy to see that any such function can be extended in a natural way to all the singular points of $V$ at which $V$ is irreducible.

**1. THEOREM.** *If a holomorphic variety $V$ is irreducible at a point $A \in \mathfrak{S}(V)$, then any weakly holomorphic function $f$ on $V$ can be assigned a unique value $f(A)$ at $A$ so that the extended function is continuous on $A \cup \mathfrak{R}(V)$.*

*Proof.* Since the desired theorem is a local result, it can be supposed that the variety $V$ is irreducible and is represented as a finite branched holomorphic covering $\pi: V \to U$ of some order $\nu$ over an open subset $U \subseteq \mathbb{C}^n$, and that $A = \pi^{-1}(\pi(A))$. The image $W = \pi(\mathfrak{S}(V)) \subset U$ of the singular locus of $V$ is then a proper holomorphic subvariety of $U$. If $f$ is weakly holomorphic on $V$, then from Theorem C5 applied to the restriction of $f$ to the finite branched holomorphic covering $\pi: V - \pi^{-1}(W) \to U - W$ it follows that there is a monic polynomial $P(X) = P(Z; X) \in {}_n\mathcal{O}_{U-W}[X]$ of degree $\nu$ such that $P(\pi(Z); f(Z)) = 0$ whenever $Z \in V - \pi^{-1}(W)$. Indeed the roots of $P(Z; X)$ are precisely the values of $f$ at the points of $\pi^{-1}(Z)$ whenever $Z \in U - W$, and since $f$ is weakly holomorphic these values are locally bounded in $V$. The coefficients of the polynomial $P(X)$ are consequently locally bounded in $U$ and so extend to holomorphic functions throughout $U$. The polynomial $P(X)$ can thus be viewed as an element $P(X) \in {}_n\mathcal{O}_U[X]$, and since $V$ is irreducible, the identity $P(\pi(Z); f(Z)) = 0$ must hold for all points $Z \in V - \mathfrak{S}(V)$. Now to prove the desired result it is sufficient just to show that all $\nu$ roots of the polynomial $P(\pi(A); X)$ coincide; for if $f(A)$ is set equal to that root, then since the roots of a polynomial are continuous functions of the coefficients, in the sense readily made precise by using Lemma I-C3, the values of $f(Z)$ are as near $f(A)$ as desired for all points $Z \in V - \mathfrak{S}(V)$ sufficiently near $A$. If the polynomial $P(\pi(A); X)$ has $k > 1$ distinct roots $X_1, \dots, X_k$, then choose disjoint open neighborhoods $\Delta(X_j, \varepsilon) \subseteq \mathbb{C}$ around these points and note by applying Lemma I-C3 again that when $Z$ is restricted to a sufficiently small connected open neighborhood $U_0$ of $A$ in $\mathbb{C}^n$, the roots of $P(Z; X)$ remain in the disjoint union $\bigcup_j \Delta(X_j; \varepsilon)$. Therefore $\{Z \in V - \mathfrak{S}(V) : \pi(Z) \in U_0$ and $f(Z) \in \Delta(X_j; \varepsilon)\}$ for $1 \leqq j \leqq k$ are nonempty disjoint open subsets of $V$, the union of which is $\pi^{-1}(U_0) \cap (V - \mathfrak{S}(V))$. But since $V$ is irreducible at $A$, the latter set must be connected when $U_0$ is sufficiently small, as is clear from Theorem E19. That is thus a contradiction, and the proof is therewith completed.

This observation then leads immediately to the following property of the rings of germs of weakly holomorphic functions on a holomorphic variety.

**2. THEOREM.** *If the holomorphic variety $V$ is irreducible at a point $A \in V$, then the ring ${}_V\hat{\mathcal{O}}_A$ is a local ring.*

*Proof.* If $A$ is a regular point of $V$, then ${}_V\hat{\mathcal{O}}_A = {}_V\mathcal{O}_A$ and the desired result has already been established. In any case the inclusion ${}_V\mathcal{O}_A \to {}_V\hat{\mathcal{O}}_A$ exhibits ${}_V\hat{\mathcal{O}}_A$ as a finitely generated module over the ring ${}_V\mathcal{O}_A$, as in Theorem Q4, and since ${}_V\mathcal{O}_A$ is Noetherian, the ring ${}_V\hat{\mathcal{O}}_A$ is necessarily also Noetherian. Furthermore if $V$ is irreducible at $A$, then by Theorem 1 any germ $\mathbf{f} \in {}_V\hat{\mathcal{O}}_A$ can be assigned a unique value $\mathbf{f}(A)$ at $A$ so that any representative function will be continuous on $A \cup \mathfrak{R}(V)$ near $A$. If $\mathbf{f}(A) \neq 0$ and $f$ is a representative function, then $1/f$ must be a bounded holomorphic function on $\mathfrak{R}(V)$ near $A$ and hence $1/\mathbf{f} \in {}_V\hat{\mathcal{O}}_A$. Conversely if $1/\mathbf{f} = \mathbf{g} \in {}_V\hat{\mathcal{O}}_A$, then $f(Z)g(Z) = 1$ on $\mathfrak{S}(V)$ near $A$ for any representative function $f$ and $g$; hence, $f(A) \neq 0$. The nonunits of ${}_V\hat{\mathcal{O}}_A$ are therefore precisely those germs $\mathbf{f} \in {}_V\hat{\mathcal{O}}_A$ such that $\mathbf{f}(A) = 0$ and hence obviously form an ideal in ${}_V\hat{\mathcal{O}}_A$. That suffices to conclude the proof.

The principal result of the present section is the theorem that the local ring ${}_V\hat{\mathcal{O}}_A$ of germs of weakly holomorphic functions, at a point $A$ at which the holomorphic variety $V$ is irreducible, is isomorphic as an algebra over the complex numbers to the local ring of another germ of a holomorphic variety. That in a sense fully describes the structure of the local ring ${}_V\hat{\mathcal{O}}_A$ and leads of course to the further analysis of this auxiliary variety. This result is actually a special case of a somewhat more general theorem, characterizing the local rings of germs of holomorphic varieties by extending the arguments used in proving Theorem B14. The general characterization has other uses as well, as will be seen later.

**3. THEOREM (Seidenberg's theorem).** *If $\mathfrak{R}$ is an algebra over the complex numbers and has the structure of a local ring with no nilpotent elements, and if there is an algebra homomorphism $\phi: {}_{\mathbf{V}}\mathcal{O} \to \mathfrak{R}$ that exhibits $\mathfrak{R}$ as a finite module over ${}_{\mathbf{V}}\mathcal{O}$ for some germ $\mathbf{V}$ of a holomorphic variety, then the algebra $\mathfrak{R}$ is isomorphic to the local ring of a germ of a holomorphic variety.*

*Proof.* Suppose that $\mathfrak{R}$ is an algebra satisfying the hypotheses as in the statement of the theorem. As a preliminary simplification, recall that for any germ $\mathbf{V}$ of a holomorphic variety there is a finite germ of a holomorphic mapping $\pi: \mathbf{V} \to \mathbb{C}^k$ for some $k$, and the induced homomorphism $\pi^*: {}_k\mathcal{O} \to {}_{\mathbf{V}}\mathcal{O}$ exhibits ${}_{\mathbf{V}}\mathcal{O}$ as a finite module over the ring ${}_k\mathcal{O}$. The composition $\phi \circ \pi^* : {}_k\mathcal{O} \to \mathcal{R}$ then exhibits $\mathcal{R}$ as a finite module over the ring ${}_k\mathcal{O}$. Thus without loss of generality it can be assumed that the germ **V** in the hypothesis of the theorem is a regular germ of a variety of dimension $k$, or equivalently that there is an algebra homomorphism $\phi: {}_k\mathcal{O} \to \mathcal{R}$ that exhibits $\mathcal{R}$ as a finite module over ${}_k\mathcal{O}$. This variant of the hypothesis will be used in the remainder of the proof.

Let $\mathfrak{m}$ be the maximal ideal in the local ring $\mathcal{R}$ and ${}_k\mathfrak{m}$ be the maximal ideal in ${}_k\mathcal{O}$. Any ring homomorphism $\phi: {}_k\mathcal{O} \to \mathcal{R}$ clearly takes units in ${}_k\mathcal{O}$ to units in $\mathcal{R}$, but of greater moment is the assertion that under the additional hypotheses here the homomorphism $\phi$ also takes nonunits in ${}_k\mathcal{O}$ to nonunits in $\mathcal{R}$. It may be recalled that this observation was also basic to the proof of Theorem B14, which was somewhat simpler in that both rings involved were assumed to be local rings of holomorphic varieties. To demonstrate this assertion in the present case suppose to the contrary that there is a nonunit $\mathbf{f} \in {}_k\mathfrak{m} \subseteq {}_k\mathcal{O}$ such that $\phi(\mathbf{f}) \notin \mathfrak{m}$; there must then be an element $r \in \mathcal{R}$ for which $r \cdot \phi(\mathbf{f}) = 1$. Since the homomorphism $\phi: {}_k\mathcal{O} \to \mathcal{R}$ by assumption exhibits $\mathcal{R}$ as a finite module over ${}_k\mathcal{O}$, any element $r \in \mathcal{R}$ is necessarily integral over the subring $\phi({}_k\mathcal{O}) \subseteq \mathcal{R}$. Hence there must be an identity of the form $r^\nu + \phi(\mathbf{f}_1) \cdot r^{\nu-1} + \cdots + \phi(\mathbf{f}_\nu) = 0$ for some integer $\nu \geq 1$ and some elements $\mathbf{f}_j \in {}_k\mathcal{O}$. Upon multiplying this identity by $\phi(\mathbf{f})^\nu$, it follows that $0 = 1 + \phi(\mathbf{f}\mathbf{f}_1) + \cdots + \phi(\mathbf{f}^\nu \mathbf{f}_\nu) = \phi(1 + \mathbf{f}\mathbf{f}_1 + \cdots + \mathbf{f}^\nu \mathbf{f}_\nu)$ and hence that $1 + \mathbf{f}\mathbf{f}_1 + \cdots + \mathbf{f}^\nu \mathbf{f}_\nu \in \ker \phi \subseteq {}_k\mathfrak{m}$. But that is clearly impossible, since $\mathbf{f} \in {}_k\mathfrak{m}$ but $1 \notin {}_k\mathfrak{m}$. It thus follows as desired that $\phi({}_k\mathfrak{m}) \subseteq \mathfrak{m}$.

As yet another preliminary observation, note that for any element $r \in \mathcal{R}$ there must exist some complex constant $a \in \mathbb{C}$ such that $r - a \in \mathfrak{m}$. Indeed since $r$ is necessarily integral over the subring $\phi({}_k\mathcal{O}) \subseteq \mathcal{R}$, there must again be an identity of the form $r^\nu + \phi(\mathbf{f}_1) \cdot r^{\nu-1} + \cdots + \phi(\mathbf{f}_\nu) = 0$ for some integer $\nu \geq 1$ and some elements $\mathbf{f}_j \in {}_k\mathcal{O}$. For each of these germs $\mathbf{f}_j$ of holomorphic functions there is of course a complex constant $c_j \in \mathbb{C}$ such that $\mathbf{f}_j - c_j \in {}_k\mathfrak{m}$, so since $\phi(\mathbf{f}_j) - c_j = \phi(\mathbf{f}_j - c_j) \in \phi({}_k\mathfrak{m}) \subseteq \mathfrak{m}$, it follows that $r^\nu + c_1 r^{\nu-1} + \cdots + c_\nu \in \mathfrak{m}$. If $a_1, \ldots, a_\nu$ are the complex roots of the polynomial $X^\nu + c_1 X^{\nu-1} + \cdots + c_\nu$, this last inclusion can be rewritten $(r - a_1) \ldots (r - a_\nu) \in \mathfrak{m}$, and since $\mathfrak{m}$ is a prime ideal, at least one of these factors $(r - a_j) \in \mathfrak{m}$, yielding the desired result.

With these preliminary observations established, select some elements $r_j \in \mathcal{R}$ so that $1, r_1, \ldots, r_n$ generate $\mathcal{R}$ as a module over the ring ${}_k\mathcal{O}$. In view of the result of the preceding paragraph it can be assumed that $r_j \in \mathfrak{m}$. The homomorphism $\phi: {}_k\mathcal{O} \to \mathcal{R}$ can be extended to a homomorphism $\Phi: {}_k\mathcal{O}[z_{k+1}, \ldots, z_{k+n}] \to \mathcal{R}$ from the formal polynomial algebra over ${}_k\mathcal{O}$ in indeterminates $z_{k+1}, \ldots, z_{k+n}$ to the algebra $\mathcal{R}$ by setting $\Phi(z_{k+j}) = r_j$. The homomorphism $\Phi$ is evidently surjective, so if the ideal $\mathfrak{A} \subseteq {}_k\mathcal{O}[z_{k+1}, \ldots, z_{k+n}]$ is the kernel of $\Phi$, it follows that

$$\mathcal{R} \cong {}_k\mathcal{O}[z_{k+1}, \ldots, z_{k+n}]/\mathfrak{A} \tag{1}$$

The polynomial ring ${}_k\mathcal{O}[z_{k+1}, \ldots, z_{k+n}]$ can be viewed as a subring of the ring ${}_{k+n}\mathcal{O}$ of holomorphic functions of $k + n$ variables, and the ideal $\mathfrak{A} \subseteq {}_k\mathcal{O}[z_{k+1}, \ldots, z_{k+n}]$ can be extended to an ideal $\tilde{\mathfrak{A}} = {}_{k+n}\mathcal{O} \cdot \mathfrak{A} \subseteq {}_{k+n}\mathcal{O}$. The natural inclusion ${}_k\mathcal{O}[z_{k+1}, \ldots, z_{k+n}] \to {}_{k+n}\mathcal{O}$ thus carries the inclusion $\mathfrak{A} \to \tilde{\mathfrak{A}}$ and so induces an algebra homomorphism

$$\theta: {}_k\mathcal{O}[z_{k+1}, \ldots, z_{k+n}]/\mathfrak{A} \to {}_{k+n}\mathcal{O}/\tilde{\mathfrak{A}} \tag{2}$$

In order to complete the proof it is sufficient just to show that $\theta$ is an isomorphism; for then from (1) and (2) it follows that $\mathcal{R} \cong {}_{k+n}\mathcal{O}/\tilde{\mathfrak{A}}$, and since by hypothesis $\mathcal{R}$ has no nilpotent elements, the ideal $\tilde{\mathfrak{A}}$ must be a radical ideal, so by Hilbert's zero-theorem $\tilde{\mathfrak{A}} = \text{id}\, \mathbf{W}$ for some germ $\mathbf{W}$ of a holomorphic variety and hence $\mathcal{R} \cong {}_{\mathbf{W}}\mathcal{O}$ as desired.

To show first that $\theta$ is surjective, begin by noting that since each element $r_j \in \mathcal{R}$ is integral over the subring $\phi({}_k\mathcal{O}) \subseteq \mathcal{R}$, there must be a monic polynomial $\tilde{\mathbf{p}}_j(X) = X^{\nu_j} + \phi(\mathbf{f}_{j1})X^{\nu_j - 1} + \cdots + \phi(\mathbf{f}_{j\nu_j}) \in \phi({}_k\mathcal{O})[X]$ such that $\tilde{\mathbf{p}}_j(r_j) = 0$. The polynomial $\mathbf{p}_j(z_{k+j}) = z_{k+j}^{\nu_j} + \mathbf{f}_{j1} z_{k+j}^{\nu_j - 1} + \cdots + \mathbf{f}_{j\nu_j} \in {}_k\mathcal{O}[z_{k+j}]$ then has the property that $\Phi(\mathbf{p}_j(z_{k+j})) = \tilde{\mathbf{p}}_j(r_j) = 0$, so that $\mathbf{p}_j(z_{k+j}) \in {}_k\mathcal{O}[z_{k+j}] \cap \mathfrak{A} \subseteq {}_k\mathcal{O}[z_{k+1}, \ldots, z_n]$. The polynomial $\mathbf{p}_j(z_{k+j})$ can also be viewed as a germ of a holomorphic function of the $k + 1$ variables $z_1, \ldots, z_k, z_{k+j}$, and as such it is regular in the variable $z_{k+j}$. The Weierstrass preparation theorem shows that $\mathbf{p}_j(z_{k+j}) = \mathbf{u}_j \cdot \mathbf{q}_j(z_{k+j})$ where $\mathbf{u}_j$ is a unit in ${}_{k+n}\mathcal{O}$ and $\mathbf{q}_j(z_{k+j}) \in {}_k\mathcal{O}[z_{k+j}]$ is a Weierstrass polynomial, and of course $\mathbf{q}_j(z_{k+j}) = \mathbf{u}_j^{-1}\mathbf{p}_j(z_{k+j}) \in {}_{k+n}\mathcal{O} \cdot \mathfrak{A} = \tilde{\mathfrak{A}}$. Now by repeated application of the Weierstrass division theorem it is clear that any germ $\mathbf{f} \in {}_{k+n}\mathcal{O}$ can be written $\mathbf{f} = \sum_{j=1}^n \mathbf{f}_j \mathbf{q}_j(z_{k+j}) + \mathbf{r}$ for some polynomial $\mathbf{r} \in {}_k\mathcal{O}[z_{k+1}, \ldots, z_{k+n}]$. Indeed it is possible to write $\mathbf{f} = \mathbf{f}_n \mathbf{q}_n(z_{k+n}) + \mathbf{r}_n$ where $\mathbf{r}_n \in {}_{k+n-1}\mathcal{O}[z_{k+n}]$, to divide each coefficient of $\mathbf{r}_n$ by $\mathbf{f}_{n-1}$ similarly, and so on. That means of course that the germ $\mathbf{f}$ can be written as a sum of an element in $\tilde{\mathfrak{A}}$ and a polynomial in ${}_k\mathcal{O}[z_{k+1}, \ldots, z_{k+n}]$, so that ${}_{k+n}\mathcal{O} = \tilde{\mathfrak{A}} + {}_k\mathcal{O}[z_{k+1}, \ldots, z_{k+n}]$, and that is clearly enough to imply that the homomorphism (2) is surjective.

Finally to show that $\theta$ is injective, consider a germ $\mathbf{p} \in {}_k\mathcal{O}[z_{k+1}, \ldots, z_{k+n}] \subseteq {}_{k+n}\mathcal{O}$ that represents an element in the kernel of $\theta$, or equivalently a germ $\mathbf{p} \in {}_k\mathcal{O}[z_{k+1}, \ldots, z_{k+n}] \cap \tilde{\mathfrak{A}}$. Let $\mathbf{q}_j \in {}_k\mathcal{O}[z_{k+1}, \ldots, z_{k+n}] \subseteq {}_{k+n}\mathcal{O}$ be any finite set of generators of the ideal $\mathfrak{A} \subseteq {}_k\mathcal{O}[z_{k+1}, \ldots, z_{k+n}]$, and note that these germs must also be generators of the ideal $\tilde{\mathfrak{A}} \subseteq {}_{k+n}\mathcal{O}$. Therefore $\mathbf{p} = \sum_j \mathbf{f}_j \mathbf{q}_j$ for some germs $\mathbf{f}_j \in {}_{k+n}\mathcal{O}$. Now for any integer $\nu > 0$ write the power series $\mathbf{f}_j$ as a sum $\mathbf{f}_j = \mathbf{f}_j' + \mathbf{f}_j''$ where $\mathbf{f}_j' \in {}_k\mathcal{O}[z_{k+1}, \ldots, z_{k+n}]$ is a polynomial in the variables $z_{k+1}, \ldots, z_{k+n}$ of degree less than $\nu$ and each monomial in the power series $\mathbf{f}_j''$ has degree at least $\nu$ in the variables $z_{k+1}, \ldots, z_{k+n}$. Note that

$$\sum_j \mathbf{f}_j'' \mathbf{q}_j = \mathbf{p} - \sum_j \mathbf{f}_j' \mathbf{q}_j \in {}_k\mathcal{O}[z_{k+1}, \ldots, z_{k+n}] \tag{3}$$

so the power series on the left-hand side of (3) must actually be a polynomial in the variables $z_{k+1}, \ldots, z_{k+n}$. Moreover since each monomial in the power series $\mathbf{f}_j''$ for each index $j$ is of degree at least $\nu$ in these variables, the same must be true for each monomial in the polynomial (3). Since $\Phi(z_{k+j}) = r_{k+j} \in \mathfrak{m}$, it follows that the image of the polynomial (3) under the homomorphism $\Phi$ must be contained in the ideal $\mathfrak{m}^\nu \subseteq \mathcal{R}$. On the other hand, $\sum_j \mathbf{f}_j' \mathbf{q}_j \in \mathfrak{A}$, the kernel of the homomorphism $\Phi$, so the image of the polynomial (3) under the homomorphism $\Phi$ is just $\Phi(\mathbf{p})$. Thus altogether

$\phi(\mathbf{p}) \in \mathfrak{m}^{\nu}$, and since that must be the case for every integer $\nu > 0$, it follows immediately from Nakayama's lemma that $\Phi(\mathbf{p}) = 0$. But then $\mathbf{p} \in \mathfrak{A}$ represents the zero element of the ring ${}_k\mathcal{O}[z_{k+1}, \ldots, z_{k+n}]/\mathfrak{A}$, so that $\Phi$ is injective as desired and the entire proof is thereby concluded.

**4. THEOREM.** *To each irreducible germ of a holomorphic variety* $\mathbf{V}$ *there is associated a unique germ of a holomorphic variety* $\hat{\mathbf{V}}$ *such that the local ring* ${}_{\hat{\mathbf{V}}}\mathcal{O}$ *of* $\hat{\mathbf{V}}$ *is isomorphic to the ring* ${}_{\mathbf{V}}\hat{\mathcal{O}}$ *of germs of weakly holomorphic functions on* $\mathbf{V}$ *as a complex algebra.*

*Proof.* If $\mathbf{V}$ is irreducible, then the ring ${}_{\mathbf{V}}\hat{\mathcal{O}}$ of germs of weakly holomorphic functions on $\mathbf{V}$ is a local ring by Theorem 2, and the natural inclusion ${}_{\mathbf{V}}\mathcal{O} \to {}_{\mathbf{V}}\hat{\mathcal{O}}$ exhibits ${}_{\mathbf{V}}\hat{\mathcal{O}}$ as a finite module over ${}_{\mathbf{V}}\mathcal{O}$ by Theorem Q4. It then follows from Seidenberg's theorem that the algebra ${}_{\mathbf{V}}\hat{\mathcal{O}}$ is isomorphic to the local ring ${}_{\hat{\mathbf{V}}}\mathcal{O}$ of some holomorphic variety $\hat{\mathbf{V}}$. The uniqueness of $\hat{\mathbf{V}}$ is a consequence of Corollary B17, thereby concluding the proof.

**5. DEFINITION.** *The* **normalization** *of an irreducible germ* $\mathbf{V}$ *of a holomorphic variety is that germ* $\hat{\mathbf{V}}$ *of a holomorphic variety for which* ${}_{\hat{\mathbf{V}}}\mathcal{O} = {}_{\mathbf{V}}\hat{\mathcal{O}}$. *The normalization of a reducible germ of a holomorphic variety is the disjoint union of the normalizations of the irreducible components.*

The structure of the ring ${}_{\mathbf{V}}\hat{\mathcal{O}}$ of germs of weakly holomorphic functions on the germ $\mathbf{V}$ of a holomorphic variety is to some extent settled by Theorem 4, since it follows from equation Q(1) that ${}_{\mathbf{V}}\hat{\mathcal{O}} = \bigoplus_j {}_{\mathbf{V}_j}\hat{\mathcal{O}}$ if $\mathbf{V}_j$ are the irreducible components of $\mathbf{V}$. A further result about the structure of ${}_{\mathbf{V}}\hat{\mathcal{O}}$, justifying the terminology introduced in the preceding definition, together with a description of the basic geometric relation between a germ $\mathbf{V}$ and its normalization $\hat{\mathbf{V}}$, is given in the following observation; it should be pointed out that this observation is often taken as the defining property of the normalization $\hat{\mathbf{V}}$.

**6. THEOREM.** *The normalization* $\hat{\mathbf{V}}$ *of an irreducible germ* $\mathbf{V}$ *of a holomorphic variety is the germ of a normal holomorphic variety, and there is a finite branched holomorphic covering* $\pi: \hat{\mathbf{V}} \to \mathbf{V}$ *of branching order one.*

*Proof.* There is of course the natural inclusion relation ${}_{\mathbf{V}}\mathcal{O} \subseteq {}_{\mathbf{V}}\hat{\mathcal{O}} \subseteq {}_{\mathbf{V}}\mathcal{M}$, where ${}_{\mathbf{V}}\mathcal{M}$ is the quotient field of the integral domain ${}_{\mathbf{V}}\mathcal{O}$ associated to the irreducible germ $\mathbf{V}$. It is clear from this that ${}_{\mathbf{V}}\hat{\mathcal{O}}$ is also an integral domain, and that ${}_{\mathbf{V}}\mathcal{M}$ can also be viewed as the quotient field of ${}_{\mathbf{V}}\hat{\mathcal{O}}$. Any element of ${}_{\mathbf{V}}\mathcal{M}$ that is integral over ${}_{\mathbf{V}}\hat{\mathcal{O}}$ must also be integral over ${}_{\mathbf{V}}\mathcal{O}$, since ${}_{\mathbf{V}}\hat{\mathcal{O}}$ is a finite module over ${}_{\mathbf{V}}\mathcal{O}$ by Theorem Q4, and hence must be contained in ${}_{\mathbf{V}}\hat{\mathcal{O}}$ by Theorem Q5. Thus the ring ${}_{\mathbf{V}}\hat{\mathcal{O}}$ is integrally closed in its quotient field. In view of the isomorphism ${}_{\hat{\mathbf{V}}}\mathcal{O} \cong {}_{\mathbf{V}}\hat{\mathcal{O}}$ it follows that the local ring ${}_{\hat{\mathbf{V}}}\mathcal{O}$ is integrally closed in its quotient field as well. Thus any variety representing the germ $\hat{\mathbf{V}}$ is normal at the base point, and hence in view of Theorem Q17, any sufficiently small open neighborhood of the base point will be a normal variety representing the germ $\hat{\mathbf{V}}$. Next note that by Theorem B14 the natural homomorphism $\pi^*: {}_{\mathbf{V}}\mathcal{O} \to {}_{\hat{\mathbf{V}}}\mathcal{O}$, the composition of the inclusion mapping ${}_{\mathbf{V}}\mathcal{O} \to {}_{\mathbf{V}}\hat{\mathcal{O}}$ and the isomorphism ${}_{\mathbf{V}}\hat{\mathcal{O}} \to {}_{\hat{\mathbf{V}}}\mathcal{O}$, is induced by some holomorphic mapping $\pi: \hat{\mathbf{V}} \to \mathbf{V}$. Since the homomorphism $\pi^*$ exhibits ${}_{\mathbf{V}}\hat{\mathcal{O}}$ as a finite module over ${}_{\mathbf{V}}\mathcal{O}$, it follows from Theorem E11 that the mapping $\pi$ is finite, and since moreover the homomorphism $\pi^*$ is injective, it follows from Corollary E9 that the mapping $\pi$ is surjective. Both germs $\mathbf{V}$ and $\hat{\mathbf{V}}$ are irreducible, $\mathbf{V}$ by hypothesis and $\hat{\mathbf{V}}$ since its local ring ${}_{\hat{\mathbf{V}}}\mathcal{O} \cong {}_{\mathbf{V}}\hat{\mathcal{O}}$ is as already noted an integral domain, so that in fact, as in Theorem E7, the mapping $\pi: \hat{\mathbf{V}} \to \mathbf{V}$ is the germ of a finite branched holomorphic covering, which must be of some pure order $v \geqq 1$. To conclude the proof it is only necessary to show that $v = 1$. If $v > 1$, there must be some germ $\mathbf{f} \in {}_{\hat{\mathbf{V}}}\mathcal{O}$ that separates the sheets of this covering and hence that cannot be contained in the subring $\pi^*({}_{\mathbf{V}}\mathcal{O}) \subseteq {}_{\hat{\mathbf{V}}}\mathcal{O}$. For any nonzero germ $\mathbf{d} \in {}_{\mathbf{V}}\mathcal{O}$ the product $\pi^*(\mathbf{d})\cdot\mathbf{f}$ also separates the sheets of this covering and hence also cannot be contained in the subring $\pi^*({}_{\mathbf{V}}\mathcal{O}) \subseteq {}_{\hat{\mathbf{V}}}\mathcal{O}$. Under the isomorphism ${}_{\hat{\mathbf{V}}}\mathcal{O} \cong {}_{\mathbf{V}}\hat{\mathcal{O}}$ the germ $\mathbf{f}$ must then correspond to a germ $\hat{\mathbf{f}} \in {}_{\mathbf{V}}\hat{\mathcal{O}}$ with the property that $\mathbf{d}\hat{\mathbf{f}} \notin {}_{\mathbf{V}}\mathcal{O}$ for any nonzero germ $\mathbf{d}$; but that is impossible, since there are by Theorem Q2 nonzero universal denominators in ${}_{\mathbf{V}}\mathcal{O}$. That contradiction means that $v = 1$ as desired, and the proof is thereby completed.

In this context it is convenient to introduce the following additional equivalence relation among germs of holomorphic varieties.

**7. DEFINITION.** *Two germs of holomorphic varieties* $\mathbf{V}$, $\mathbf{W}$ *are called* **weakly equivalent** *if they have arbitrarily small representatives* $V$, $W$ *for which there are thin holomorphic subvarieties* $V_1 \subset V$, $W_1 \subset W$ *such that the complements* $V - V_1$, $W - W_1$ *are biholomorphically equivalent varieties.*

Recall that a thin holomorphic subvariety $V_1 \subset V$ not only is a proper subvariety of $V$ but also meets each irreducible component of $V$ in a proper subvariety of that component; the complement $V - V_1$ is thus a dense open subset of $V$. If $V$ and $W$ are holomorphic varieties representing weakly equivalent germs $\mathbf{V}$ and $\mathbf{W}$, so that $\mathbf{V} = \mathbf{V}_A$ for some point $A \in V$ and $\mathbf{W} = \mathbf{W}_B$ for some point $B \in W$, and if $V_1 \subset V$ and $W_1 \subset W$ are thin subvarieties such that $V - V_1$ and $W - W_1$ are biholomorphic, then the definition also requires that there must be arbitrarily small open neighborhoods $V_A$ of $A$ in $V$ and $W_B$ of $B$ in $W$ such that $V_A \cap (V - V_1)$ and $W_B \cap (W - W_1)$ are biholomorphic. It may be recalled from the discussion in section N that if $V$ is a holomorphic modification of $W$ with exceptional locus $E \cap V$ and center a single point $C \in W$, then $V - E$ and $W - C$ are biholomorphic, but for no point $A \in E$ can there be arbitrarily small open neighborhoods $V_A$ of $A$ in $V$ and $W_C$ of $C$ in $W$ for which $V_A \cap (V - E)$ and $W_C \cap (W - C)$ are biholomorphic. Thus it is important to keep in mind the point that Definition 7 really refers to a local situation, as distinct from the more global situations arising in holomorphic modifications. It is clear that weak equivalence of germs of holomorphic varieties is actually an equivalence relation in the usual sense, and that it is a weaker equivalence relation than that of equivalence of germs of holomorphic varieties as discussed in section B and hence can be viewed as an equivalence relation between germs of holomorphic varieties rather than just between representative subvarieties. The relevance of this definition to the discussion here is as indicated in the following result.

**8. THEOREM.** *Two germs of holomorphic varieties are weakly equivalent precisely when they have the same normalization.*

*Proof.* It is evident from Definition 7 that two reducible germs of holomorphic varieties are weakly equivalent precisely when their irreducible components are correspondingly weakly equivalent in pairs, since removing the singular locus from a variety leaves the irreducible components as disjoint varieties, so it is only necessary to prove the theorem in the special case of irreducible germs of varieties. For that purpose note first as an immediate consequence of Theorem 6 that any irreducible germ of a holomorphic variety is weakly equivalent to its normalization. It is perhaps worth noting in passing that weak equivalence of germs of varieties is to some extent just the natural extension of the relation of one germ being a finite branched holomorphic covering of branching order one over another germ to an equivalence relation. On the one hand, it is obvious that any two irreducible germs of holomorphic varieties having the same normalizations are weakly equivalent to each other. On the other hand, if two irreducible germs of holomorphic varieties are weakly equivalent, then their normalizations are also weakly equivalent, so to conclude the proof it is sufficient just to show that any two normal germs that are weakly equivalent are actually equivalent germs of varieties.

Suppose then that $V_j$ are holomorphic subvarieties of bounded open neighborhoods $U_j$ of the origin in some complex spaces $\mathbb{C}^{n_j}$, that as varieties they are normal, and that the germs of $V_1$ and $V_2$ at the origin represent weakly equivalent germs of holomorphic varieties. After shrinking the neighborhoods $U_j$ suitably, there will be thin holomorphic subvarieties $W_j \subset V_j$ and a biholomorphic mapping $F: V_1 - W_1 \to V_2 - W_2$. The coefficient functions of the mapping $F$ are then uniformly bounded holomorphic functions on $V_1 - W_1$, since their values lie in the bounded subset $U_2 \subseteq \mathbb{C}^{n_2}$, and it follows from Theorem Q8 that these functions extend to holomorphic functions on all of $W_1$; that provides an extension of the given mapping $F$ to a holomorphic mapping $\tilde{F}: V_1 \to V_2$. The same argument applied to the inverse mapping $F^{-1}$ yields a holomorphic mapping $\tilde{G}: V_2 \to V_1$. Since the compositions $\tilde{F} \circ \tilde{G}$ and $\tilde{G} \circ \tilde{F}$ are the identity on dense open subsets of their domains, they must be the identity everywhere; that means that $\tilde{F}$ is a biholomorphic mapping as desired, and the proof is thereby concluded.

**9. THEOREM.** *Two germs of holomorphic varieties are weakly equivalent precisely when their rings of germs of weakly holomorphic functions are isomorphic as complex algebras.*

*Proof.* Consider first the case of irreducible germs of holomorphic varieties. Since the ring of germs of weakly holomorphic functions on the germ of a variety is isomorphic as a complex algebra to the ring of germs of holomorphic functions on the normalization of that germ, to say that two germs of varieties have isomorphic algebras of weakly holomorphic functions is by Corollary B17 the same as saying that the two germs have the same normalization. The desired result in this case then follows immediately from the preceding theorem. Next if $\mathbf{V} = \bigcup_j \mathbf{V}_j$ is a reducible germ of a holomorphic variety with irreducible components $\mathbf{V}_j$, then as already noted ${}_{\mathbf{V}}\tilde{\mathcal{O}} = \bigoplus_j {}_{\mathbf{V}_j}\tilde{\mathcal{O}}$. It is quite easy to see that the subrings ${}_{\mathbf{V}_j}\tilde{\mathcal{O}} \subseteq {}_{\mathbf{V}}\tilde{\mathcal{O}}$ are really intrinsically determined. Indeed the germs $\mathbf{f} \in {}_{\mathbf{V}}\tilde{\mathcal{O}}$ that are identically one on some $\mathbf{V}_j \cap \mathfrak{R}(\mathbf{V})$ and identically zero on the rest of $\mathfrak{R}(\mathbf{V})$ are uniquely characterized by the properties that $\mathbf{f}^2 = \mathbf{f}$ and that $\mathbf{f}$ cannot be written in a nontrivial way as a sum of other germs $\mathbf{g} \in {}_{\mathbf{V}}\tilde{\mathcal{O}}$ also satisfying the condition that $\mathbf{g}^2 = \mathbf{g}$. The subring ${}_{\mathbf{V}_j}\tilde{\mathcal{O}}$ is then isomorphic to the image of ${}_{\mathbf{V}}\tilde{\mathcal{O}}$ under the mapping $\mathbf{f}^*: {}_{\mathbf{V}}\tilde{\mathcal{O}} \to {}_{\mathbf{V}}\tilde{\mathcal{O}}$ that sends a germ $\mathbf{g} \in {}_{\mathbf{V}}\tilde{\mathcal{O}}$ to the product $\mathbf{fg} \in {}_{\mathbf{V}}\tilde{\mathcal{O}}$. Thus the rings of germs of weakly holomorphic functions on two reducible germs of holomorphic varieties are isomorphic as algebras precisely when the rings of germs of weakly holomorphic functions on the irreducible components are isomorphic in corresponding pairs. But by the first part of the proof that is the same as saying that the irreducible components are weakly equivalent in pairs; hence, the two germs of varieties must have the same normalization and so by Theorem 8 are weakly equivalent. That suffices to conclude the proof.

It has been pointed out several times that a finite branched holomorphic covering of order one is not necessarily a biholomorphic mapping, and the discussion here may help to clarify this situation somewhat further. The equivalence relation of Definition 7 is as noted the natural extension of the relation of being a finite branched holomorphic covering of order one to an equivalence relation. The ring of germs of weakly holomorphic functions is an algebraic invariant characterizing this equivalence relation. Such functions do arise fairly naturally in connection with the analysis of the extent to which Riemann's removable singularities theorem holds on general holomorphic varieties. That this equivalence relation is really quite basic is demonstrated even more strongly though by the observation that the ring of germs of meromorphic functions is also a characteristic invariant for it. Perhaps the easiest way to demonstrate this is by using some properties of valuations, following Hironaka. Recall that a **valuation** (with value group the group $\mathbb{Z}$ of integers, as will be assumed throughout this discussion) on a field $\mathcal{M}$ is a mapping $v: \mathcal{M}^* \to \mathbb{Z}$ from the set $\mathcal{M}^*$ of nonzero elements of the field $\mathcal{M}$ to the group $\mathbb{Z}$ of integers such that

$$\begin{cases} \text{(i)} \quad v(fg) = v(f) + v(g) & \text{for any } f, g \in \mathcal{M}^* \\ \text{(ii)} \quad v(f+g) \geqq \min(v(f), v(g)) & \text{for any } f, g \in \mathcal{M}^* \end{cases} \tag{4}$$

The field ${}_{\mathbf{V}}\mathcal{M}$ of germs of meromorphic functions on any irreducible germ $\mathbf{V}$ of a holomorphic variety has a great many valuations, enough indeed to provide an intrinsic characterization of the subring ${}_{\mathbf{V}}\mathcal{O} \subseteq {}_{\mathbf{V}}\mathcal{M}$ as follows.

**10. LEMMA.** *For any germ* $\mathbf{V}$ *of a normal holomorphic variety,* ${}_{\mathbf{V}}\mathcal{O} = \{\mathbf{f} \in {}_{\mathbf{V}}\mathcal{M} : v(\mathbf{f}) \geqq 0$ for every valuation $v$ on ${}_{\mathbf{V}}\mathcal{M}\}$.

*Proof.* Consider first an arbitrary valuation $v$ of the field ${}_{\mathbf{V}}\mathcal{M}$. If $f \in {}_{\mathbf{V}}\mathcal{O} \subset {}_{\mathbf{V}}\mathcal{M}$ is the germ of a nonvanishing holomorphic function, it is possible for any integer $n \neq 0$ to write $\mathbf{f} = \mathbf{g}^n$ where $\mathbf{g} \in {}_{\mathbf{V}}\mathcal{O}$. It then follows from property (i) of the characterization (4) of valuations that $v(\mathbf{f}) = nv(\mathbf{g})$, so that the integer $v(\mathbf{f}) \in \mathbb{Z}$ is divisible by an arbitrary nonzero integer $n$, and that of course can happen only if $v(\mathbf{f}) = 0$. Any germ $\mathbf{f} \in {}_{\mathbf{V}}\mathcal{O}$ can be written as the sum $\mathbf{f} = \mathbf{f}_1 + \mathbf{f}_2$ of two germs of nonvanishing holomorphic functions—for instance, by taking for $\mathbf{f}_1$ a nonzero constant $c \neq \mathbf{f}(0)$ and putting $\mathbf{f}_2 = \mathbf{f} - c$. It then follows from property (ii) of the characterization (4) that $v(\mathbf{f}) \geqq \min(v(\mathbf{f}_1), v(\mathbf{f}_2)) = 0$, since $v(\mathbf{f}_1) = v(\mathbf{f}_2) = 0$ as just noted. Thus $v(\mathbf{f}) \geqq 0$ for every $\mathbf{f} \in {}_{\mathbf{V}}\mathcal{O}$.

To conclude the proof it is necessary to show conversely that if $\mathbf{f} \in {}_{\mathbf{V}}\mathcal{M}$ is not holomorphic, there is some valuation $v$ on ${}_{\mathbf{V}}\mathcal{M}$ for which $v(\mathbf{f}) < 0$, and that can be achieved by constructing a class of valuations on ${}_{\mathbf{V}}\mathcal{M}$ as follows. Choose a normal and irreducible variety $V$ so that the germ of $V$ at a point $O \in V$ is the given germ $\mathbf{V}$. The regular part of $V$ is a connected complex manifold $\mathfrak{R}(V)$, and just as in the discussion of divisors in section O, to each meromorphic function not identically zero on $\mathfrak{R}(V)$ there can be associated its divisor on $\mathfrak{R}(V)$. Fix an irreducible subvariety $W \subset \mathfrak{R}(V)$ of codimension one containing the point $O$ in its closure, and for any meromorphic function $f$ not identically zero on $\mathfrak{R}(V)$ let $v_W(f)$ be the coefficient with which the subvariety $W$ appears in the divisor of $f$. That clearly determines a valuation on the field of meromorphic functions on $\mathfrak{R}(V)$. For any connected open neighborhood $U$ of the point $O$ in $V$ the restriction of $W$ to $U \cap \mathfrak{R}(V)$ determines similarly a valuation on the field of meromorphic functions in that subset, and gives the same value to the restriction to $U \cap \mathfrak{R}(V)$ of a meromorphic function in $\mathfrak{R}(V)$ as does the initial valuation. There is thus determined a valuation $v_{\mathbf{W}}$ on the field ${}_{\mathbf{V}}\mathcal{M}$, depending of course only on the germ $\mathbf{W}$ of the subvariety at the point $O$ in the obvious sense. Now consider a germ $\mathbf{f} \in {}_{\mathbf{V}}\mathcal{M}$ that is not holomorphic and let $f$ be a meromorphic function in an open neighborhood $U$ of the point $O$ in $V$ representing that germ. The point $O$ must then be a singular point of the meromorphic function $f$. It follows from Theorem Q12 that $\dim \mathfrak{S}(V) \leqq \dim V - 2$, and with this in mind it further follows from Theorem Q15 that the restriction of $f$ to $U \cap \mathfrak{R}(V)$ cannot be holomorphic. Indeed that must be so for arbitrarily small neighborhoods $U$, since $f$ is not holomorphic at $O$. This is the point at which the normality of $V$ is really needed. There must thus be some subvariety $W \subset U \cap \mathfrak{R}(V)$ of codimension one contained in the polar locus of this restriction $f|(U \cap \mathfrak{R}(V))$, and for that subvariety $v_W(f) < 0$. Moreover that must as noted be the case for any open neighborhood $U$ of the point $O$ in $V$. These subvarieties $W$ form the restriction to $U \cap \mathfrak{R}(V)$ of the singular locus of the function $f$, a holomorphic subvariety of $U$, so there are only finitely many of them if $U$ is sufficiently small. At least one of these subvarieties must then contain the point $O$ in its closure, and that subvariety determines a valuation $v_{\mathbf{W}}$ on ${}_{\mathbf{V}}\mathcal{M}$ for which $v_{\mathbf{W}}(\mathbf{f}) < 0$ as desired. That suffices to conclude the proof of the theorem.

This observation can then be used to establish the next result, also following Hironaka.

**11. THEOREM.** *Two germs of holomorphic varieties are weakly equivalent precisely when their rings of germs of meromorphic functions are isomorphic as complex algebras.*

*Proof.* First consider the case of two irreducible germs $\mathbf{V}$ and $\mathbf{W}$ of holomorphic varieties. On the one hand, if $\mathbf{V}$ and $\mathbf{W}$ are weakly equivalent, then by Theorem 9 the rings ${}_{\mathbf{V}}\hat{\mathcal{O}}$ and ${}_{\mathbf{W}}\hat{\mathcal{O}}$ are isomorphic as complex algebras. But as already noted ${}_{\mathbf{V}}\mathcal{M}$ can be described as the quotient field of the integral domain ${}_{\mathbf{V}}\hat{\mathcal{O}}$, and similarly of course for ${}_{\mathbf{W}}\mathcal{M}$; hence, ${}_{\mathbf{V}}\mathcal{M}$ and ${}_{\mathbf{W}}\mathcal{M}$ are also isomorphic as complex algebras. On the other hand, suppose that the fields ${}_{\mathbf{V}}\mathcal{M}$ and ${}_{\mathbf{W}}\mathcal{M}$ are isomorphic as algebras. Now $\mathbf{V}$ is weakly equivalent to its normalization $\hat{\mathbf{V}}$ by Theorem 8 and the fields ${}_{\mathbf{V}}\mathcal{M}$ and ${}_{\hat{\mathbf{V}}}\mathcal{M}$ are then isomorphic as complex algebras by the first part of the proof, and similarly of course for $\mathbf{W}$. Thus it is possible for the remainder of the proof to replace $\mathbf{V}$ and $\mathbf{W}$ by their normalizations, or equivalently just to assume that $\mathbf{V}$ and $\mathbf{W}$ are themselves normal. The subrings ${}_{\mathbf{V}}\mathcal{O} \subset {}_{\mathbf{V}}\mathcal{M}$ and ${}_{\mathbf{W}}\mathcal{O} \subset {}_{\mathbf{W}}\mathcal{M}$ are then intrinsically determined as in Lemma 10, so the isomorphism between ${}_{\mathbf{V}}\mathcal{M}$ and ${}_{\mathbf{W}}\mathcal{M}$ must determine an isomorphism between ${}_{\mathbf{V}}\mathcal{O}$ and ${}_{\mathbf{W}}\mathcal{O}$. But then $\mathbf{V}$ and $\mathbf{W}$ are equivalent germs of holomorphic varieties by Corollary B17 and so are of course weakly equivalent as well, as desired. Next the case of reducible varieties can be handled just as in the proof of Theorem 9. If $\mathbf{V} = \bigcup_j \mathbf{V}_j$ is a reducible germ with irreducible components $\mathbf{V}_j$, then as noted earlier, ${}_{\mathbf{V}}\mathcal{M} = \bigoplus_j {}_{\mathbf{V}_j}\mathcal{M}$, and the subfields ${}_{\mathbf{V}_j}\mathcal{M} \subseteq {}_{\mathbf{V}}\mathcal{M}$ are intrinsically determined. Indeed the germs $\mathbf{f} \in {}_{\mathbf{V}}\hat{\mathcal{O}} \subset {}_{\mathbf{V}}\mathcal{M}$ that are identically one on some $\mathbf{V}_j \cap \mathfrak{R}(V)$ and identically zero elsewhere on $\mathfrak{R}(\mathbf{V})$ are uniquely characterized by the properties that $\mathbf{f}^2 = \mathbf{f}$ and that $\mathbf{f}$ cannot be written in a nontrivial way as a sum of other germs $\mathbf{g} \in {}_{\mathbf{V}}\mathcal{M}$ also satisfying the identity $\mathbf{g}^2 = \mathbf{g}$. The subfield ${}_{\mathbf{V}_j}\mathcal{M}$ is then isomorphic to the image of ${}_{\mathbf{V}}\mathcal{M}$ under the mapping $\mathbf{f}^*\colon {}_{\mathbf{V}}\mathcal{M} \to {}_{\mathbf{V}}\mathcal{M}$ that sends a germ $\mathbf{g} \in {}_{\mathbf{V}}\mathcal{M}$ to the product $\mathbf{fg} \in {}_{\mathbf{V}}\mathcal{M}$. Thus the rings of germs of meromorphic functions on two reducible germs of holomorphic varieties are isomorphic as algebras precisely when the fields of germs of meromorphic functions on the irreducible components are isomorphic in corresponding pairs. But by the first part of the proof, that is equivalent to the condition that the irreducible components are weakly equivalent in pairs; hence, the two germs of varieties must have the same normalization and so by Theorem 8 are weakly equivalent. That suffices to conclude the proof of the theorem.

It is perhaps worth pointing out as a simple consequence of the preceding theorem that when applied to not necessarily normal varieties, the condition of Lemma 10 provides an intrinsic characterization of the ring of weakly holomorphic functions in the field of germs of meromorphic functions on an irreducible germ of a holomorphic variety. It is also an evident consequence of the preceding theorem that there is in general no intrinsic characterization of the ring of holomorphic functions just in terms of the field of germs of meromorphic functions.

The results established thus far provide an interesting approach to the problem of classifying germs of holomorphic varieties. A first step is classification under the relation of weak equivalence, which is algebraically the same as classifying the fields of germs of meromorphic functions on irreducible germs of holomorphic varieties. That is admittedly a monumental task, but nonetheless one that is at least approachable. Indeed the local parametrization theorem, Theorem D10, shows among other things that any germ of an irreducible variety of dimension $n$ is weakly equivalent to the germ of a subvariety at the origin in $\mathbb{C}^{n+1}$ defined as the zero locus of some Weierstrass polynomial $\mathbf{p}(z_{n+1}) \in {}_n\mathcal{O}[z_{n+1}]$. Thus all the fields that can arise in this classification are necessarily of the form ${}_{n+1}\mathcal{O}/\mathfrak{p}$ for prime ideals of the form $\mathfrak{p} = {}_{n+1}\mathcal{O} \cdot \mathbf{p}(z_{n+1})$ and hence are algebraic extensions of the field ${}_n\mathcal{M}$ generated by a single element. Having found any such field, the local ring of the normal germ with that field of germs of meromorphic functions is intrinsically determined by Lemma 10, thus leading geometrically to the classification of all germs of normal varieties. It is perhaps worth repeating again the observation that as a consequence of Theorem 8 classification by weak equivalence and classification by ordinary equivalence are the same thing for germs of normal varieties. The next step is the classification of all germs of holomorphic varieties having a given normalization **V**. It follows from Theorem 6 that for any such germ **W** there is a finite branched holomorphic covering $\pi: \mathbf{V} \to \mathbf{W}$ of branching order one. The induced homomorphism $\pi^*: {}_{\mathbf{W}}\mathcal{O} \to {}_{\mathbf{V}}\mathcal{O}$ imbeds ${}_{\mathbf{W}}\mathcal{O}$ isomorphically as a subring of the local ring ${}_{\mathbf{V}}\mathcal{O}$, so the algebraic problem is just that of determining and classifying all the subrings $\mathcal{R} \subseteq {}_{\mathbf{V}}\mathcal{O}$ that can so arise. These subrings can be characterized purely algebraically by the properties that (i) $\sqrt{\mathfrak{A}} = {}_{\mathbf{V}}\mathfrak{m}$ where $\mathfrak{A} = {}_{\mathbf{V}}\mathcal{O}\cdot(\mathcal{R} \cap {}_{\mathbf{V}}\mathfrak{m})$, and (ii) $\mathcal{R} = \bigcap_{\nu=1}^{\infty} (\mathcal{R} + {}_{\mathbf{V}}\mathfrak{m}^{\nu})$. There is not much point in giving the proof of this assertion here, but a proof and an illustration of how this characterization can be used can be found in *Lectures on Analytic Varieties: Finite Analytic Mappings* (Mathematical Notes, vol. 14, Princeton University Press, 1974).

It should also at least be mentioned that Seidenberg's theorem has several other applications rather similar to the one discussed here. There are a number of interesting rings lying between the local ring ${}_{\mathbf{V}}\mathcal{O}$ of a germ **V** of a holomorphic variety and the ring ${}_{\mathbf{V}}\hat{\mathcal{O}}$ of germs of weakly holomorphic functions on **V**. First and foremost is the ring ${}_{\mathbf{V}}\tilde{\mathcal{O}}$ of germs of continuous weakly holomorphic functions, for which of course ${}_{\mathbf{V}}\mathcal{O} \subseteq {}_{\mathbf{V}}\tilde{\mathcal{O}} \subseteq {}_{\mathbf{V}}\hat{\mathcal{O}}$. If the variety $V$ representing **V** is irreducible at all points in an open neighborhood of the base point, a condition that is stronger than being irreducible just at the base point it may be recalled, then it follows from Theorem 1 that ${}_{\mathbf{V}}\tilde{\mathcal{O}} = {}_{\mathbf{V}}\hat{\mathcal{O}}$. The ring ${}_{\mathbf{V}}\tilde{\mathcal{O}}$ is always a local ring, for the argument of Theorem 2 can be applied to the ring ${}_{\mathbf{V}}\tilde{\mathcal{O}}$ for an arbitrary germ **V** since by definition all germs in ${}_{\mathbf{V}}\tilde{\mathcal{O}}$ are represented by continuous functions. Moreover as a subring of ${}_{\mathbf{V}}\hat{\mathcal{O}}$ the ring ${}_{\mathbf{V}}\tilde{\mathcal{O}}$ is also a finitely generated module over ${}_{\mathbf{V}}\mathcal{O}$. It then follows immediately from Seidenberg's theorem that to any germ **V** of a holomorphic variety there is associated a unique germ $\tilde{\mathbf{V}}$ of a holomorphic variety such that the local ring ${}_{\tilde{\mathbf{V}}}\mathcal{O}$ is isomorphic to the ring ${}_{\mathbf{V}}\tilde{\mathcal{O}}$ as a complex algebra; the germ $\tilde{\mathbf{V}}$ is sometimes called the **maximalization** of **V**. The inclusions ${}_{\mathbf{V}}\mathcal{O} \subseteq {}_{\mathbf{V}}\tilde{\mathcal{O}} \subseteq {}_{\mathbf{V}}\hat{\mathcal{O}}$ imply the existence of finite branched holomorphic covers $\hat{\pi}: \hat{\mathbf{V}} \to \tilde{\mathbf{V}}$ and $\tilde{\pi}: \tilde{\mathbf{V}} \to \mathbf{V}$ of branching order one; the composition $\pi = \tilde{\pi} \circ \hat{\pi}: \hat{\mathbf{V}} \to \mathbf{V}$ is the mapping already considered here. All three germs are thus weakly equivalent, and since $\tilde{\pi}$ preserves the continuous holomorphic functions, it is not difficult to verify that it is a homeomorphism. Correspondingly it is possible to introduce still more rings to which Seidenberg's theorem can be applied. For instance, it is possible to consider subvarieties in $\mathbb{C}^n$ and the rings of weakly holomorphic functions that are not just continuous on the subvariety but can even be represented as the restrictions of functions of some differentiability class in the underlying real coordinates. A further discussion, particularly of maximalization, and references to the reasonably extensive literature on these topics can be found in G. Fischer, *Complex Analytic Geometry* (Lecture Notes in Math., vol. 538, Springer-Verlag, 1976).

Finally there is a global form of the normalization of a holomorphic variety, a rather simple but important and frequently used result, as follows.

**12. THEOREM.** *To any holomorphic variety $V$ there is associated a unique normal holomorphic variety $\hat{V}$ for which there is a finite branched holomorphic covering $\pi: \hat{V} \to V$ of branching order one.*

*Proof.* First, to establish the uniqueness assertion, recall from Theorem Q15 that any variety $V$ is normal outside a holomorphic subvariety $W$ of the singular locus of $V$. If $\hat{V}_j$ are normal varieties for which there are finite branched holomorphic coverings $\pi_j: \hat{V}_j \to V$ of branching order one, then as an almost immediate consequence of Theorem Q10 it follows that the restrictions $\pi_j | \hat{V}_j - \pi_j^{-1}(W): \hat{V}_j - \pi_j^{-1}(W) \to V - W$ are biholomorphic mappings; consequently the various varieties $\hat{V}_j - \pi_j^{-1}(W)$ are canonically biholomorphic to one another. Moreover it is apparent that this biholomorphic equivalence establishes a weak equivalence between the germs of the various varieties $\hat{V}_j - \pi_j^{-1}(W)$ at corresponding points, for the corresponding points are determined by a point in $V$ and an irreducible component of $V$ at that point. But by Theorem 8 this means that the biholomorphic mappings between the various varieties $\hat{V}_j - \pi_j^{-1}(W)$ extend to locally biholomorphic mappings at each point of $\hat{V}_j$; hence, the varieties $\hat{V}_j$ are also biholomorphic to one another as desired. With the uniqueness thus established, the existence follows readily from the local results already established. Indeed it follows immediately from Theorem 6 that there is a covering of $V$ by open sets $U_\alpha$, for each of which there are a normal holomorphic variety $\hat{U}_\alpha$ and a finite branched holomorphic covering $\pi_\alpha: \hat{U}_\alpha \to U_\alpha$ of branching order one. The uniqueness result then implies that the varieties $\pi_\alpha^{-1}(U_\alpha \cap U_\beta)$ and $\pi_\beta^{-1}(U_\alpha \cap U_\beta)$ are canonically biholomorphic, so that the collection of varieties $\{\hat{U}_\alpha\}$ can be viewed as the coordinate neighborhoods describing a variety $\hat{V}$ having the desired properties, and the proof is thereby concluded.

**13. DEFINITION.** *The* **normalization** *of a holomorphic variety $V$ is that normal holomorphic variety $\hat{V}$ for which there exists a finite branched holomorphic covering $\pi: \hat{V} \to V$ of branching order one.*

The function theory on a variety and on its normalization are very closely related, indeed are identical if only weakly holomorphic or meromorphic functions are considered, so for some purposes it is really enough to restrict to normal varieties. In some cases it is convenient to have an explicit statement of this assertion, in the following slightly more general form.

**14. THEOREM.** *If $\pi: V \to W$ is a finite branched holomorphic covering of branching order one, then the natural induced mappings $\pi^*: {}_W\hat{\mathcal{O}}_W \to {}_V\hat{\mathcal{O}}_V$ and $\pi^*: {}_W\mathcal{M}_W \to {}_V\mathcal{M}_V$ are isomorphisms.*

*Proof.* There are thin holomorphic subvarieties $V_1 \subset V$ and $W_1 \subset W$ such that the restriction

$$\pi | V - V_1 : V - V_1 \to W - W_1$$

is a biholomorphic mapping, and it can be assumed that $V_1$ and $W_1$ contain the singular loci of their respective varieties. The weakly holomorphic functions on $V$

can then be identified with the holomorphic functions on $V - V_1$ that are locally bounded on $V$, and similarly for $W$, so that it is evident that the induced mapping $\pi^*: {}_W\hat{\mathcal{O}}_W \rightarrow {}_V\hat{\mathcal{O}}_V$ is an isomorphism. The meromorphic functions on $V$ can be viewed as those meromorphic functions on $V - V_1$ that can be represented in a neighborhood of any point of $V$ as a quotient of weakly holomorphic functions, and it is evident from this and the first part of the proof that the induced mapping $\pi^*: {}_W\mathcal{M}_W \rightarrow {}_V\mathcal{M}_V$ is also an isomorphism, thereby concluding the proof.

A simple example of the usefulness of the normalization of a holomorphic variety is the following extension theorem for meromorphic functions, the version of Theorem O7 for arbitrary varieties that completes the parallelism between extension theorems for meromorphic functions on regular varieties in section O and on general varieties in section P.

**15. THEOREM.** *If $V$ is a holomorphic variety of pure dimension $n$, $W$ is an irreducible subvariety of $V$ with* $\dim W = n - 1$, *$f$ is a meromorphic function on $V - W$, and there is a point $A \in W$ such that $f$ extends to a meromorphic function in an open neighborhood of $A$ in $V$, then $f$ extends to a meromorphic function on $V$.*

*Proof.* In view of Theorem P8 it is only necessary to show that $f$ extends to a meromorphic function on each irreducible component of $V$; so consider some such component $V_1$. If $W$ is not entirely contained in $V_1$, then $V_1 \cap W$ must be a proper holomorphic subvariety of $W$ and hence $\dim V_1 \cap W \leqq n - 2$. Since $f$ is meromorphic on $V_1 - V_1 \cap W$, it must extend to a meromorphic function on all of $V_1$ by Levi's theorem, Theorem P10. It can therefore be supposed for the remainder of the proof that $W \subseteq V_1$ and hence in particular that the point $A$ of the hypothesis is contained in $V_1$. The restriction of the function $f$ to the regular points of $V_1$ is a meromorphic function on the complement of the irreducible holomorphic subvariety $W \cap \mathfrak{R}(V_1)$ of the complex manifold $\mathfrak{R}(V_1)$, and since $f$ is meromorphic in an open neighborhood of the point $A$ in $V_1$, it must in particular extend as a meromorphic function to those points of $W \cap \mathfrak{R}(V_1)$ in that neighborhood. It then follows from Theorem O7, which evidently holds for arbitrary complex manifolds as well as for just open subsets of $\mathbb{C}^n$, that $f$ extends to a meromorphic function on all of $\mathfrak{R}(V_1)$. In view of Theorem 14 it can be supposed that $V_1$ is a normal variety, and hence by Theorem Q12 that $\dim \mathfrak{S}(V_1) \leqq n - 2$. Another application of Levi's theorem, Theorem P10, then shows that $f$ extends to a meromorphic function on all of $V$, thereby concluding the proof.

# Bibliography


1. Abhyanker, S. S. Local Analytic Geometry. Academic Press; New York, 1964.
2. Andreotti, A., and Stoll, W. Analytic and Algebraic Dependence of Meromorphic Functions. Lecture Notes in Math., 234. Springer-Verlag; Heidelberg, 1970.
3. Baily, W. L. Several Complex Variables. University of Chicago Press; Chicago, 1957.
4. Bănică, C., and Stănăşilă, O. Algebraic Methods in the Global Theory of Complex Spaces. J. Wiley; London, 1976.
5. Beals, R., and Greiner, P. Calculus on Heisenberg Manifolds. Annals of Math. Studies, 119. Princeton University Press; Princeton, NJ, 1988.
6. Behnke, H., and Thullen, P. Theorie der Funktionen mehrerer komplexer Veränderlichen. (2 Aufl.) Ergebnisse Math; 51. Springer-Verlag; Heidelberg, 1970.
7. Bergman, S. The Kernel Function and Conformal Mapping. Math. Surveys, American Math. Society; Providence, RI, 1950.
8. Bergman, S. Sur les Fonctions Orthogonales de Plusieurs Variables Complexes avec Applications à la Théorie des Fonctions Analytiques. Interscience; New York. 1941.
9. Bers, L. Introduction to Several Complex Variables. New York University Press; New York, 1964.
10. Bochner, S., and Martin, W. T. Several Complex Variables. Princeton University Press; Princeton, NJ, 1948.
11. Cartan, H. Séminaire Henri Cartan. 1951/52: Fonctions Analytiques de plusiers variables complexes. 1953/54: Fonctions automorphes et espaces analytiques. 1960/61: Familles d'espaces complexes et fondements de la géométrie analytique. W. A. Benjamin; New York, 1967.
12. Cazacu, C. A. Theorie der Funktionen mehrerer komplexer Veränderlichen. VEB Deutscher Vorlag der Wissenschaften; Berlin, 1975.
13. Chern, S. S. Complex Manifolds Without Potential Theory. Van Nostrand; Princeton, NJ, 1962.
14. Coleff, N. R., and Herrera, M. E. Les courants résiduel associés à une forme méromorphe. Lecture Notes in Math., 633. Springer-Verlag; Heidelberg, 1978.
15. Ehrenpreis, L. Fourier Analysis in Several Complex Variables. Wiley-Interscience; New York, 1970.
16. Fischer, G. Complex Analytic Geometry. Lecture Notes in Math., 538. Springer-Verlag; Heidelberg, 1976.
17. Folland, G., and Kohn, J. J. The Neumann Problem for the Cauchy-Riemann Complex. Annals of Math. Studies, 75. Princeton University Press; Princeton, NJ, 1972.

18. Fornaess, J. E. Recent Developments in Several Complex Variables. Annals of Math. Studies, 100. Princeton University Press; Princeton, NJ, 1981.
19. Fornaess, J. E., and Stensønes, B. Lectures on Counterexamples in Several Complex Variables. Math. Notes, 33. Princeton University Press; Princeton, NJ, 1987.
20. Fuks, B. A. Introduction to the Theory of Analytic Functions of Several Complex Variables. American Math. Society; Providence, RI, 1963.
21. Gamelin, T. W. Uniform Algebras. Prentice-Hall; Englewood Cliffs, NJ, 1969.
22. Grauert, H., and Frtizsche, K. Several Complex Variables. Graduate Texts in Math., 38. Springer-Verlag; New York, 1976.
23. Grauert, H., and Remmert, R. Analytische Stellenalgebren. Grundlehren Math. Wissenschaften, 176. Springer-Verlag; Heidelberg, 1971.
24. Grauert, H., and Remmert, R. Theorie der Steinschen Räumen. Grundlehren Math. Wissenschaften, 236. Springer-Verlag; Heidelberg, 1979.
25. Greiner, P., and Stein, E. M. Estimates for the $\partial$-Neumann Problem. Math. Notes, 19. Princeton University Press; Princeton, NJ, 1977.
26. Griffiths, P. A. Entire Holomorphic Mappings in One and Several Variables. Annals of Math. Studies, 85. Princeton University Press; Princeton, NJ, 1976.
27. Gunning, R. C. Lectures on Complex Analytic Varieties: The Local Parametrization Theorem. Mathematical Notes, 10. Princeton University Press; Princeton, NJ, 1970.
28. Gunning, R. C. Lectures on Complex Analytic Varieties: Finite Analytic Mappings. Math. Notes, 14. Princeton University Press; Princeton, NJ, 1974.
29. Gunning, R. C., and Rossi, H. Analytic Functions of Several Complex Variables. Prentice-Hall; Englewood Cliffs, NJ, 1965.
30. Henkin, G. M., and Leiterer, J. Theory of Functions on Complex Manifolds. Birkhäuser; Boston, 1984.
31. Hervé, M. Several Complex Variables, Local Theory. Oxford University Press; Oxford, 1963.
32. Hörmander, L. An Introduction to Complex Analysis in Several Variables. Van Nostrand; Princeton, NJ, 1966.
33. Hua, L. K. Harmonic Analysis of Functions of Several Complex Variables in the Classical Domains. Math. Monographs, 6. American Mathematical Society; Providence, RI, 1963.
34. Kaup, L., and Kaup, B. Holomorphic Functions of Several Variables. Walter de Gruyter; Berlin, 1983.
35. Kobayashi, S. Hyperbolic Manifolds and Holomorphic Mappings. Marcel Dekker; New York, 1970.
36. Krantz, Steven G. Function Theory of Several Complex Variables. J. Wiley; New York, 1982.
37. Laufer, H. B. Normal Two-Dimensional Singularities. Annals of Math. Studies, 71. Princeton University Press; Princeton, NJ, 1971.
38. Lelong, P. Fonctionelles analytiques et fonctions entières ($n$ variables). Séminaire de Math. supérieure, 28. Les Presses de L'Université de Montreal; Montreal, 1968.
39. Lelong, P. Fonctions Plurisousharmoniques et Formes Différentielles Positives. Gordon and Breach; Paris, 1968.
40. Malgrange, B. Lectures on the Theory of Functions of Several Variables. Tata Institute; Bombay, 1958.
41. Milnor, J. Singular Points on Complex Hypersurfaces. Annals of Math. Studies, 61. Princeton University Press; Princeton, NJ, 1968.
42. Narasimhan, R. Analysis on Real and Complex Manifolds. Masson; Paris, 1973.
43. Narasimhan, R. Introduction to the Theory of Analytic Spaces. Lecture Notes in Math., 25. Springer-Verlag; Heidelberg, 1966.

44. Narasimhan, R. Several Complex Variables. University of Chicago Press; Chicago, 1971.
45. Oka, K. Sur les Fonctions Analytiques de Plusieurs Variables. Iwanami Shoten; Tokyo, 1961.
46. Osgood, W. F. Lehrbuch der Funktionentheorie, Bd II. Chelsea; New York, 1965.
47. Pflug, R. P. Holomorphiegebiete, pseudokonvexe Gebiete, und das Levi-Problem. Lecture Notes in Math., 432. Springer-Verlag; Heidelberg, 1975.
48. Range, R. M. Holomorphic Functions and Integral Representations in Several Complex Variables. Graduate Texts in Mathematics, 108. Springer-Verlag; New York, 1986.
49. Rothstein, W. Vorlesungen über Einführung in die Funktionentheorie mehrerer komplexer Veränderlichen, Bd. I, II. Aschendorffsche Verlagsbuchhandlung; Münster, 1965.
50. Rudin, W. Function Theory in the Unit Ball of $\mathbb{C}^n$. Grundlehren der Math. Wissenschaften, 241. Springer-Verlag; Berlin, 1980.
51. Rudin, W. Function Theory in Polydiscs. W. A. Benjamin; New York, 1969.
52. Rudin, W. Lectures on the Edge-of-the-Wedge Theorem. CBMS 6. American Mathematical Society; Providence, RI, 1970.
53. Siegel, C. L. Analytic Functions of Several Complex Variables. Institute for Advanced Study; Princeton, NJ, 1948.
54. Siu, Y-T, and Trautmann, G. Gap-Sheaves and Extension of Coherent Analytic Sheaves. Lecture Notes in Math., 172. Springer-Verlag; Heidelberg, 1971.
55. Stein, E. M. Boundary Behavior of Holomorphic Functions of Several Complex Variables. Math. Notes, 11. Princeton University Press; Princeton, NJ, 1972.
56. Vladimirov, V. S. Les Fonctions de Plusieurs Variables Complexes (et leur application à la théorie quantique des champs). Dunod; Paris, 1967.
57. Vladimirov, V. S. Methods of the Theory of Functions of Many Complex Variables. M.I.T. Press; Cambridge, MA, 1966.
58. Wells, R. O., Jr. Differential Analysis on Complex Manifolds. Prentice-Hall; Englewood Cliffs, NJ, 1973.
59. Wermer, J. Banach Algebras and Several Complex Variables. Graduate Texts in Math., 35. Springer-Verlag; New York, 1976.
60. Whitney, H. Complex Analytic Varieties. Addison-Wesley; Reading, MA, 1972.

# Index

Milton Keynes UK
Ingram Content Group UK Ltd.
UKHW040103071024
449327UK00019B/773

9 780367 450793